Restoration and History

Routledge Studies in Modern History

Restoration and History

The Search for a Usable Environmental Past

Edited by Marcus Hall

Routledge
Taylor & Francis Group
New York   London

First published 2010
by Routledge
270 Madison Ave, New York, NY 10016

Simultaneously published in the UK
by Routledge
2 Park Square, Milton Park, Abingdon, Oxon OX14 4RN

Routledge is an imprint of the Taylor & Francis Group, an informa business

© 2010 Taylor & Francis

Typeset in Sabon by IBT Global.
Printed and bound in the United States of America on acid-free paper by IBT Global.

Library of Congress Cataloging-in-Publication Data
   Restoration and history : the search for a usable environmental past / edited by Marcus Hall.
      p. cm.—(Routledge studies in modern history ; 8)
   "Simultaneously published in the UK"—T.p. verso.
   Papers from a meeting of an interdisciplinary group of ecologists, geographers, anthropologists, sociologists, historians, and philosophers held July 2006 in Zurich, Switzerland.
   Includes bibliographical references and index.
   1. Restoration ecology—Congresses.   2. Landscape ecology—Congresses.   3. Historic preservation—Congresses.   4. Historic sites—Conservation and restoration—Congresses.   5. Ecology—History—Congresses.   6. Landscape—History—Congresses.   7. Restoration ecology—Case studies—Congresses.
   8. Landscape ecology—Case studies—Congresses.   9. Historic preservation—Case studies—Congresses.   10. Historic sites—Conservation and restoration—Case studies—Congresses.   I. Hall, Marcus, 1959–
   QH541.15.R45R485 2010
   333.71'53—dc22
   2009029812

ISBN10: 0-415-87176-X (hbk)
ISBN10: 0-203-86037-3 (ebk)

ISBN13: 978-0-415-87176-1 (hbk)
ISBN13: 978-0-203-86037-3 (ebk)

For Alex, Aldo, and Elena

Contents

Figures

Tables

Acknowledgements

Large quantities of emotional, logistical, intellectual, and financial support are required for the production of any book, and this one consumed more than its share of all of them. Every author of the following chapters deserves much praise, enduring as she or he did, repeated requests to rethink, recraft, and rewrite. The range of expertise represented in these pages is proof that interdisciplinary collaboration is possible and desirable, and, when pursuing good environmental restoration, crucial. The Institute of Environmental Sciences at the University of Zurich was the principal sponsor of the initial meeting that brought all contributors together, with the University of Zurich's Botanical Gardens providing the venue. As many participants as authors also attended that 2006 meeting, sharing ideas and offering constructive criticism. To all of these institutions and individuals, together with their supporting institutions, thanks are due. To name names, Bernhard Schmid provided unfailing enthusiasm for the project; Christine Müller offered a stimulating keynote; Dan Tamir and Elena Conti volunteered onsite organizational support; Christian Göldi toured us through Dietikon's Limmat restoration site to demonstrate that words can produce results; ever-present Isabel Schöchli and Lilli Strasser made sure that everyone kept smiling even after returning home. Routledge editors and associates, to include Laura Stearns, Nicholas Mendoza, and Terence Johnson, saw value in the book and guided it to efficient completion. Two anonymous reviewers posed a range of valuable suggestions, most of which found traction in these chapters. And Fabrizio Frascaroli, Keya Howard, and Nuru Kitara provided indexing polish. Restoration is indeed a group project, and a heartfelt thanks goes to all for joining me in the search for a usable environmental past.

Part I
Introduction

1 Introduction

Tempo and Mode in Restoration

Marcus Hall

Give a man a seed and he will grow a tree; teach a man to restore and he will save the planet. But do we restore by growing trees? The carbon-sinkers would have us bring back whole rainforests to reduce global warming as the extra foliage would consume a good deal of carbon dioxide. But these newly planted forests would certainly be different than those that formerly grew on the land. Planting a new forest might restore the earlier function of taking up carbon, but there would be plenty of other functions no longer performed, be they utilitarian or aesthetic. Canopy structures, biodiversity levels, flower aromas, and bird behaviors all may well be different in a replacement rainforest, even after the most conscientious of restoration programs. Other characteristics or functions of the new forest, meanwhile, might be enhanced beyond those found in the old forest, satisfying biocentrists or anthropocentrists, to include improved soil retaining, timber growing, wildlife propagating, and scenery viewing. One might set out to "restore a forest" but some functions will be diminished or lost at the same time that others are recovered or multiplied. Deforested forests can never be brought back, if for the simple reason that those forests are now gone. With these limitations in mind, should we plant new forests to be as much as possible like their former counterparts, or else plant designer forests to satisfy our day's most pressing needs? Should we attempt to re-create forests, or simply create them? Should the past determine our restorative methods, and should the present influence our restorative goals? Can we predict the future of restoration?

This book explores how a consideration of *time* can improve the practice of environmental restoration. Historic ecosystems can serve as models for our restorative goals if we can just describe such ecosystems. Here, the historian's craft, from archival to palaeoarchaeological, can help reveal the composition of former ecosystems: yet, exposing every last detail of an earlier ecosystem will be impossible by any historic method, even if we have numerous old photographs of a site and detailed oral histories of its changes. At the very least, one needs to decide which of the many former snapshots to emulate. Even if an ecosystem's snapshot is not a restoration practitioner's goal, which instead centers on revealing former ecological processes for relaunching former ecological trajectories, historic information can still be useful for describing recurring events, such as the frequency of forest fires or early lists of species. Ecosystems are historic entities that depend on what happened before to become what they are now. Despite attempts by last century's ecologists to discover and describe "true" climaxes and accurate successional communities, ecosystems—like human systems—never exactly repeat themselves. Sites with nearly identical rainfall, temperature, soil, and sunlight may nurture very different plant and animal communities. The past can reveal process, but some of this process may not be made to reappear by any method—natural or human—so that restoring for former process also has its limitations.

Baselines, also called reference states, are useful measures in evolving ecosystems, but here again, a crucial challenge to relying on environmental baselines is selecting and describing such states. In the Americas, flora and fauna changed dramatically after Columbus made his famous voyages so that 1492 is an excellent baseline for those who want to bring back former conditions. After this date, novel species and human immigrants would start transforming landscapes of North and South America. But there were other dramatic turning points in human-induced change, be these the eighteenth century's agricultural revolution, the nineteenth century's industrial revolution, or even the Pleistocene's arrival of *Homo sapiens* to North America some 13,000 years ago. Each of these turning points has been justified as an appropriate restorative goal. For restorationists working in the Old World of Europe, Asia, or Africa, one can envision very different baselines that might depend on human population thresholds, for example, or else settlement events or technological and agricultural breakthroughs, such as the 1910 Haber–Bosch invention of capturing atmospheric nitrogen for adding to fertilizer. Europeans may therefore hold more recent or else more distant restoration goals then their colleagues overseas. Perhaps Americans (and Australians) simply confront less ambiguity in establishing restoration goals, and need not worry about the dilemmas posed by Europe's increasingly popular pursuit of "renaturing"—sometimes called "new naturing"—which can be understood as the process of returning appropriate nature to a site. When Germans *renature* their canalized streams and rivers, as by removing dykes and reinserting meanders, are they assuming fundamentally different historic baselines than when Canadians *restore* their own rivers, as by removing alien species or reintroducing natives? What conditions should be brought back, and do such conditions represent new natures or better pasts—or is an ecosystem's former "health" and "integrity" more important than the physical re-creation of a baseline? Can non-native species belong in properly restored sites? Can *re-wilding* be a legitimate goal in Europe, or is this a Holy Grail better pursued in the New World?

These were the questions posed to an interdisciplinary group of ecologists, geographers, anthropologists, sociologists, historians, and philosophers who met in Zurich, Switzerland, in July, 2006. Their collective answer is given in the following pages, and it is not a unified answer. If restoration in its barest form is retrieving a previous condition, as David Lowenthal points out in Chapter 2, we realize that there have been many earlier conditions and many ways of identifying those conditions. Time's role in restoration reveals that there is also a past of restoration, as well as past assumptions about restoration, and such assumptions have implications for current restoration practitioners that are social and political and not just ecological. It was our meeting's conviction that many amateur and professional restorationists set out to bring back original natural conditions without thinking very hard about their own notions of "original" or "natural." Before we ever reinsert meanders in a dyked river, reflood a drained wetland, reintroduce a native species, weed or cull an invasive plant or insect, or allow a lightning-ignited forest fire to continue burning, we must appreciate that the past and our understanding of it are crucial to successful restoration. Our governments are willing to spend billions on restoration projects—in Florida's Everglades, along the Rhine River, across the South China Sea—without acknowledging that many former ecosystems existed on these sites, that there have been many former ways of understanding such systems, and that former generations have already wrestled with repairing elements of them. The following pages aim to reveal how the consideration of restoration's temporal dimensions can improve our practice of it.

The organizational structure in the upcoming six parts is broadly chronological and topical, by taking the past of restoration as a way of highlighting its current challenges and of outlining its future possibilities. The first part considers *restoration in history*, by presenting several historical case studies, while the second part turns to *history in restoration*, by surmising the role of historical thinking in restoration projects. The third and fourth parts address restorationists' perennial questions about *target states* and then *initial states*: before one sets out to bring back certain conditions (be they wild or humanized or something else), one must first decide on which sites to work on. The restorative process requires selecting conditions one hopes to produce, as well selecting places that need to be restored. In the fifth part, authors offer tentative answers to questions raised in preceding chapters, while acknowledging restoration's complexity and outlining the progress made in understanding it. Implementing good restoration is the subject of the sixth part and conclusion: these last chapters consider the "uses of history" for identifying better restorative practices and improving land management.

For sake of clarifying our arguments, "rewilding" in the following pages is the human endeavor of bringing ecosystems toward untouched conditions, usually ones like those found in pre-degraded pasts; "regardening" is the effort of reproducing desirable humanized conditions and humanized natures; "renaturing" denotes activities pursued to make new natures or better natures, which depend on inherited assumptions and human relationships. Thus, rewilding counteracts the human propensity to degrade, regardening reinforces the human ability to improve, while renaturing positions humans alongside the non-human world for the benefit of both. Distinctions between these restorative types are necessarily messy, and one of the book's larger goals is to explore how these somewhat unconventional terms can contribute to better practices of restoration. Here I review the major themes taken up later in greater detail.

In Chapter 2, David Lowenthal contextualizes the temporal role of restoration by underscoring that time's passage can be cyclical as well as linear. Elements of natural systems and human systems can be viewed as cycling backward as well as marching irreversibly forward: if *restoration* manifests time's cycle, *sustainability* manifests time's arrow. Both pursuits are vital to the way we set out to promote and preserve life on earth. Whether one aims to restore or sustain, Lowenthal believes that we must acknowledge a past ecosystem's future, prehistory's tendency toward diversity, and restoration's value for informing natural and human history.

James Feldman takes us to the shores of Lake Superior and the Apostle Islands to show how forest rewilding and spontaneous regrowth stemmed more from last century's economic and political circumstances than from hands-on restorers in the field. While enlarging the concept of "wildness," Feldman's message is that *restoring* involves much more than ecological issues. Timo Myllyntaus then brings us to Europe and the restoration of Finland's coniferous forests. His study sets up the book's recurring transatlantic (and transpacific) contrasts, so that pre-settlement, pre-industrial, and pre-Pleistocene baselines take on more complicated meanings when located in Europe. Myllyntaus notes that in the nineteenth century, his country's forest managers restored with cultural values in mind (such as productivity) and only in the late twentieth century did they aim to create natural or "near-natural" states, despite sixty centuries of previous human inhabitation. Feldman's story suggests that time's passing may be the only real way to reinstate wild conditions, while Myllyntaus's contribution forces us to reconsider the nature of "original" states.

Mairi Stewart and Althea Davies turn to Britain in order to expose historical assumptions of restorationists. They find that mythic truths are often more important than objective field data for providing information about what formerly grew on the land. Good-meaning restorationists currently plant trees over barren lands where forests never grew, at least in human memory. Stewart and Davies suggest that while land managers need not replicate precise historical conditions, such managers at least need to acknowledge human values—not historic landscapes—as the main inspiration for their projects. Davies, and then Nicola Whitehouse, point to evidence from the deeper past, stemming from pollen deposits in soggy heaths and fossilized insects in prehistoric bogs, to demonstrate that biotic communities varied widely in the intervening centuries, so that restoration goals should not rely on single vegetational states. Depending on climatic patterns, geological and biological processes, as well as human land uses, vegetation has shifted in the northern part of the British Isles from mixed woodlands to wet bogs and back again. It seems that any one single former ecosystemic state should not be favored. But if ecological processes are the restorationist's goal, how is one supposed to reproduce "moving trajectories" that include a potential for producing various vegetational states? Restorers may need to plan on centuries, if not millennia, for regenerating wild heaths and bogs, each one uniquely adapted to its site.

Closed canopy forests may *not* have been the norm in medieval Europe, despite popular opinion to contrary. Frans Vera argues in his contribution that the myth of unbroken deciduous forests in pre-agricultural, mainland Europe (and eastern North America) must be reconsidered in light of evidence for extensive grazing by former wild herbivores and other large ungulates. Etymology as well as ecology, says Vera, provide evidence for shifting assumptions about vegetational baselines, so that many of today's land managers in these temperate areas mistakenly aim to reestablish thick forests instead of woodlands laced with meadows and pastures: if "original-natural" conditions are desired in restored sites, Vera argues that managers should reintroduce grazing analogues (such as Heck Cattle) instead of planting trees. While critics counter that central Europe's medieval ecosystem was itself very different from still earlier, Ice-Age states, one can see that restorationists working on any landscape must be well informed about historic assumptions they have inherited.

Even if former natures are ultimately unknowable and former natural processes can never be exactly repeated, an appreciation for time on the land is still fundamental to those hoping to reinstate natural conditions. Our very notions of *natural*, after all, are derived from those who came before us. In the early twentieth-century, "wild gardens" and "wildflower gardens," for example, became fashionable in the United States and Britain even though such garden plots hardly represented untouched renditions of pristine nature. Rather, these wild gardens reflected the day's idealized notions of wildness, which at the time included colorful flowers in pleasing arrangements that depended on the gardener's tastes. Such gardeners borrowed from (and reacted to) former gardeners, so that even today's "native plant gardens" continue to reflect biases handed down from earlier gardeners. Restorationists always stand on the shoulders of former restorationists. If the aim of restoration is *assisting the recovery of an ecosystem that has been degraded, damaged, or destroyed* (a definition adopted by the Society of Ecological Restoration International), we automatically depend on our forerunners to understand what is meant by *ecosystem*, and how one interprets or measures *degraded*, *damaged*, and *destroyed*.

An ongoing debate for those who assist ecosystems to recover is the extent to which restorationists are simply gardeners with a deep concern for biodiversity and native species. After all, wild gardeners, native plant gardeners, ecological engineers, and naturalistic landscape designers also restore elements of former landscapes. These designers may largely be re-gardening so as to maintain or reinstate previous humanized landscapes. In his chapter, Chris Smout helps distinguish regardeners from rewilders, pointing out that the latter generally avoid designating any historic snapshot as their goal while often working on larger-scale projects. He also notes that in Britain many botanists classify non-native plants as so-called *archaeo-phytes* or *neophytes*, the latter arriving in the British Isles after 1500. A British alien plant's degree of belonging apparently depends on its date of arrival, rather than on its ecological role in the ecosystem—or else on its usefulness to people or their psychological attachment to it. A rewilder in the Scottish Highlands may therefore be able to rely on history, after all, for identifying target flora. Smout believes that nature and culture, both, are crucial to the pursuit of restoration. Both undergo changes through time, and both are understood by us through our changing perceptions of them.

The third part focuses on the challenges of planning restoration, espe-cially through selecting target states. Anita Guerrini and Jenifer Dugan introduce us to a typical degraded site along coastal California that is in need of repair. They find natural as well as human remnants worth sav-ing and enhancing, and search the historical record for clues about former states that might be reestablished: they conclude that it is ecologists and his-torians working together who can produce the best results. Ian Rotherham and Keith Harrison next aim to reveal the condition of a British fens before it was drained, employing palynological and archival analyses for reveal-ing restoration's best goals for the site; their theme is that a rich wetland heritage was lost and should now be recreated. Jan Dizard cautions that restorers must not be merely striving for historical fidelity, as the results will always fall short: one can never predict how an ecosystem would be today if it had never been altered by the human hand. Instead, restorers should rely on historical study and its insights for learning about human mistakes, and for finding ways to avoid repeating them. For Dizard, those who repair the earth should ultimately marvel at nature's ways even more than lament nature's losses.

Restorationists spend a great deal of energy thinking about the natural systems they want to create, but they don't spend enough effort examining their reasons for choosing the site they work on. Should we concentrate on rehabilitating landfills and mining quarries, or else on weedy prairies and drained wetlands? Should we dedicate ourselves to a site near home or to a remote site—or should we concentrate intensively on a small area or superficially on a large area? One site may be deemed more degraded than another, and so more worthy of restoration, but there are many definitions of degradation: soil erosion may be more—or less—serious than infesta-tion by alien species. Human values and human biases inevitably factor into any restoration project. Soil erosion, biodiversity loss, and diminished productivity, moreover, can be entirely natural processes, so that even pristine sites exhibiting these traits may be judged worthy of restorative management.

David Sprague and Nobusuke Iwasaki in the fourth part take up this issue of degradation, and identifying initial states, by asserting that Japanese rice paddies can be critical natural spaces. Re-flooding and re-wetting abandoned rice paddies can bring benefit to wetland creatures as well as human societies. While so many wetlands in Europe were once despised and so were drained, Japan's wetlands were once revered but were then abandoned, and so dried up. Now restorationists around the world set out to bring back their watery landscapes, but for different reasons. Concepts and assumptions, as well as politics and power, determine what gets restored. David Tomblin looks at traditional restoration in North America's Indian lands to discover that White Mountain Apache regardeners and Western rewilders may hold irreconcilable differences. Renewing Indian lands integrates the human element; Western rewilders scrupulously exclude that element.

Next, David Casagrande and Miguel Vasquez juxtapose two very different restoration projects in the United States to show that collective, not personal, decisions are usually key to determining what gets restored. They see promise in *renaturing*, which they interpret as reinstating healthy human and natural relationships, appropriate for lower-income urbanites in Connecticut as well as the Hopi of Arizona. Historical fidelity is integral to these renaturation projects, but so is a community's needs and its ability to participate in both process and product. According to Casagrande and Vasquez, deciding on the target state, as well as the place to be targeted, should be a community process that is negotiated and democratically agreed upon. Lynne Westphal, Paul Gobster, and Matthias Gross then do the Herculean task of categorizing renaturing activities, which span from holistic to partial goals according to whether one tries to bring back a species, a habitat, or a cultural landscape. Their concern is large urban areas in Europe and the United States so that, in their case, historic replacing and narrow rewilding are not options. But here again, historical process in both conception and implementation is crucial to each of their restoration models.

Having considered history *of* restoration and history *in* restoration, and then turning *to what* and *from what* to restore, the last two parts are more concerned with the future of the field. Instead of relying on their own judgments about restoration's target states and initial states, Kathy Hodder and James Bullock ask the interest groups themselves for answers to these questions. They dwell on the potentially enigmatic practice of rewilding, whereby human action is meant to remove the results of human action. Deciding on goals and sites, on species and features, becomes a matter of consensus. Here, rewilding is the greatest good for the greatest number over the longest period of time. Jozef Keulartz and then David Kidner in turn offer philosophical responses to the answer of the human role in restoration. Lacking public surveys, Keulartz summarizes a host of theoretical issues that have arisen concerning the restorative endeavor, while reminding us that we have no choice but to interpret the world through metaphors, and that restoration is a supreme metaphor. The fields of art, engineering, and medicine all provide restorers with ways of understanding degraded and idealized systems, so that converting the former to the latter depends on the restorer's particular background. For his part, Kidner emphasizes that restoration necessarily relies on non-rational thinking to utilize intuition and passion instead of inflexible definitions. There is a psychology of restoration that makes process as important as product, and both are tied to a restorer's relations with other people. One sets out to assist an ecosystem to recover, but *assistance*, *ecosystem*, and *recovery* are all mediated by humans pondering other humans.

Daniel McCool considers three American river restoration projects to show how rewilding, regardening, and renaturing have been adopted to different sites. His is a pragmatic approach that views a river's future needs along a background of political contingencies. Power relationships—between person and person, between individuals and organizations, between local and distant interests, between culture and nature—all form the riparian system on which decisions are made. Eileen Crist then leads us into the future-present of high technology applied to restoration. Creating anew takes on a literal meaning with cloning at the genetic and organismal level: if science can allow us to replicate endangered species, Crist wonders about the implications of a science that can replicate endangered ecosystems.

As a prelude to a conclusion, Josh Donlan and Harry Greene reflect on the aftermath of their controversial Pleistocene Rewilding proposal. They predict distinct advantages to ecosystems in reproducing prehistoric evolutionary processes through the introduction of African lions, cheetahs, and elephants to serve as megafauna analogues for extinct North American cheetahs and mastadons. What Donlan and Greene discovered in their reading public was either immense enthusiasm or else instinctual repulsion for their proposal. These ecologists have been forced to confront the social aftermath of their biological reasoning, which assumes that nature's purest state was the one before humans entered it. Here is surely an extreme view of restoration from which other restoration projects can be judged. Yet it turns out that analogues of extinct species are already being introduced within and beyond North America for reproducing biological processes, so that Donlan and Greene's rewilding scheme is not as fantastical as their responders may at first assume. Pleistocene Rewilding forces us to think again about restoration's optimal target and initial states, about the species that make up those states.

Winding up our query is Eric Higgs' call to restore dirt under our fingernails. Higgs undertakes the critical task of grounding our insights, reminding us that there are real consequences for why, how, and what we restore.

Rampant pollution, pesky invasives, marred scenery, and environmental injustice are not going away, so that assisting the recovery of damaged environments is one of our most crucial pursuits.

Our goal here is not to criticize restoration or to belabor the assumptions on which it depends, much less to offer policy recommendations, but to ask harder questions so that we can better understand and improve this endeavor we all embrace. Growing trees is not always restoration. But growing trees will anchor soil and absorb carbon, and if we can incorporate the historic record while discovering our core values, we can select appropriate tree species, plant them in the right places, and nurture them along with other biota and ourselves, to begin to restore. Ultimately, by *re-storying* nature we can offer richer histories of past environments, and so identify better ways of bringing them back.

Part I
Restoration in History

2 Reflections on Humpty–Dumpty Ecology

David Lowenthal

"O! call back yesterday, bid time return," cries Salisbury in Shakespeare's *Richard II*, bearing the king dire news. Yearning for restoration is age-old. So is faith it will come to pass. "Every city and village and field will be restored, just as it was," foretold a fourth-century ecclesiastic.[1] From divine fulfillment, restoration devolved into human agency. "Not a thing in the past has not left its memories," mused H. G. Wells. "Some day we may learn to gather in that forgotten gossamer, . . . weave its strands together again, until the whole past is restored to us."[2]

Like myriad *re-* words (repair, revert, renew, renaissance, reform), restoration implies going back to an earlier condition, often the pristine original. The previous is held better—healthier, safer, purer, truer, more enduring, beautiful, or authentic—than what now exists. Whether with lost or stolen property, damaged paintings, deteriorated health, reputations damaged by accusation or slander, security from danger, or undermined trust and relationships, the aim is "retrieval of an original favored condition," a status quo ante.[3] As *re-member* and *re-concile* suggest, the purpose is essentially therapeutic: to recoup physical health, to redress a grievance, to mend social wounds. Restoration rebinds what has been *dis-membered* or sundered by *dis-cord*.[4] To restore is to make whole again, in plain defiance of "All the king's horses, and all the king's men [who] couldn't put Humpty together again." "One can no more restore an area of natural beauty—or a painting . . . —to its original state than one can turn women into the little girls they once were."[5] The futility of such efforts made restoration a byword, in English Restoration comedy, for transgressive farce.[6]

RESTORATION UBIQUITOUS AND INNATE

Restorations are legion: forests, rivers, gardens, governments, buildings, furniture, sculpture, paintings, music, medicine, bygone ideologies and dynasties—Confucian precepts, Ten Commandments, the Augustan, Stuart, Bourbon, Meiji restorations. Image restoration ranges from digital repair of blurred or degraded photographs to public relations repair of tarnished reputations of countries or corporations, priests or presidents.[7] Yet restoration realms are seldom viewed in concert. While "literature from . . . history, anthropology, and philosophy tackles issues relating to restoration goals," an ecologist finds "little cross-referencing among these disciplines or with the ecological literature."[8] Few would guess from the journal's title that *Contemporary Esthetics and Restorative Practice* concerns teeth, or that dentistry debates over aggressive versus conservative restoration, and stability versus aesthetics, mirror other restoration realms.[9] *Restoring Nature* is subtitled *Perspectives from the Social Sciences and Humanities*, its authors are mainly philosophers and social scientists, but its topics are entirely environmental.[10]

Paul Eggert's *Securing the Past* compares painting, sculpture, and building conservation with textual editing. This "first concerted effort to examine together the linked philosophies" of building and art restoration with literary works notes that editors do not physically alter what they restore, whereas historic house, painting, and sculpture conservators irreparably change objects. But all share traditions of misguided confidence in definitive and non-intrusive restoration.[11] Absent from Eggert's purview, however, is any mention of ecology. In response to Marcus Hall's plea for more conversations between restorationists of nature and of culture,[12] I here discuss their intertwined histories.

Restoration is instinctive. Young children exhibit faith in restorative powers that rejoin things broken, rejuvenate the old, bring the dead back to life. Nothing irretrievably wears out, no act is irreversible.[13] To restore someone (or something) to what it was before harm was done is not only achievable but obligatory. The child feels responsible for causing the injury and must make amends—reparation.[14] Gradually we curb our own and others' restorative powers. Yet the urge to recover remains compelling: residues of restorative faith suffuse thought, speech, and behavior. Like the Victorian poet, we call back yesterday, but it never returns. "Backward, turn backward, O Time, in your flight, / Make me a child again just for to-night!"[15]

To shed adulthood's shackles was Thomas Traherne's seventeenth-century dream, escaping "the dirty devices of this world" to replenish the "sweet and curious apprehensions" of childhood's restorative innocence:

> All things were spotless and pure and glorious . . . I knew not that there were any sins, or complaints, or laws. . . . Everything was at rest, free and immortal. I knew nothing of sickness or death. . . . All time was eternity, and a perpetual Sabbath. . . . Boys and girls tumbling in the street, and playing, were moving jewels. I knew not that they were born or should die. But all things abided eternally . . . The city seemed to stand in Eden.[16]

Traherne's dream mirrors his clerical contemporary Thomas Burnet's accolade to "Providence; which loves to recover what was lost or decayed, . . .

and what was originally good and happy, to make it so again." Like Burnet's restored primordial Earth, "smooth, regular, and uniform, [with] not a wrinkle, scar or fracture in all its body,"[17] Traherne's childhood Eden is a divinely ordered paradise. Restored to Traherne is no unadulterated world but a prelapsarian realm not of nature, rather *spiritually* natural.

Traherne and Burnet followed Renaissance and Reformation zeal to restore ancient pasts. Renaissance humanists sought to reverse the medieval retrogression that had obliterated the Classical legacy. Religious reformers sought to cleanse the church of the Satanic corruption that had defiled Christianity's original innocence. Both evils must be expunged to retrieve the pure source of inspiration, Christ and the classics. Restoring the Golden Age—the primitive church, the classical vision—were humanists' and Protestants' parallel and often conjoined aims.[18] Their restorations were likened to archaeology and medicine.[19] Just as antiquarians pieced together imperial Rome from vestiges of temples and statuary, scholars collated remnants of classical "unearthed fragments." Reuniting such fragments was an act of healing. Rediscovering Quintilian's previously "mangled and mutilated" works restored him "to his original dress and dignity and . . . sound health."[20] Humanists were doctors restoring lacerated heroes— ancient exiled texts—to honor and safety.[21] In his life-work of restoring theology, Erasmus, unwell, hoped the physician Paracelsus might "restore me also."[22]

TIME'S CYCLES AND ARROWS IN TERRESTRIAL AND HUMAN HISTORY

We sense time both as circle and as arrow. Time's circles are the ceaselessly recurrent cycles of nature's constancies: the waxing and waning, ebbing and flowing of diurnal, lunar, and seasonal rhythms and planetary orbits, along with our everyday breathing and heartbeats, sleeping and waking. Time's cycles generate thoughts of restoration.

Time's arrow subverts and denies restoration. The arrow flies only once from the irrecoverable past toward the foreign future, never again the same. The targets of time's arrow are the contingent events and sporadic vagaries of human and natural history, a temporal dimension distinct from natural law's clockwork time. The interplay of circle and arrow continually shapes our lives. Habitual customs—lawlike, regular, predictable—interact with the uncertainties of history's directional, singular events. As biology conjoins repetitive analogy with ancestral homology, so society overlays rhythmic stability with novel disturbances.[23]

Past and future long seemed much alike, however. In traditional societies, lived time was more circle than arrow, both natural and human annals overwhelmingly repetitive. There was no new thing under the sun (Ecclesiastes 1:9). The fixity of divine and natural law and the enduring sameness of human nature showed what had happened before would happen again. Resurrection and re-enactment suffused religious creeds. Secular chronicles were repetitive: like Earth reborn every new year's day, each ruler's inaugural year restarted the calendar at year one. (Not until the eighteenth century was regnal dating replaced by the global chronology that subjects us all to time's arrow.) Politics too were recurrently repetitive. Monarchy led to tyranny, then to aristocracy, oligarchy, democracy, and anarchy, followed by restoration to monarchy.

Time's cycle dominated geology into the nineteenth century. James Hutton limned Earth's saga, hugely lengthened by him, "as a stately series of strictly repeating events, the making and remaking of continents as regular as the revolution of planets." Continental uplift recurrently restored matter washed away from the land. Hutton's terrestrial uplift, erosion, deposition, back to uplift mirrored the political monarchy–aristocracy–oligarchy–democracy–anarchy–monarchy cycle.[24]

Charles Lyell depicted Earth cycling in eternal steady state, every destruction renewed. Singular events—floods, earthquakes, comets—were trifling local aberrations amid overarching uniformity. Lyell posited wholesale restoration. When global warming resumed, "then might those genera of animals return, of which the memorials are preserved in the ancient rocks." Out-aging today's Pleistocene rewilders, Lyell fancied "the huge iguanodon might reappear in the woods, and the ichthyosaur in the sea, while the pterodactyl might flit again through the umbrageous groves of tree-ferns."[25] Only after 1860 did fossil and artifactual finds, Darwinian evolution, and Marshian ecological history persuade Lyell, with utmost reluctance, to admit progress in animate life, to include mankind in the history of nature, and to give up time's restorative terrestrial cycle for time's directional arrow.[26]

RESTORATION DENIED: GOSPELS, ENLIGHTENMENT, EVOLUTION, ENTROPY

Time's arrow took early flight in Judeo-Christian annals. The Old Testament chronicled a contingent history following one-time Creation and Adam's Fall; the New Testament inserted into secular annals several unique sacred events, the birth, life, death, and resurrection of the son of God, and His eventual Second Coming. Judaism and Christianity posited a singular flow of time, events stemming from divine or human will happening once and only once. But such temporal consciousness was absent from the habitual regimen of ordinary life, rare even in scholarly thought, and remained uncommon until the eighteenth century.

Enlightenment savants began to see human affairs not as repetitive or cyclical but infused by ongoing improvement. Innovation, once a threat to settled order, became a welcome harbinger of progress. Far from being impious, secular advance fulfilled scriptural commands: God left the world unfinished for man, made in His image, to perfect. Occasional setbacks might occur, but not wholesale regress; progress was cumulative, each generation building on prior advances.

The shift from restorative-laden circle to directional arrow transformed Western societies from the late eighteenth century. Stasis gave way to linear progress; preordained *life cycles* became self-fashioned *life courses*.[27] New World Manifest Destiny bore out foretold Enlightenment improvement, alike attractive and assured.

Natural history too was transformed. Cyclic regularities were not wholly abandoned. But they were interrupted, accelerated, or retarded, warped into new trajectories by episodic catastrophes—asteroid impacts, tsunamis, heat-occluding volcanic dust. These one-off dislocations reshaped continents, leaving in their wake novel landforms, plants, and animals—incontrovertible proof of time's arrow. Unique and unrepeatable natural history did not conflict with biblical faith, which likewise embraced a unique and unrepeatable history. The Christian saga was indeed a major stimulus to awareness of nature's contingent events. Earth like human history progressively shed perennial regularities for sporadic singularities.[28]

Evolutionary biology reinforced the shift from circle to arrow. Competitive selection was, to be sure, constantly recurrent, but it led to ever-novel conditions, continually extinguishing old and engendering new life-forms. Nothing struck (indeed, dismayed) Darwin more forcibly than the absence, amid the teeming variety of extant species, of virtually all previous ones. His ineluctable doleful conclusion was that "not one living species will transmit its unaltered likeness to a distant futurity."[29] Darwin's "intolerable thought that [man] and all other sentient beings are doomed to complete annihilation" was memorably echoed, the very year *On the Origin of Species* appeared, in Edward FitzGerald's *Rubaiyat*:

One thing is certain, that life flies;
One thing is certain and the Rest is Lies;
The Flower that once has blown for ever dies.[30]

Even more sobering than biological carnage was the running down of the entire universe, notably the impending heat death of the sun. "Within a finite period of time to come," Kelvin warned of entropy's grim implacability, "the earth [would become] unfit for the habitation of man." Indeed, none of its inhabitants could "continue to enjoy the light and heat essential to their life, for many million years longer"—in Helmholtz's accredited estimate, about 17 million.[31] Fears of the sun's non-reappearance underlay much primitive myth. However remote, these new forecasts sorely undermined faith in nature's everlasting stability. The withdrawal of the sun become, with Freud, a classic symptom of paranoia. Humanity's age-old terror lest daylight not be restored remains part of today's mind-set.[32]

RESTORATION REDUX: NOSTALGIC
REACTION TO HISTORY'S TERRORS

The anxieties unleashed by time's arrow went beyond mordant aware-ness of evolutionary extinctions and the second law of thermodynamics. Especially distressing was the lapse of cyclical rhythms in human affairs, the replacement of clockwork regularities by volatile uncertainties. "I shot an arrow into the air / It fell to earth, I knew not where."[33] Longfellow's couplet said what many felt: time's straying arrow left outcomes uncer-tain, unforeseeable, bewildering—*wildness* meant *confusion*. The arrow's trajectory mainly exposed mankind's crimes and errors. From the French Revolution on, accelerating historical change fuelled fears of society spin-ning out of control. "The series of events comes swifter and swifter," wrote Carlyle, "velocity increasing as the square of time."[34] Writers from Ten-nyson and Hardy and Ruskin to Spengler and Wells likened the decline of the West to the death of the universe. Fin-de-siècle forecasts of universal winter, reprieved by Rutherford's 1904 discovery of the sun's radioactive energy, resurfaced as fiery fate by nuclear fission. The Bomb and its dread progeny have since made doom-laden prognoses common coin.[35]

Such fears intensified nostalgia for earlier times, for restoration to a blessed state when change was slow, cyclic, or imperceptible. Haunted by Mircea Eliade's "terror of history,"[36] men averted their gaze from tempo-ral chaos in vain yet imperishable hope of restoring intelligible, depend-able certitudes. Against revolution made fearsome in regicidal England and France, Stuart and Bourbon monarchical restoration promised time-honored security.[37]

RESTORING WHAT FEELS NATURAL, RESTORING NATURE

Regime restorers eagerly deployed analogies with nature. Edmund Burke termed nature's stability the antidote to the novel turmoils of the French Revolution. Ingrained custom was the "natural" moral propensity: states-men should emulate nature's cyclical constancy. "By preserving the method of nature in the conduct of the state, in what we improve we are never wholly new, in what we retain we are never wholly obsolete." In a com-monwealth respectful of tradition, no radical break disturbed life's regular rhythms, and subjects rested content with ancient authority. Restorers in seventeenth-century England had "regenerated the deficient part of the old constitution through the parts which were not impaired. They kept these old parts exactly as they were, that the part recovered might be suited to them."[38] Burke's aim to minimize change in human affairs strikingly fore-shadows the tectonic gradualism of Lyell, for whom slow and insensible mental progress likewise mirrored nature's own way.[39]

To repel convulsive innovation, traditional custom was extolled as "natural." The agenda was not, however, restoration to Hobbes's reviled state of nature, stripped of civic improvement. Domesticating raw nature—building shelters against climatic extremes, cooking food to make it clean and safe, storing grain and oil and livestock against dearth or famine—was "natural." Restoring the fabric of nature itself gained favor in Victorian reaction to industrial blight and urban squalor. But like biblical Eden, English "nature" thus redeemed was agrarian or emparked or gardenesque, thoroughly humanized, intensely managed. Bucolic nature was perfected by cultivation and control, idealized in the canvases of Claude and Poussin, the verses of Goldsmith and Wordsworth.

First to exalt untouched nature as Edenic restoration were late nineteenth century Americans, aghast at their land's lost purity and vanishing wilderness.[40] Their icon was Longfellow's forest primeval, not a Wordsworthian garden. "One is nearer God's heart in a garden / Than anywhere else on earth," penned Dorothy Frances Gurney in England; across the Atlantic her "garden" literally morphed into "forest."[41]

The wilderness cult transformed symbols of triumphant conquest into emblems of horrendous despoliation. Once the logger's ax and the hewn stump had stood for civilized progress; they now betokened the rape of nature.[42] In New York's Central Park the landscape architect Frederick Law Olmsted "planted trees to look like 'natural scenery' with such success that those who accepted "the scenery as 'natural', objected to cutting the trees he had planned to cull."[43] Olmsted's success in concealing traces of human intervention heralded landscapers whose task, Hall writes, was "to create just the right look of human-free nature in each national park."[44]

Restoration as rewilding featured a 1908 best-seller set in Tennessee's Cumberland Gap. Fouled by soulless loggers, a once crystal-clear stream is now "black as soot" and choked with sawdust. Tree-felling, "the cruel deadly work of civilization, [meant] a buzzing monster . . . biting a savage way through a log, that screamed with pain as the brutal thing tore through its vitals." Our hero, a mining engineer turned nature-lover, vows to restore Lonesome Cove:

> "I'll tear down those mining shacks, . . . stock the river with bass again. And I'll plant young poplars to cover the sight of every bit of uptorn earth . . . I'll bury every bottle and tin can in the Cove. I'll take away every sign of civilization . . ."
>
> "And leave old Mother Nature to cover up the scars," says his fiancée, June.
>
> "So that Lonesome Cove will be just as it was."
>
> "Just as it was in the beginning," echoed June.
>
> "And shall be to the end."[45]

Restoration redeems all: corporate greed vanquished, machine-age poisons excised, nature left to heal, Edenic plenitude in everlasting tranquility.[46]

Restoration ecology's redemptive bent reflects biblical tradition. Repair "to the same state again" accords with Burnet's "methods of Providence." In the rapturous second Golden Age of his *Sacred Theory of the Earth* (1691), all lands will be "restored to the same posture they had at the beginning . . . before any disorder came into the natural or moral world."[47] Two centuries later, another English cleric took heart that St. John's Revelation placed "the restoration of man and the restoration of nature . . . side by side."[48] Rebutting Lynn White's 1967 assault against Judeo-Christian environmental abuse—"Christianity is the most anthropocentric religion the world has seen"—restoration theologians urge us to bring the created world as close as possible to that perfect restoration for which God has destined it.[49]

DECAY AND RESTORATION IN THE ARTS

Restoration in architecture, sculpture, painting, and music likewise reflects changing notions of originality, authenticity, purity, and sustainability. Around 1800 the visual arts began to exchange classical completeness for romantic fragmentation—letting, even helping, things decay. Before, antique sculptures were seldom exhibited or marketed unless restored, however dubiously; that heads and limbs badly mismatched torsos mattered little. After, breakage and mutilation proclaimed a work's age and fidelity to its origins. Contradicting restoration, marble limbs were hacked off, coins and canvases artificially aged, buildings artfully disarrayed, furniture distressed. Picturesque ruination, natural and contrived, patinated buildings and gardens with broken stones, moss, and lichen. The cult of ruins extended to literature and to life itself, memoirs typically termed "Fragments," suicide invested with the pathos of genius truncated.[50]

Eternal verity, not mortal decay, inspired Victorian church restorers. Dilapidated medieval structures were restored not to how they had been, but how they *should* have been. Thousands of French and English cathedrals and churches were antiquated, with nineteenth-century materials and technical skills, back to "pure" Gothic. Aesthetic piety replaced ancient builders' errors and imperfections with idealized archaisms. "To restore a building," proclaimed Viollet-le-Duc, "is to reinstate it in a condition of completeness that could never have existed at any given time."[51]

Scrape bred Anti-Scrape, the stance immortalized by John Ruskin and William Morris that forbade all intervention. Renovation violated the original. "Restoration is impossible," protested historian-archivist Francis Palgrave.

> You cannot grind old bones new. You may repeat the outward form (though rarely with minute accuracy), but you cannot the material, the

bedding and laying, and above all the tooling . . . There is an anachronism in every stone. . . . The sensation of sham is invincible.[52]

Medieval buildings were "monuments of a bygone art, created by bygone manners, that modern art cannot meddle with without destroying."[53] Restoration was not only implausible but impious. Old edifices like living beings deserved daily care, not artificial rejuvenation. Venerable buildings, like Burke's ancient institutions, should be altered as little as possible.[54]

Revealingly, ecclesiastical restorers often reiterated these precepts and were shocked when shown to have contravened them. George Gilbert Scott sought "the least possible displacement of old stone," removing only "features which have actually been destroyed, [because] an original detail [however] decayed and mutilated [was] infinitely more valuable than the most skilful attempt at its restoration." Notorious for replacing surviving Norman with neo-Gothic, Scott confessed to slippage between belief and behavior. Having "restored" a dilapidated fifteenth-century chapel at Wakefield, Scott was later "filled with wonder how I ever was induced to consent to it at all, as it was contrary my own principles. I think of this with the utmost shame and chagrin."[55] The architect G. E. Street likewise flouted his own precept that "in dealing with old buildings . . . we cannot be wrong in letting well alone." He reproached restorers of Burgos Cathedral and St Mark's, Venice, for the same intrusive meddling that led Street himself to replace a fourteenth-century choir arm of Dublin's Christ Church Cathedral with a poor pastiche of the original.[56] Yet neither Scott nor Street were hypocrites. They believed they were restoring the true past, while manufacturing simulacra.

Anti-scrape tenets long ruled European art and architecture. When forced to intervene lest a building collapse or a painting perish, conservators stressed fidelity to its original state. Restoration was a last resort. Honoring the initial structure meant expunging traces of previous ill-conceived restorations, focusing all attention on what was original. Unavoidable replacements were in contrasting textures and colors glaringly distinct from the old. But these blatant disjunctions destroyed aesthetic unity and dimmed the aura of antiquity.

Hence some later restorers opted to stress "original" *aesthetic* qualities, stripping off marks of age and wear, while others emphasized *venerableness*, antiquating paintings with patinas of varnish, buildings with lichen. To retrieve supposed initial intent or impressions, many Old Masters emerged from restoration with varnish removed and colors gleaming as if new-made. Either way, aesthetic taste overruled anti-scrape purism. Replacements chosen to match, not clash with, original elements, were detectable as new only by close inspection of tooling on stonework, dates on stained glass, tints slightly differing.

The restoration history of Giotto's fourteenth-century *Life and Miracles of St. Francis*, in Santa Croce, Florence, typifies changing criteria. Partly destroyed and whitewashed over in the eighteenth century, the frescoes were restored in the mid-nineteenth century by Gaetano Bianchi, who repainted Giotto's scenes, interpolating pseudo-Giottoesque figures in lacunae. Mid-twentieth century restorers condemned Bianchi's "forgeries," expunged by Leonetto Tintori to highlight the remnant original. Critics "praised the recovery of the 'true', albeit fragmentary state" for revealing Giotto's "original intent." But the 1970s repudiated Tintori's purist restoration as "optically disruptive" and exalted aesthetic unity. Following Cesare Brandi's dictum for "a visible dialogue between past and present," restorers recaptured the frescoes' original significance as "Franciscan stories with deep Christian meaning."[57]

Restoration increasingly embraces artifacts' total history. Worth inheres not only in original materials, forms, and intentions, but also in the attritions of time and the interventions of collectors, curators, conservators, even thieves and forgers. Valued objects continually accrue new meanings and values, altering and shedding older ones. Prizing the palimpsest means respecting previous restorations, historically significant for, often visually essential to, the surviving original. James Wyatt's long-reviled (and later replaced) restoration at Hereford Cathedral would now be seen as "integral to the building's history, no more to be demolished than the medieval masonry behind" it.[58] "What de-restoration achieves" may be less informative, notes a sculpture historian, than what can be learned from the restoration itself.[59]

Restoration conflicts ceaselessly embroil conservators, curators, and the public. Promoters celebrated the 1990s' renovation of the Sistine Chapel and Michelangelo's *Last Judgment* as a "Glorious Restoration," freeing the frescoes of five centuries' accumulated grime, crude repainting, and earlier restorers' darkened glue size, and revealing Michelangelo's coloristic genius. Critics termed it a Chernobyl-like disaster, voiding the frescoes of divine inspiration and destroying them forever.[60] For restorers, removing the veil of time, stripping away disfiguring history, recovering "full chromatic effects" dispelled the nineteenth-century myth that Michelangelo was "a black and melancholy artist." Critics, to the contrary, believed the expunged veil contained Michelangelo's *a secco* shadow-and-chiaroscuro finishing, the darkening intended, his art tonal not chromatic.[61]

Difficulties of determining intention—the restoration may have revealed Michelangelo's first creative burst at the expense of removing his second thoughts—afflict most restorations. Creators change intentions as they go on, and often again after finishing. "Any artist's intention," warns an art historian, is "a complex and shifting compound of conscious and unconscious aspirations, adjustments, re-definitions, acts of chance and evasions."[62] Authorial aims are further complicated by patrons, clients, collaborators, publishers, editors, creditors. With their interventions the creator may or may not agree, but must nonetheless contend, lest his work remain unfinished, unsung, or unsold.[63]

But injunctions about original structures, original intent, original any-thing are fast eroding. Rigid restoration strictures become ever less tenable. That no assemblage, no structure, no image can be returned to an original or any previous state is ever more evident. Restorations at best approximate or suggest what once was. And every restoration is filtered through and tinctured by irremediably modern minds.[64] Anachronism is unavoidable.

Early music restoration offers a case in point. Enraptured by rediscov-ered melodic marvels, twentieth-century devotees retrieved original scores, instruments, acoustics, performance modes, audience habits. Some claimed to perform pre-classical music as it had been played, others as it should have been. Whether authentic notation, original instruments, composers' intentions, or listeners' expectations mattered most was hotly debated. But period recitals could not replicate past circumstances. Vocal restorations, for example, would require modern castrati, gelded in childhood, and boy sopranos with voices still unbroken at sixteen, yet mature enough both to master early music complexity and to "enter . . . into the spiritual world of their forebears five centuries ago."[65] By the 1980s, restoration to any *Ur* condition came to seem impossible, restoring in toto preposterous. How-ever faithful a recital to sixteenth- or seventeenth-century origins, modern ears, accustomed to modern sounds and unused to the older ones, necessar-ily hear music differently. One might reconstruct, revive, rebuild, but not retrieve the musical past. "Restoration" is now understood as *revitalizing* early music, for its own sake and for enhancing the mainstream musical canon it led to.[66]

DIVERGENT THEORY AND PRACTICE
VIS-À-VIS NATURE AND CULTURE

Restorationists in every domain borrow one another's language. Aldo Leo-pold's maxim for conserving biota, never to "discard seemingly useless parts, [for] to keep every wheel and cog is the first precaution of intelli-gent tinkering,"[67] echoed Burke's constitutional restorers who "kept these old parts exactly as they were" and Scott's "least possible displacement of old stone." Ecologists' metaphors moved from homeostatic self-regulating machines in engineering and cybernetics to art and aesthetics, thence, in tune with evolving images of man as nature's steward rather than master, to medicine and health care.[68]

As guardians of culture shifted from remaking things whole, to rever-ing original fragments, to recreating palimpsests, analogous impulses led environmental stewards from regenerating gardens, to restoring degraded landscapes, to rewilding, and to processual concerns. Insights from art and architecture served cautionary functions. "Just as faked art is less valuable than authentic art," warned an environmental philospher, "faked nature is less valuable than original nature."[69] The Sierra Club's David Brower famously likened Bureau of Reclamation plans for damming the Grand Canyon to improve tourist access to "flood[ing] the Sistine Chapel so tourists can get nearer to the ceiling."[70] How far can or should damaged landscapes be returned to their "original" state is termed the "Sistine Chapel Debate."[71]

Yet these parallels are far from equivalent. The heritage of nature is perceived and appraised differently from that of culture, arousing unlike restoration aims. Restorers of culture revert either to a specific moment—of inception or creation, of peak prowess or beauty or fame, of some iconic person or event—or to an agglomeration of epochs or lifespans. Neither suits nature restoration: no natural assemblage had either creator or specific moment of creation, and all keep changing. Although they are palimpsests, their diachronic historicity is invisible to all but expert eyes. Paintings, plays, creeds, codes of law get restored to whenever is felt to have been most new and fresh, most effective, most intelligible, most admired, most sacred. None of these criteria apply to ecological restoration, for most favored times of nature antedate human existence. Instead, landscapes are restored back to when least deranged by human action, most ecologically diverse, or most systemically stable and sustainable.

Cultural legacy is prized mainly as individual items created over finite periods of time—Lascaux, Stonehenge, Parthenon, Chartres, Monticello, Gettysburg, Mona Lisa, a First Folio—each made noteworthy by some specific person, event, or quality. Contrariwise, nature's legacies are mainly aggregates engendered over eons; single plants or animals matter less than clusters, swarms, or herds, species, genera, or ecosystems. Exceptions abound: Niagara Falls and Yosemite are more valued for their singular scenery than as examples of generic waterfalls and canyons. But restoring culture stresses unique particulars, restoring nature composite amalgams.[72] Indeed, some deplore dwelling on specific endangered species; these doomed relics deflect concern from the paramount need to restore natural selection in general, currently imperiled by human "unintelligent design," as the vital engine of evolutionary diversity.[73] And unlike artifacts and works of art, restoring particular species or discrete reserves impinges beyond their boundaries. River restorers must consider downstream impacts; reintroducing wolves to national parks must deal with distant ranchers' fears; transgenic restorers of pathogen-devastated American elm and chestnut must address species alteration, ecosystem imbalance, and food-chain worries.[74]

Nature restorers currently find more fitting analogies with medicine and health care, integrating ecosystem management with individual and community healing. Just as a prosthetic leg aims "to rehabilitate the function of leg rather than to recompose original flesh and bones," the restoration ecologist's concern is ecosystem function, not composition—"not just certain species, communities, or habitats, but all natural and anthropogenic flow processes."[75] And an urban ecologist claims "people . . . relate to the analogy of restoring the human body and are intrigued by the similarities between the two processes." Both are systems with many interacting parts; "environment is not something to be passively fixed like a car but instead is actively healed like the human body." Radical ecosystem intervention is major surgery, raking up stream and shore debris recalls aspirins for headaches, band-aids for cuts and bruises.[76] Environmental psychologists credit psychic health restoration to recuperation in unstressful natural surroundings. And those so restored in turn promote efforts to restore nature.[77]

Like physicians, ecologists say, they give nature a helping hand. Ecological restoration is likened to setting a broken bone: the healer resets the trajectory, but nature does most of work.[78] But a fundamental distinction remains. To ecologists "nature itself is the best restorer of all," as John Cairns remarked anent the speedy revival of sinuosity in the no longer channeled Kissimmee River, and the return of fish species to the unpolluted tidal Thames[79] Few physicians so glorify non-intervention. "Curing sick watersheds," noted a botanist, "is not unlike the responsibilities demanded of doctors."[80] Nor is it entirely like it. Dying landscapes are held to require terminal care like dying people; ecosystem health even figures in medical schooling.[81] But human death is terminal, whereas landscapes endure, albeit in altered and transformed states.

Artifacts and institutions are created by human agency and designed for human purpose; nature has no design. Restoring a work of art alters an existing artifact; restoring a natural environment turns it into an artifact. Hence restoration's "big lie": "Artifactual restored nature is . . . fundamentally different from natural objects and systems."[82] To be sure, most restorations aim to repair already degraded artifactual nature; no place on Earth remains untouched by human agency. But we readily conceive predisturbance nature and exalt nature freed from human impress; artifacts cannot be so conceived. Nothing in the arts resembles the guilt-ridden duty of restitution that drives ecological restoration.[83] Geriatric medicine offers no parallel to the spiritual epiphany of ecological restoration.[84]

Immaculate purity, once divine and saintly, now idealizes nature more than culture. Nature is widely held to deserve autonomy, to be unshackled, left to its own devices. That nature can and should repair itself—an idle fancy for humans dependent on agriculture, architecture, antibiotics, reservoirs, sewage systems—is scientifically untenable. Ecology half a century ago abandoned equilibrium models that equated non-interference with environmental health and stable climaxes, yet many ecologists still elevate nature over culture, deploring humanity's imprint as retrogression from the untouched fundament. In the very book that launched UNESCO's cultural landscapes program, essay after essay terms culture a menace to nature and ranks anthropogenic below pristine landscapes—even when agreeing that none *are* pristine. A prime criterion for cultural World Heritage designation is "harmony with nature."[85] Public faith in the beneficent stability of untamed nature persists, in denial of all experience of natural disasters. Hence the contrariness of gardeners who denounce exotics while happily cultivating them, and of rewilding devotees who prefer nature restored with no sullying trace of human agency.

Stewards of nature and culture alike know that human meddling is ubiquitous and unavoidable, but they react in opposite ways. The cultural custodian accepts intervention as normal and necessary, the custodian of nature feels it reprehensible and masks or conforms it to natural outcomes. Culture is ours to tamper with; nature, increasingly, is not. Buildings and works of art are, to be sure, impermanent creations that many feel should be left to age and perish at some natural tempo. But like physicians, most cultural heritage managers now eschew the hands-off stance as untrue to history and unacceptable to their clients. In contrast, even remedial disturbance to nature seems distressing, notably to Americans.[86]

Consider public reactions to the Tower of Pisa, to Avebury, and to Yellowstone's Old Faithful geyser, each recently newsworthy for restoration to supposed stability. Few demurred against lifting the Pisa campanile, whose collapse was imminent, back to its historic incline; though nature caused it to lean, the tower is a human artifact. Underpinning and uprighting precariously tilted sarsens in Avebury's prehistoric stone circle likewise evoked widespread approval.[87] With Yellowstone it was quite the contrary. Public outrage met a Procter & Gamble television ad that a dose of their laxative made Old Faithful's erratic eruptions 'regular' again. Nature was *ipso facto* 'faithful', butting in sacrilegious.[88]

Yet in ecology, as in art, what restorers do is often at odds with what they think or say. River restoration practitioners and stakeholders in 36 countries were asked what restoration meant.[89] Four out of five opted for Cairns's strict canonical definition, "complete structural and functional return to a predisturbance state."[90] But what they actually did embraced not just *return* and *recovery* but *improving*, even *creating*. "Most activities the restoration community undertakes are actually recovery, rehabilitation or enhancement," in accord with a British ecologist's "opportunity to create something new and more valuable than what was there originally."[91] While declaring adherence to "the most . . . widely accepted definition in the restoration literature," they regularly transgressed it.[92] They willingly manipulate, but—like Victorian architectural restorers—are reluctant to admit that they do so. "We bury our efforts beneath an ecological cover," concludes Eric Higgs, "and pretty quickly a landscape that depends on or originates in extensive human contrivance becomes naturalized."[93] Hence ecologists' charges that landscape restoration is "gardening dressed up with jargon to simulate ecology," "quite literally, agriculture in reverse," or "a fiction" best renamed landscape architecture.[94] "Restoration is fencing, planting, fertilizing, tilling, and weeding the wildland garden: succession, bioremediation, reforestation, afforestation, fire control, prescribed burning, crowd control, biological control, . . . and much more."[95]

Much more means restoring not just ecosystems but "the human communities that sustain and are sustained by" them. And to realize that "ecosystem values will shift over time as they have been doing throughout history," restorers must work with "kindred, intellectual adventurers" in the social sciences, arts, and humanities.[96]

EMBRYONIC RESTORATION

Christian theologians envision restoration not as completed deed but ongoing preparation for future redemption. Anticipatory restoration appeals to visionaries concerned about potential apocalypse. We cannot know what future generations may want, but we can anticipate what they may need to recover from global calamity. To regenerate a viable ecology or a workable society, access to records, ideas, and techniques of civilized history could be crucial. The earliest known encyclopedic time capsule was begun in seventh-century China. Fearing imminent cataclysm, Buddhists at Cloud Dwelling Monastery near Beijing inscribed the tenets and history of their faith on stone tablets sealed up in caves for the instruction and salvation of possible survivors. Monastic successors added more inscribed tablets, numbering almost ten thousand by the twelfth century, for humanity's potential rehabilitation. "After the days of doom, the [tablets] would emerge from the earth," to edify future people.[97]

Feeding remote posterity is the aim of the latest prologue to restoration, the Svalbard seed bank in arctic Norway. Seeds of three million plants and animal DNA samples will be stored in sub-freezing safety, so that survivors of global catastrophe might restart agriculture.[98] Others envisage the moon as a fail-safe DNA storage locker. Lacking air, the moon is erosion-free and far enough from Earth to escape contamination from nuclear fallout.[99] Safer and more durable still is deep space. The science-fiction-inspired Alliance to Rescue Civilization plans to launch a rocket packed with data about Earth and its inhabitants into deep space, for far future retrieval in some remote galaxy.[100] Restoration as potential and incipient, as seed not fruit, as data not deed, is ongoing and cumulative, inviting incremental enrichment by future generations. No museum exhibit fixed in amber, envisaged restoration draws on creative drive along with conserving instinct.

Lauding nature's cyclical restorations, Lyell rejected them in human affairs. He ridiculed ancient mythic reiterations wherein "the same individual men were doomed to be re-born, . . . the same arts were to be invented, and the same cities built and destroyed." In lieu of physical restoration Lyell celebrated rediscovery. Delving into Earth's past, he borrowed a celebrated aphorism from Barthold Niebuhr's *History of Rome*: "He who calls what has vanished back again into being, enjoys a bliss like that of creating."[101] Indeed, since reliving the past was a power the ancients themselves had lacked, "to restore great things is sometimes not only a harder but a nobler task than to have introduced them."[102]

NOTES

1. Bishop Nemesius of Emesa, *On the Nature of Man*, quoted in G. J. Whitrow, *The Nature of Time* (London: Penguin, 1975), 17.
2. H. G. Wells, *The Dream* (London: Collins, 1929), 236.
3. R. A. Duff, "Restorative punishment and punitive restoration," in Lode Walgrave, ed., *Restorative Justice and the Law* (Cullompton, England: Willan, 2002), 82–100 at 84.
4. Brian A. Weiner, *Sins of the Parents: The Politics of National Apologies in the United States* (Philadelphia: Temple University Press, 2005), 116; Charles S. Maier, "Overcoming the past? Narrative and negotiation, remembering and reparation: issues at the interface of history and the law," in John Torpey, ed., *Politics and the Past: On Repairing Historical Injustices* (Lanham, MD: Rowman and Littlefield, 2003), 295–304.
5. Midas Dekkers, *The Way of All Flesh: A Celebration of Decay* (London: Harvill, 2000), 94.
6. Steven McElroy, "The comedy is Restoration, but the sex is timeless" (review of David Grimm, "Measure for Pleasure"), *New York Times*, 7 Mar. 2006, B3.
7. William L. Benoit, *Accounts, Excuses, and Apologies: A Theory of Image Restoration Strategies* (Albany, NY: SUNY Press, 1995); Brett A. Miller, *Divine Apology: The Discourse of Religious Image Restoration* (Westport: Praeger 2002); Thomas L. Pedigo, *Restoration Manual: A Workbook for Restoring Fallen Ministers and Religious Leaders*, 5th edn. (Colorado Springs: Winning Edge, 2007); James Kauffman, "When sorry is not enough: Archbishop Cardinal Bernard Law's image restoration strategies in the statement on sexual abuse of munors by clergy," *Public Relations Review* 34:3 (Sept. 2008): 58–62; Joseph R. Blaney and William L. Benoit, *The Clinton Scandals and the Politics of Image Restoration* (New York: Praeger 2001).
8. Richard J. Hobbs, "Setting effective and realistic restoration goals: key directions for research," *Restoration Ecology* 15 (2007): 354–57 at 356.
9. Gordon J. Christensen, "What has happened to conservative tooth restorations?" and idem, "Longevity versus esthetics: the great restorative debate," *Journal of the American Dental Association* 136 (2005): 1436–37 and 138 (2007): 1013–15.
10. Paul H. Gobster and R. Bruce Hull, eds., *Restoring Nature: Perspectives from the Social Sciences and Humanities* (Washington, DC: Island Press, 2000).
11. Paul Eggert, *Securing the Past: Conservation in Art, Architecture and Literature* (Cambridge: Cambridge University Press, 2009), i, 9, 139, 154.
12. Marcus Hall, *Earth Repair: A Transatlantic History of Environmental Restoration* (Charlottesville: University of Virginia Press, 2005), 245.
13. Jean Piaget, *The Child's Conception of the World* [1929] (Paterson, NJ: Littlefield & Adams, 1960), 361–67; Virginia Slaughter, Raquel Jaakola, and Susan Carey, "Constructing a coherent theory: children's biological understanding of life and death," in Michael Siegal and Candida Peterson, eds., *Children's Understanding of Biology and Health* (Cambridge: Cambridge University Press, 1999), 71–96.
14. Brandon Hamber, "Narrowing the micro and the macro: a psychological perspective on reparations in societies in transition," in Pablo de Greiff, ed., *The Handbook of Reparations* (Oxford: Oxford University Press, 2005), 560–88 at 562–63.
15. Elizabeth Akers Allen, "Rock Me to Sleep, Mother" (1859) (Boston, 1883).

16. Thomas Traherne, "The Third Century" [c.1660], in his *Centuries, Poems, and Thanksgivings*, 2 v. (Oxford: Clarendon, 1958), 1: 110–68 at 110–12.
17. Thomas Burnet, *The Sacred Theory of the Earth* [1691] (Carbondale: University of Illinois Press, 1965), 53, 64.
18. Anthony Kemp, *The Estrangement of the Past* (London: Oxford University Press, 1991).
19. Thomas M. Greene, *The Light in Troy: Imitation and Discovery in Renaissance Poetry* (New Haven: Yale University Press, 1982), 92.
20. Poggio Bracciolini to Guarino of Verona (1446), in *Petrarch's Letters to Classical Authors* (Chicago: University of Chicago Press, 1910), 93.
21. A. Bartlett Giamatti, "Hippolytus among the exiles: the romance of early humanism," in his *Exile and Change in Renaissance Literature* (New Haven: Yale University Press, 1984), 12–32 at 24, 26.
22. Desiderius Erasmus to Theophrastus Paracelsus, March 1527, in Johan Huizinga, *Erasmus and the Age of Reformation* [1924] (New York: Harper & Row, 1957), 242–43.
23. Stephen Jay Gould, *Time's Arrow, Time's Cycle: Myth and Metaphor in the Discovery of Geological Time* (Cambridge: Harvard University Press, 1987), 196–98.
24. James Hutton, *Theory of the Earth with Proofs and Illustrations* (Edinburgh, 1795); Gould, *Time's Arrow, Time's Cycle*, 77–79, 129.
25. Charles Lyell, *Principles of Geology, Being an Attempt to Explain the Former Changes of the Earth's Surface by Reference to Causes Now in Operation* (London, 1830), 1: 75–76, 141–42, 165–66, 473; Gould, *Time's Arrow, Time's Cycle*, 105–45. Whereas Lyell relied on nature to resurrect extinct megafauna, current rewilding requires human agency to transplant extant proxy species from other continents (or from zoos).
26. Charles Lyell, *The Geological Evidences of the Antiquity of Man* (London, 1863); Gould, *Time's Arrow*, 168. George P. Marsh's *Man and Nature* (New York, 1864) forced Lyell to abandon his view that human impact on nature was negligible (Lyell to Marsh, Sept. 22, 1865, cited in my *George Perkins Marsh, Prophet of Conservation* (Seattle: [University of Washington Press, 2000], 302).
27. John Demos, *Circle and Lines: The Shape of Life in Early America* (Cambridge: Harvard University Press, 2004), 61–77.
28. Martin J. S. Rudwick, *Bursting the Limits of Time: The Reconstruction of Geohistory in an Age of Revolution* (Chicago: University of Chicago Press, 2005), 188–93, 642–51.
29. Charles Darwin, *On the Origin of Species by Means of Natural Selection*. [1859] (Oxford University Press, 1998), 395. See Fiona J. Stafford, *The Last of the Race: The Growth of a Myth from Milton to Darwin* (Oxford: Clarendon Press, 1994), 292, 304.
30. Francis Darwin, ed., *The Life and Letters of Charles Darwin, Including an Autobiographical Chapter* [1876/1887] (reprint, Chestnut Hill, MA: Adamant Media Corp., 2001), 282; Edward FitzGerald, *Rubáiyàt of Omar Khayyám* [1859], stanza 26 (Charlottesville: University of Virginia Press, 1997), 184.
31. William Thomson (later Lord Kelvin), "On the age of the sun's heat" (1862), quoted in Stafford, *Last of the Race*, 304; Hermann von Helmholtz, "Observations on the sun's store of force" (1854), cited in P. F. Strawson, *Analysis and Metaphysics: An Introduction to Philosophy* (London: Oxford University Press, 1992), 222. See Gillian Beer, "'The death of the sun': Victorian solar physics and solar myth," in J. B. Bullen, ed., *The Sun Is God: Painting, Literature, and Mythology in the Nineteenth Century* (Oxford: Clarendon Press, 1989), 159–80.
32. Paul T. Davies, *The Last Three Minutes: Conjectures about the Ultimate Fate of the Universe* (New York: Basic Books, 1994), 9–13; "Rotation of Earth

plunges entire North American continent into darkness," *The Onion*, 27 Feb. 2006, p. 1.

33. Henry Wadsworth longfellow, "The Arrow and the Song," [1845], in *The Poetical Works of Henry Wadsworth Longfellow, with Bibliographical and Critical Notes*, Riverside Edition (Boston and New York: Houghton, Mifflin, 1890), I, 234.

34. Thomas Carlyle, "Shooting Niagara: and after?" [1867], in his *Critical and Miscellaneous Essays*, 3 vols. (London, 1887–1888), 3: 590.

35. Jerome Hamilton Buckley, *The Triumph of Time: A Study of the Victorian Concepts of Time, History, Progress, and Decadence* (Cambridge: Harvard University Press, 1967), 55–70; Beer, "Death of the sun," 171–73.

36. Mircea Eliade, *The Myth of the Eternal Return* (New York: Bollingen/Pantheon, 1954).

37. Geoffrey Cubitt, "The political uses of seventeenth-century English history in Bourbon Restoration France," *Historical Journal* 50:1 (2007): 73–95.

38. Edmund Burke, *Reflections on the Revolution in France* [1790] (Stanford: Stanford University Press, 2001), 184–85, 181, 170.

39. Lyell, *Principles of Geology* (1830), 1: 72.

40. Carolyn Merchant, *Reinventing Eden: The Fate of Nature in Western Culture* (New York: Routledge, 2003).

41. Sign on tree in Mianus River Gorge Preserve, Bedford, NY, quoted in James Duncan and Nancy Duncan, "Aestheticization of the politics of landscape preservation," *Annals of the Association of American Geographers* 91 (2001): 405.

42. Thomas R. Cox et al., *This Well-Wooded Land: Americans and Their Forests from Colonial Times to the Present* (Lincoln: University of Nebraska Press, 1985), 144–47; Nicolai Cikovsky, "'The Ravages of the Axe': the meaning of the tree stump in nineteenth-century American art," *Art Bulletin* 61 (1971): 611–26 at 613.

43. Anne Whiston Spirn, "Constructing Nature: The legacy of Frederick Law Olmsted," in William Cronon, ed., *Uncommon Ground: Toward Reinventing Nature* (New York: W. W. Norton, 1995), 111–12.

44. Hall, *Earth Repair*, 141–43.

45. John Fox, Jr., *The Trail of the Lonesome Pine* (New York: Grosset & Dunlap, 1908), 201–2.

46. A generation later, historical restoration reified Fox's fiction: in 1940 Daniel Boone's famed Wilderness Road was reborn as Cumberland Gap National Historical Park.

47. Burnet, *Sacred Theory of the Earth*, 376, 257.

48. Regius professor of divinity Brooke Foss Westcott, *The Gospel of Life: Thoughts Introductory to the Study of Christian Doctrine* (London, 1892), 243.

49. Lynn White, Jr., "The historical roots of our ecologic crisis," *Science* 155:3767 (1967): 1203–7 at 1205; Douglas J. Moo, "Nature in the new creation: New Testament eschatology and the environment," *Journal of the Evangelical Theological Society* 49 (2006): 449–88 at 467–82.

50. I detail this transition in *The Past Is a Foreign Country* (Cambridge: Cambridge University Press, 1985), 145–82, and in "The value of age and decay," in W. E. Krumbein et al., eds., *Durability and Change: The Science, Responsibility, and Cost of Sustaining Cultural Heritage* (London: John Wiley, 1994), 39–49.

51. Eugène-Emmanuel Viollet-le-Duc, "On restoration," quoted in Eggert, *Securing the Past*, 54.

52. Francis Palgrave to Dawson Turner, 19 July 1847; Francis Palgrave, *History of Normandy and England* (1851); both in my *Past Is a Foreign Country*, 278.

53. William Morris, "Repair not restoration" (1877), in Stephan Tschudi-Madsen, *Restoration and Anti-Restoration: A Study in English Restoration Philosophy*, 2nd edn. (Oslo: Universitetsforlaget, 1976), Annex VI.

54. John Ruskin, *Modern* Painters (1846) and *Seven Lamps of Architecture* (1849), cited in my *Past Is a Foreign Country*, 164–68, 278–80. To be sure, Burke *commanded* restoration, whereas Ruskin and Morris *condemned* it; but they treasured alike a past perfected by venerating its integrity.

55. Giles Gilbert Scott, *Plea for the Faithful Restoration of Our Ancient Churches* (1850), and idem, *Recollections* (1879), quoted in my *Past Is a Foreign Country*, 326.

56. George Edmund Street, "Destructive restoration on the Continent" (1857), "Report to the S.P.A.B." (1880–1886), and *Some Account of Gothic Architecture in Spain* (1865), quoted in my *Past Is a Foreign Country*, 151, 278, 327.

57. Cathleen Sara Hoeniger, "Aesthetic unity or conservation honesty? Four generations of wall-painting restorers in Italy and the changing approaches to loss, 1850–1970," in Andrew Oddy and Sandra Smith, eds., *Past Practice–Future Prospects*, British Museum Occasional Paper No. 145 (London: British Museum Press, 2001), 115–22 at 119–21, quoting Umberto Baldini (1978), Bruce Cole (1976), and Cesari Brandi (1963).

58. Philip Wilkinson, "Restoration: a dynamic process," in *Restoration: The Story Continues* (London: English Heritage, 2004), 5–23 at 22.

59. Jerry Podany, "Restoring what wasn't there: reconsideration of the eighteenth-century restorations to the Lansdowne *Herakles* in the collection of the J. Paul Getty Museum," in Andrew Oddy, ed., *Restoration: Is It Acceptable?* British Museum Occasional Paper 99 (London: British Museum Press, 1994), 9–16 at 15.

60. Carlo Pietrangeli et al., *The Sistine Chapel: A Glorious Restoration* (New York: Abrams, 1999); Peter Layne Arguimbau, blog 5 Oct. 2006; Richard Serrin, "Michelangelo and the destruction of the Sistine Chapel," lecture, Classical Design Foundation, Jan. 5, 2006, incorporating his "Lies and misdemeanors: Gianluigi Colalucci's Sistine Chapel revisited." "Glorious restoration" patently recalls the biblically prophesied return of the Israelites and the ensuing millennial kingdom and England's Stuart restoration; see John Morrill, "The later Stuarts: a glorious restoration?"*History Today* 38:7 (July 1988): 8–16. "Glorious" also designates the 1990s' restoration of Mission San Xavier del Bac in Tucson, Arizona, 2000s' renovation of Mobile, Alabama's 1850s' Cathedral-Basilica of the Immaculate Conception, the renovated seventeenth-century Nether Auchendrane House in Ayrshire, Scotland, and several architectural heritage projects in Ulster.

61. James Beck, with Michael Daley, *Art Restoration: The Culture, the Business and the Scandal* (New York: Norton, 1996), 88–100; Eggert, *Securing the Past*, 90–93.

62. Martin Kemp, "Looking at Leonardo's *Last Supper*," in Peter Booth et al., eds., *Appearance, Opinion, Change: Evaluating the Look of Paintings* (London: U.K. Institute for Conservation, 1990), 14–21 at 18.

63. Hence the difficulty of deciding the *Ur*-text for D. H. Lawrence's *Sons and Lovers* and for Theodore Dreiser's *Sister Carrie*, both original publications much altered by author-sanctioned editorial intercession (Eggert, *Securing the Past*, 192–94).

64. Sergio Palazzi, "Restoration: dealing with a ghost," in Andrew Oddy and Sara Carroll, eds., *Reversibility–Does It Exist?* British Museum Occasional Paper 135 (London: British Museum Press, 1999), 175–79 at 176.

65. Richard Taruskin, "The pastness of the present and the presence of the past," in Nicholas Kenyon, ed., *Authenticity and Early Music* (Oxford: Oxford

University Press, 1988), 137–207; John Butt, *Playing with History: The Historical Approach to Musical Performance* (Cambridge: Cambridge University Press, 2003), 43; David Lowenthal, "From harmony of the spheres to national anthem: reflections on musical heritage," *GeoJournal* 65 (2006): 3–15 at 8–9.

66. Stan Godlovitch, "Performance authenticity: possible, practical, virtuous," in Salim Kemal and Ivan Gaskell, eds., *Performance and Authenticity in the Arts* (Cambridge: Cambridge University Press, 1999), 154–74.

67. Aldo Leopold, "The Round River," in his *A Sand County Almanac, with Other Essays on Conservation from* Round River [1949/1953] (New York: Oxford University Press, 1966), 177.

68. Jozef Keulartz, "Using metaphors in restoring nature," *Nature and Culture* 2 (2007): 27–48.

69. Robert Elliot, *Faking Nature: The Ethics of Environmental Restoration* [1982] (London: Routledge, 1997), vii.

70. Sierra Club advertisements, June 1966, quoted in Hall, *Earth Repair*, 1–2, 13–15.

71. Peter Losin, "Faking nature—a review," *Restoration & Management Notes* 4 (1986): 55; "The Sistine Chapel debate: Peter Losin replies," *Restoration & Management Notes* 6 (1988): 6.

72. On these differences see my "Natural and cultural heritage," in Kenneth R. Olwig and David Lowenthal, eds., *The Nature of Cultural Heritage and the Culture of Natural Heritage: Northern Perspectives on a Contested Patrimony* (London: Routledge, 2006), 79–90.

73. Stephen M. Meyer, *The End of the Wild* (Cambridge: M.I.T. Press, 2006).

74. Mark C. Buckley and Elizabeth E. Crone, "Negative off-site impacts of ecological restoration: understanding and addressing the conflict," *Conservation Biology* 22:5 (2008): 1118–24; S. A. Merkle et al., "Restoration of threatened species: a noble cause for transgenic trees," *Tree Genetics & Genomes* 3:2 (April 2007): 111–18. Restoration ecology is in this sense at odds with the emerging field of reintroduction biology's focus on single species, usually charismatic large vertebrates (Philip J. Seddon, Doug P. Armstrong, and Richard F. Maloney, "Developing the science of reintroduction biology," *Conservation Biology* 21:2 [2007]: 303–12).

75. Young Choi, "Restoration ecology to the future: a call for new paradigm," *Restoration Ecology* 15 (2007): 351–53 at 352; Zev Naveh, *Transdisciplinary Challenges in Landscape Ecology and Restoration Ecology—An Anthology* (Dordrecht: Springer 2007).

76. Valentin Schaefer, "Science, stewardship, and spirituality: the human body as a model for ecological restoration," *Restoration Ecology* 14:1 (2006): 1–3.

77. Stephen Kaplan, "The restorative benefits of nature: toward an integrative framework," *Journal of Environmental Psychology* 15 (1995): 169–82; Terry Hartig and Henk Staats, "The need for psychological restoration as a determinant of environmental preferences," *Journal of Environmental Psychology* 26 (2006): 215–26; Agnes E. van den Berg, Terry Hartig, and Henk Staats, "Preference for nature in urbanized societies: stress, restoration, and the pursuit of sustainability," *Journal of Social Issues* (63:1) 2007 79–96; Terry Hartig, Florian G. Kaiser, and Peter A. Bowler, "Psychological restoration in nature as a positive motivation for ecological behavior, *Environment & Behavior* 33:4 (July 2001): 590–607. However, those interviewed were by and large urban North Americans and northern Europeans predisposed to idealize nature to begin with.

78. William Throop and Rebecca Purdom, "Wilderness restoration: the paradox of public participation," *Restoration Ecology* 14 (2006): 493–99 at 497

(citing Holmes Rolston III, *Conserving Natural Value* [New York: Columbia University Press, 1994]).
79. John Cairns, Jr., "Restoring damaged aquatic ecosystems," *Journal of Social, Political & Economic Studies* 31:1 (2006): 53–74 at 55–56.
80. Walter P. Cottam (1958) quoted in Hall, *Earth Repair*, 125.
81. Linda J. Krisjanson and Richard J. Hobbs, "Degrading landscapes: lessons from palliative care," *Ecosystem Health* 7:4 (Dec. 2001): 203–13; David J. Rapport et al., "Strange bedfellows: ecosystem health in the medical curriculum," *Ecosystem Health* 7:3 (Sept 2001): 155–62.
82. Eric Katz, "The big lie: human restoration of nature," *Research in Philosophy and Technology* 12 (1992): 231–41 at 235.
83. Elliot, *Faking Nature*, 111–12. Ecological restoration is a "ritual of atonement for living in a culture that is responsible for causing morally unacceptable environmental degradation, . . . imbu[ing] the practitioner with optimism and a sense of expiation." The hope of redemption makes "ecological restoration . . . especially attractive to those whose cultural roots stem from the Protestant Reformation" (Andre F. Clewell and James Aronson, "Motivations for the restoration of ecosystems," *Conservation Biology* 20 [2006]: 420–28 at 423).
84. "Mainstream doctors are turned off by geriatrics. The Old Crock . . . has diabetes . . . " (Felix Silverstone, quoted in Atul Gawande, "Annals of medicine: the way we age now," *New Yorker*, 30 Apr. 2007, 53).
85. Bernd von Droste, Harald Plachter, and Mechtild Rössler, eds., *Cultural Landscapes of Universal Value—Components of a Global Strategy* (Jena: Gustav Fischer Verlag/UNESCO, 1995).
86. Jan E. Dizard, *Going Wild: Hunting, Animals Rights, and the Contested Meaning of Nature*, rev. ed. (Amherst: University of Massachusetts Press, 1999); David Lowenthal, "Environment as heritage," in Kate Flint and Howard Morphy, eds., *Culture, Landscape, and the Environment: The Linacre Lectures 1997* (Oxford: Oxford University Press, 2000), 198–217; Hall, *Earth Repair*, 11–13, 138–49, 195–99.
87. Simon de Bruxelles, "Ancient stones to regain true standing," *The Times* (London), 8 Apr. 2003; Richard Scott, "The accidental rainforest, the leaning tower of Pisa, and making the most of opportunity," in Ian D. Rotherham, ed., *Loving the Aliens??!!? Ecology, History and Management of Exotic Plants and Animals: Issues for Nature Conservation*, Special Ser. no. 4, *Journal of Practical Ecology and Conservation* (June 2005), 83–84.
88. Ed Fotherington, "Stick to prunes," *Audubon* 6 (2003): 1; Andy Opel, "Corporate culture keeps nature regular: the 'super citizen,' the media, and the 'Metamucil and Old Faithful' ad," *Capitalism, Nature, Socialism* 17(3) (Sept. 2006): 100–13 at 105–6.
89. Joseph M. Wheaton, Stephen E. Darby, David A. Sear, and Jim A. Milne, "Does scientific conjecture accurately describe restoration practice? Insight from an international river restoration survey," *Area* 38 (2006): 128–42. The website 2003–2004 survey enlisted 300 respondents.
90. John Cairns, Jr., "The status of the theoretical and applied science of restoration ecology," *The Environmental Professional* 13 (1991): 186–94 at 187.
91. Anthony D. Bradshaw, "Alternative endpoints for reclamation," in John Cairns, Jr., ed., *Rehabilitating Damaged Ecosystems*, 2 vols. (Boca Raton, Fla.: CRC Press, 1988), 2: 69–85 at 74. A recent survey of 89 global restoration projects found only 21 limited to ceasing degradation or extirpating damaging species; 28 included planting and 25 topographic remodeling (José M.Rey Benayas et al., "Enhancement of biodiversity and ecosystem services

by ecological restoration: a meta-analysis," *Science* 325:5944 [2009]: 1121–24).

92. Wheaton et al., "Does scientific conjecture accurately describe restoration practice?"

93. Eric S. Higgs, "Restoration goes wild: a reply to Throop and Purdom," *Restoration Ecology* 14 (2006): 500–3 at 502.

94. Peter Del Tredici, "Neocreationism and the illusion of ecological restoration," *Harvard Design Magazine*, no. 20 (Spring/Summer 2004): 87–89; William R. Jordan III, "Restoration, community, and wilderness," in Gobster and Hull, *Restoring Nature*, 21–36 at 27; Mark A. Davis, "'Restoration'—a misnomer?" *Science* 287 (2000): 1203.

95. Daniel Janzen, "Gardenification of wildland nature and the human footprint," *Science* 279 (1998): 1312–13.

96. Eric Higgs, "The two-culture problem: ecological restoration and the integration of knowledge," *Restoration Ecology* 13 (2005): 159–64.

97. Lothar Ledderose, "Carving sutras into stone before the catastrophe: the inscription of 1118 at Cloud Dwelling Monastery near Beijing," *Proceedings of the British Academy* 125 (2004): 381–454 at 390–92, 396. Excavated in 1957, the stones were again reburied in 1999.

98. John Seabrook, "Sowing for Apocalypse," *New Yorker*, 27 Aug. 2007: 60–71.

99. Richard Morgan, "Proposed use for Moon: storage locker for DNA," *International Herald Tribune*, 2 Aug. 2006, 3; William E. Burrows, *The Survival Imperative: Using Space to Protect Earth* (New York: Forge Books, 2006), 208–35.

100. Gregory Benford, *Deep Time: How Humanity Communicates Across Millennia* (New York: Avon, 1999), 93–127.

101. Lyell, *Principles* (1830), I: 74; Gould, *Time's Cycle*, 155; Barthold Georg Niebuhr, *The History of Rome* [1811–1812] (Philadelphia, 1835), I: 4, quoted in Linda Dowling, "Roman decadence and Victorian historiography," *Victorian Studies* 28 (1985): 579–609 at 535.

102. Erasmus to Pope Leo X, 1 Feb. 1516, letter 384, *Correspondence*, in his *Complete Works* (Toronto: University of Toronto Press, 1976), 3:221–22.

3 Spontaneous Rewilding of the Apostle Islands

James Feldman

In 1931, Harlan Kelsey, a representative of the U.S. National Park Service (NPS), traveled to the far northern tip of Wisconsin to evaluate the Apostle Islands as a potential addition to the national park system. Kelsey was not impressed with what he saw. "What must have been once a far more striking and characteristic landscape of dark coniferous original forest growth has been obliterated by the axe followed by fire. . . . The ecological conditions have been so violently disturbed that probably never could they be more than remotely reproduced." Kelsey reported that destructive logging practices of the previous half-century had robbed the area of its value as a park. But Kelsey was wrong, at least in his assessment of the islands' future. By the 1960s, the island forests had regenerated and the nation was enjoying an environmental awakening. Congress created Apostle Islands National Lakeshore in 1970. When NPS administrators published the park's first management plan in 1977, they determined that 97 percent of the park should be managed as a wilderness area. In November 2004, Congress formally designated a majority of the park as wilderness.[1]

Rewilding is typically understood in North America to mean large-scale human intervention to restore the wild qualities of the landscape. For Josh Donlan and colleagues (Chapter 26, this volume), this could mean introducing analogues of megafauna extinct since the Pleistocene. For others, it involves the restoration of top carnivores and other keystone species. But can wildness return without such direct human action? The rewilding of the Apostle Islands suggests that it can. From the 1850s until 1970, the extraction of natural resources like fish, lumber, and stone provided the basis for the local economy, with predictable impacts on island environments. Although some of the islands have been under state or federal management since the 1950s, this management never involved tree planting, road removal, invasive species control, or other techniques associated with rewilding and restoration ecology. And yet, now this place is a wilderness. In 2003, NPS managers even discovered wolf tracks on Sand Island—the area's top carnivore had returned after an absence of over fifty years, a powerful statement about the renewed wildness of the islands. How did this transition occur? What were the conditions necessary for this kind of "spontaneous" rewilding, or rewilding without direct intervention?[2]

Figure 3.1   Apostle Islands from the air with York Island in foreground. (Photograph by William Cronon)

Although resource managers never took a hands-on management approach, the return of wildness to the Apostle Islands was not accidental. Nor was it simply an ecological process. Jan Dizard (Chapter 14, this volume) suggests that restoration and rewilding projects are artifacts of the desires and decisions of specific cultural moments. Whose artifact, then, are the now wild island landscapes? What cultural, economic, and political processes led to their rewilding? As the environments of the Apostle Islands regained their wilderness characteristics in the wake of logging and other extractive activities, the state—represented at various times by agencies from several levels of government—reshaped the islands by influencing the ways that people valued and used them. The rewilding of the islands resulted from the emergence of a consumer society that valued the islands as a recreational amenity and a scientific laboratory, and from the actions of a powerful, modern state that both promoted and benefited from these new uses of island resources.

A shift from an economy based on production to one based on consumption triggered the rewilding of the islands. The islands are now valued as a site of recreation, not resource production. Tourism has been an important part of the regional economy since the 1870s, when railroads promoted the Apostle Islands as a resort destination. This early tourist economy shared island resources with other industries that relied on resource extraction. These activities were not mutually exclusive; indeed, the logging, fishing, and tourism industries complemented and reinforced each other. But over the course of the twentieth century, the loggers and commercial fishermen disappeared, replaced by anglers and kayakers. This transition had social costs, as some people (often residents of the region) lost access to resources in favor of others (often from outside the region). As the state asserted its authority to manage natural resources, it encouraged this transition from production to consumption and shaped the rewilding of the islands. Four episodes of state activity fostered the creation of the modern island wilderness: regulation of the fishery between the 1880s and 1970s; the use of rural zoning to bolster tourism in the 1930s; land acquisition for the purposes of outdoor recreation in the 1950s; and management of the islands to foster particular kinds of recreational activity in the 1970s.

The state role in promoting consumer-oriented uses of nature is perhaps most direct in fisheries management. Although commercial fishermen plied island waters as early as the 1830s, no regulation existed until the 1880s, when the Wisconsin legislature set minimum weights for sale and mesh sizes for commercial nets. In the following four decades, the state gradually tightened these regulations, establishing closed seasons, equipment restrictions, and spawning sanctuaries. The goal of these regulations was to protect and improve the commercial fishery.[3]

Regulation benefited some fishermen more than others. Large firms, wageworkers, and full-time fishermen benefited the most from regulation, because their larger capital made it easier to adapt to restrictions on equipment and to take advantage of new technologies. Part-time and itinerant fishermen suffered. Regulation also curtailed the fishing of the state's Native American population, such as the Ojibwe who lived near the Apostle Islands, whose fishing activities the state also sought to control. Although the Ojibwe consistently maintained that they had a treaty-guaranteed right to fish, state and federal courts did not at that time recognize these rights.[4]

The state's growing control over both commercial fishing and Ojibwe usufructuary rights was part of a larger process. These were the first steps toward bringing what anthropologist and political scientist James C. Scott calls "legibility" to the management of natural resources: that is, simplification for easier state management. Regulations like closed seasons and mesh sizes systematized fishing, making it easier for the state to control. A 1909 law requiring commercial fishermen to register for a license serves as a classic example of state action designed to increase legibility. With licensed fishermen, the state could more easily track and control commercial fishing.[5]

The state also used its growing authority to promote a sport fishery. In 1879, the state began to supplement the natural reproduction of game fish and to introduce exotic sport fish. The motivation for this was explicitly economic. As railroad lines reached all corners of the state, and as railroad companies promoted sporting and hunting opportunities throughout northern Wisconsin, the value of the tourist economy expanded rapidly. Sport fishing anchored this new economy of leisure. "Our lakes and rivers are also attractive to [tourists] because of their fish supply," explained one fisheries report. "This supply needs continual protection as well as reinforcement." The state also regulated the sport fishery by setting seasons on sport fish and outlawing the sale of game species. These restrictions again privileged some activities—and some people—over others. Residents of rural northern Wisconsin who depended on fishing for subsistence or market sale lost access to resources in favor of wealthy urban sportsmen.[6]

The collapse of the Lake Superior trout fishery in the 1950s brought sport and commercial fishermen into direct competition. Sportsmen blamed commercial fishermen for over-fishing, and argued that their uses deserved priority because recreation brought more money to the local economy. Commercial fishermen blamed the invasion of the exotic sea lamprey for the decline in lake trout, and believed that their economic needs should take precedence over the leisure activities of sportsmen. Under constant pressure from sportsmen's groups, the state closed the commercial trout fishery in 1962, but allowed continued sport fishing for the species.[7]

Relief from fishing pressure combined with an aggressive lamprey control program paid quick dividends, and lake trout stocks rebounded by 1970. But when the state reopened the trout fishery, it put the needs of the sport fishery first, limiting the number of commercial fishermen and the amount each could catch. The Wisconsin Conservation Department explained its position as an attempt to find "the greatest good recreationally, aesthetically, and economically." This meant "precedence in management is given to sport fishing, since it provides a greater benefit." State officials prioritized the value of the fishery as a recreational amenity rather than for commercial production.[8]

The twin drives for the simplification of resource management and the encouragement of an economy based on recreational consumption of nature can also be seen in Wisconsin's rural zoning initiatives of the 1930s and 1940s. Rural zoning was a response to the tax delinquency crisis that gripped northern Wisconsin in the 1920s. For decades, settlers had attempted to carve farms out of the stumplands left behind after the logging era. The failure of these attempts led to widespread tax delinquency. In 1929, the state legislature granted counties the right to zone land for exclusive forest, agricultural, or recreational use. This would prevent agricultural settlement on submarginal lands and relieve county governments of the obligation to provide services to isolated farm families. Zoning for recreation and tourism emerged as a central tenet of economic recovery for the region. A typical recreational area might prohibit such industrial activity as quarries or sawmills, but allow development for summer homes. One expert explained: "Recreational land means taxable wealth. A zoned area dedicated to recreation insuring a quiet, beautiful, undisturbed area in which to build a summer home will help to attract the recreation seeker. . . ."[9]

Rural zoning was a part of the same process of simplification that took place in the fisheries. Faced with the disorder and inefficiency of tax delinquency and the settlement of submarginal lands, state officials used zoning to order the landscape for easier management. The state assumed a more powerful role in determining economic activity, an authority that increased steadily over the twentieth century.

Rural zoning had important implications for the Apostle Islands. Madeline Island was designated as "agricultural and recreational" while the rest of the islands were categorized as "forestry and recreational." These islands were closed to potential settlement, to all uses other than forestry and tourism. Significantly, Madeline is the only island today not included within the national lakeshore. State and county officials had segregated the Apostles for their recreational value. This prescribed the types of activity permissible in the islands and structured the way that people valued island resources. These actions fostered rewilding, even if they did not lead to the deliberate restoration of the pre-logging forests, because they culminated in the view of the islands as a wilderness, as a place valued for a specific type of recreation.[10]

The next step in this process involved outright state acquisition and management of the islands. In the 1950s, the Apostles emerged as a premier spot for primitive outdoor recreation. In the prosperous years that followed World War II, the nation as a whole and Wisconsin in particular experienced a boom in outdoor recreation. Between 1950 and 1960, for example, visits to Wisconsin state lands jumped 243 percent. The state's recreational facilities could not meet this booming demand.[11] State officials viewed outdoor recreation as an opportunity to ameliorate the chronic economic depression that had stifled northern Wisconsin since the end of the logging era. They wanted to connect the region's lakes, forests, and rivers to the recreationally starved urbanites of Midwestern cities. Beginning in the mid-1950s, Wisconsin governor Gaylord Nelson made the improvement of parks, roads, and state forests the centerpiece of his plans for economic development and environmental stewardship.[12]

In 1952, representatives of the Wisconsin Conservation Department (WCD) and sportsmen's groups toured the Apostle Islands with the idea of acquiring land in the area for these purposes. They liked what they found: a forest recovering after the destructive logging of previous generations; a landscape seemingly tailor-made for scientific experiments in ecology and game management due to different island histories of logging, fire, and deer browse; and a wonderland for the types of outdoor recreation that were surging in popularity.

How could state officials find so much to value in the Apostles, just two decades after NPS investigators described the islands as a logging- and fire-devastated wasteland? Certainly, the trained scientists of the WCD recognized the signs of logging. Even today, large stumps and fire scars are still evident on some of the islands, as well as the ruins of logging camps and industrial machinery. But the WCD officials valued the islands for their "natural condition." Wisconsin's chief forester described Stockton Island in 1956 as "an island almost in its natural condition; no fires have destroyed the forests. It has not been logged since 1915 . . ." By this, the forester did not mean the absence of human activity, but that people had not directly shaped forest regeneration. Stockton Island had no roads, no buildings save the ruins of a few logging camps, and no active forest management. As plans for public acquisition of the islands developed, the wildness of the islands—that is, their freedom from human control—became their chief selling point. The prevailing Clementsian idea that forests inevitably and predictably matured from disturbance to a climax ecosystem increased the scientific value of the islands' "natural condition." The islands were particularly suited to the study of the relationship between deer and forest regeneration in a cutover landscape. One ecologist, who conducted research on the island forests in the late 1950s, called the Apostles a "ready made experiment for the ecologist" because of their location, their isolation from each other and the mainland, and their different histories of fire, logging, deer browsing, and agriculture.[13]

But WCD officials believed that the Apostles' greatest value lay in the opportunity they provided for wilderness recreation. Visitors to the islands had long noted their scenic beauty, their red sandstone cliffs and beaches, and the opportunities for sport fishing. The islands could also provide a type of recreation not possible in most places. As one official explained: "The value of an undeveloped area where it is possible to get away from the hustle and bustle of modern living cannot be overestimated. . . . One of Stockton Island's greatest assets is its inaccessibility. The fact that a vacation on the island requires some planning and the possibility that one may be stranded for a few extra days makes it all the more desirable."[14] In 1956, the state purchased the 10,000-acre island.[15]

Crucially, the WCD had no plans to develop or even formally manage Stockton Island. By the 1950s, forest managers had the silvicultural techniques for reforestation, and had conducted such projects elsewhere in Wisconsin. But WCD officials had neither the funding nor the inclination to pursue such a policy. Instead, Stockton Island would be retained "as a wilderness area . . . open for limited recreational use" and scientific research. It would receive no active management. After all, it was this very lack of human intervention and management that made the island so appealing in the first place.[16]

Local residents of the region had a very different idea of how the state should use its newly acquired land. They wanted a state park—with campgrounds, picnic facilities, shelters, and concessions—arguing that a park would bring in more tourists and dollars than an undeveloped wilderness area. One local official believed "it would not help the state, the county, or the islands if they are established as a pure wilderness area . . . there is already enough wilderness area in the north . . . the need is for

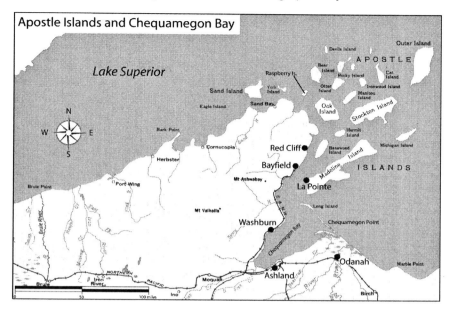

Figure 3.2 The Apostle Islands and the Chequamegon Bay. (Map by Joel Heiman)

well-developed state parks."[17] But as with fisheries management and rural zoning, state resource management privileged some uses of nature over others, and the state had no plans for this kind of development.

The creation of Apostle Islands National Lakeshore in 1970 finalized the transition from producer- to consumer-oriented uses of the islands. The policies adopted by the NPS hastened the rewilding of the islands—albeit still without direct restoration. The federal government purchased (at times using its power of eminent domain) the islands that remained in private ownership, ending the last vestiges of the production economy by stopping the remaining logging operations. NPS policies ensured that the islands would be used for a specific type of recreation—for wilderness recreation. For example, the NPS chose not to pursue plans for a scenic road along the mainland shore or an island hotel, although these measures had been included in the original park proposal. The NPS also shut down the Rocky Island Air Haven, a restaurant run by a fishing family from the shore of Rocky Island. The NPS prescribed the kinds of activities that were allowed in the islands, ensuring that island landscapes would be used and valued as a recreational wilderness.[18]

NPS land management further reinforced this perspective. The NPS removed many of the fishermen's cabins and logging camps, the remnants of the former extractive economy. The purpose of this policy was to enhance the park's wilderness character—to make the islands seem more wild—by erasing evidence of the logging, fishing, and other resource production related activities of the past. This was as close to pro-active rewilding as the NPS would come. NPS resource managers considered active management to encourage the restoration of pre-logging forest conditions—most notably for the hemlock forests that had failed to regenerate after logging and fire—but they never implemented these policies. They did, however, make the conscious decision to let the forest regenerate in ways that would enhance the seemingly pristine qualities of the islands. Surveys conducted soon after the NPS assumed control of the islands in 1970 identified twenty-eight forest clearings, ranging in size from one to twenty acres, where logging, farming, or some other activity had left an opening in the woods. Lakeshore officials chose not to maintain these clearings—other than the few in use as campsites or for other recreational purposes—despite the cultural values associated with them. These choices enhanced the wilderness qualities of the islands and met both the desires of the general public and the dictates of NPS policy. The "spontaneous" rewilding of the Apostle Islands was really quite intentional.[19]

Over the course of nearly a century, the state—first the state of Wisconsin and then the U.S. federal government—fostered the rewilding of the islands, the transition from production to consumption, and the valuation of the islands for wilderness recreation and scientific study. Why did the state do this?

First, modern American ideas about wilderness as a place devoid of human impact—"untrammeled by man," as the Wilderness Act puts it—crystallized in the mid-twentieth century. This created a self-fulfilling prophecy for island landscapes. Once people started to conceive of the islands as a recreational wilderness, the state made the Apostles conform to the wilderness ideal. This explains the WCD's refusal to build picnic areas and the NPS policies of removing cultural resources to foster the appearance of wilderness.

Second, the emergence of modern consumer society brought new ways of valuing places like the Apostle Islands. Although once valued as a source of resource production, by the 1950s most Americans valued scenic areas primarily as a recreational amenity.[20] State promotion of a tourist economy can be understood simply as a search for economic security. Although the islands still retained significant production value in the form of commercial fisheries and merchantable second growth timber, their value for recreation, as a consumer amenity, proved greater.

Finally, the shift from production to consumption was tied to the rise in state authority over the course of the twentieth century. The state gradually asserted its authority to manage natural resources, an authority that demanded legibility. State officials used closed seasons and rural zoning to simplify and order the natural world, to make it easier to manage and control. Both the state and federal governments allocated increasingly scarce resources, determining how those resources could be used to maximize desirable goals like nature protection, economic stability, and opportunities for outdoor recreation. A wilderness landscape, a place where activities are rigidly prescribed, is a legible one. The state's emphasis on the shift from production to consumption—rather than the blend of the two that had existed in the past—was the result of the need for legibility.

One of the drawbacks of focusing on state power and legibility is that it leads us to think of the state as an all-powerful, independent actor. This tendency obscures the role of people and their very human choices in shaping state action. This rewilding of the islands was in part the result of a drive toward legibility, but it was also the result of railroad executives trying to lure fare-paying passengers onto their trains, wealthy urban sportsmen looking to secure access to game fish, and Gaylord Nelson protecting his chances at reelection. That is, some groups were able to use this growing state authority for their own ends.

It is the legacy of these choices that make the Apostle Islands a worthwhile place to consider the conditions necessary for rewilding, and that offers lessons for how we understand the presence of the past in ecological restoration. Although not the product of direct, hands-on restoration efforts, the rewilding of the islands was far from accidental. Nor was it simply an ecological process. Rewilding required a set of economic, social, and political conditions: a consumer society that valued the islands for recreation and research, and a powerful state that benefited from segregating tourism from other types of economic activity. For the islands to regain their wilderness character, these ideas had to subjugate and replace an older, more locally informed way of using and valuing nature. It is not enough to pay attention only to ecological processes like evolutionary potential or predator-prey relationships. The way that people intend to use newly wild places—and which people do the using—matters greatly in the process of rewilding. This lesson is easily generalized to apply to other newly wild sites, whether the return of wildness to these places was spontaneous or deliberate.

What have these choices meant for the environments of the Apostle Islands? At least seven islands boast patches of old growth forest, stands saved from the axe by their ownership history or inaccessibility. Ecologists consider the 185-acre stand of hemlock-hardwood old growth Outer Island to be among the most important of its type in the in the Great Lakes. Other islands now contain maturing second- or third-growth forests. Although these forests certainly differ from those that clothed the islands prior to logging and farming, they nevertheless retain significant ecological value. Island forests contain remarkably few invasive exotic species, and provide critical habitat for a number of threatened or endangered plants and animals. One survey in 2000 documented 79 percent of the plants in the islands as native, among the highest percentages on any public lands in the Great Lakes region. Ecologists have used the islands to study deer browse patterns, beaver colonization, hemlock regeneration, and a host of other questions. Indeed, the varying island histories of land use, fire, deer population, and exotic species have made the Apostles a laboratory for the study of the complicated interactions between human choices and natural processes.[21]

Stories from places like the Apostle Islands suggest that rewilding can occur spontaneously, without direct planning or action. And yet, the lesson of this story is certainly not that wildness will always grow back, and that intervention and management in the name of restoring wilderness (and wildness) are not necessary. Rather, the lesson is that rewilding landscapes are the artifacts of particular ideas about and uses of nature. We need to pay attention to the political, social, economic, and ecological conditions that create these artifacts. These conditions need to be right for any rewilding project to succeed, and when these conditions are met, a wide variety of landscapes can be restored. Identifying these factors will give us a more complex and nuanced understanding of the places that we cherish and hope to protect.

NOTES

1. Harlan Kelsey to Horace M. Albright, January 20, 1931, National Archives, RG 79, Box 2822, proposed national parks, 0–32; Apostle Islands National Lakeshore, *General Management Plan: Apostle Islands National Lakeshore, Wisconsin* (Denver: U.S. Department of the Interior/National Park Service, 1989).
2. On Pleistocene Rewildling, see Josh Donlan et al., "Re-wilding North America," *Nature* 436, 913–14 (18 Aug. 2005); also, Donlan, this volume.
3. *Laws of Wisconsin, 1887* (Madison: Democrat Printing Company, 1887), ch. 520, s.2.
4. James Nevin, "Artificial Propagation versus a Close Season For the Great Lakes," *Transactions of the American Fisheries Society* 27 (1898): 25; Ronald N. Satz, Chippewa Treaty Rights: The Reserved Rights of Wisconsin's Chippewa Indians in Historical Perspective," *Transactions of the Wisconsin Academy of Sciences, Arts, and Letters* 79 (1991): 1–251.
5. James C. Scott, *Seeing Like a State: How Certain Schemes to Improve the Human Condition Have Failed* (New Haven: Yale University Press, 1998), 2–3; *Laws of Wisconsin, 1909* (Madison: Democrat Printing Company, 1909), ch. 357, s.1.
6. Fish Commissioners of the State of Wisconsin, *Thirteenth (Fourth Biennial) Report of the Fish Commissioners of the State of Wisconsin, 1889–1890* (Madison: Democrat Printing Company, 1891), 4; *Biennial Report of the Commissioners of Fisheries of Wisconsin for the Years 1907 and 1908*, 13.
7. Brian Belonger, "Lake Trout Sport Fishing in the Wisconsin Waters of Lake Superior," (Madison: Wisconsin Department of Natural Resources, 1969); Wisconsin Conservation Commission, Minutes, June 9–10, 1950, September 12, 1952, and May 14, 1959, Wisconsin Natural Resources Board Papers, Series 812, Wisconsin Historical Society, Madison, Wisconsin [hereafter, CC Minutes].
8. *Department of Natural Resources, State of Wisconsin, 1967–69 Biennial Report* (Madison: Wisconsin Department of Natural Resources, 1969), 17.

9. James Kates, *Planning a Wilderness: Regenerating the Great Lakes Cutover Region* (Minneapolis: University of Minnesota Press, 2001), 15–51; W. A. Rowlands and F. B. Trenk, *Rural Zoning Ordinances in Wisconsin* (Madison: University of Wisconsin Agricultural Experiment Station, 1936), 13.
10. Ashland County Land Use Planning Committee, *Ashland County Intensive Land Use Planning Report* (Ashland County Land Use Planning Committee, Ashland, WI, May 1, 1941), 21, 25.
11. Roman H. Koenings, "The Status of the State Parks in Wisconsin," delivered at the National Conference on State Parks, September 1960, Box 815, Folder 6, Wisconsin Conservation Department Files, Wisconsin Historical Society.
12. I. V. Fine, Ralph B. Hovind, and Philip H. Lewis, Jr., *The Lake Superior Region Recreational Potential: Preliminary Report* (Madison: Wisconsin Department of Resource Development, 1962); Thomas R. Huffman, *Protectors of the Land and Water: Environmentalism in Wisconsin, 1961–1968* (Chapel Hill: University of North Carolina Press, 1994), 9–35.
13. Wisconsin Legislative Council, Conservation Committee Minutes, January 9, 1956, Box 453, Folder 39, WCD Files; E. W. Beals and Grant Cottam, "The Forest Vegetation of the Apostle Islands," *Ecology* 4 (October, 1960): 73.
14. Burton L. Dahlberg to L. P. Voigt, May 17, 1955, Box 453, Folder 39, WCD.
15. CC Minutes, February 9 and August 12, 1955.
16. CC Minutes, February 9, 1955, and March 9, 1956; Ray M. Stroud to State Conservation Department, May 5, 1955, and L. P. Voigt to Conservation Commission, May 23, 1955, Box 453, Folder 39, WCD Files.
17. Wisconsin Legislative Council, Conservation Committee Minutes, August 24, 1956, Folder 39, Box 453, WCD Files.
18. North Central Field Committee, *Proposed Apostle Islands National Lakeshore, Bayfield and Ashland Counties, Wisconsin* (Washington, DC: United States Department of the Interior, 1965); Grace Nourse, interview by Greta Swenson, Bayfield, WI, 1981, transcript in the AINL Name File, Bayfield, Wisconsin.
19. James Feldman and Robert W. Mackreth, "Wasteland, Wilderness, or Workplace: Perceiving and Preserving the Apostle Islands," in *Protecting our Diverse Heritage: The Role of Parks, Protected Areas, and Cultural Sites*, ed. David S. Harmon et al., (Hancock, MI: George Wright Society, 2004); Robert B. Brander, *Environmental Assessment: Natural Resources Inventory and Management, Apostle Islands National Lakeshore, Wisconsin* (Bayfield, WI: Apostle Islands National Lakeshore/National Park Service, 1981): 127–29.
20. Gregory Summers, *Consuming Nature: Environmentalism in the Fox River Valley, 1850–1950* (Lawrence: University Press of Kansas, 2006).
21. Emmet J. Judziewicz and Rudy G. Koch, "Flora and Vegetation of the Apostle Islands and Madeline Island, Ashland and Bayfield Counties, Wisconsin," *The Michigan Botanist* 32 (March 1993): 43–189; Emmet J. Judziewicz, "Survey of Non-native (Exotic) Vascular Plant Species of Campgrounds and Developed Areas, at the Apostle Islands National Lakeshore," 2000, unpublished report, AINL Library, 8.

4 Changing Forests, Moving Targets in Finland

Timo Myllyntaus

For millennia, white-backed woodpecker *(Dendrocopos leucotos)* had lived in Finland's old-growth boreal forests dominated by birch and other deciduous trees before a marked shift took place. The twentieth century brought a steep decline to this species, with numbers dropping by 90 percent since the late 1950s, making this woodpecker the most endangered of our forest birds. In 1994 only eleven nests of this bird were found in the entire country. The situation at this time was even worse for carabid beetles, a subspecies of ground beetles whose numbers had also been in dramatic decline for decades.[1]

The primary threat to these endangered species was habitat disruption. This did not happen in a small, densely populated country containing minor forested regions: The rapid decrease in biodiversity took place in Finland, which by area is one of the largest, and least-densely populated countries in Europe, with only sixteen inhabitants per square kilometer. Finland is also one of the most forested countries in Europe with approximately two-thirds of its area covered by woods.[2] The country is also Europe's second-largest exporter of paper and paper board.

Forest exploitation in Finland has therefore been intensive. The process of industrialization spurred charcoal furnaces, sawmills, pulping mills, and paper mills. Because Finnish forests are managed primarily for producing timber and other paper products, the southern half of the country preserves less than 1 percent of its forest, and the northern half preserves only 10 percent, resulting in a national average of 2 percent preservation. The Finnish public, though, expects that at least 10 percent of the country's forests be maintained as old-growth. The European Commission has set similar conservation goals. As a result, Finland is now implementing the so-called *Natura 2000 programme*, which has been part of the EU agenda for nature conservation since the 1990s.[3]

Most Finnish forests, especially in the south, are therefore less than 150 years old and hardly "old-growth." Virgin forests have been slashed and burned or cut for various purposes, or destroyed in forest fires. Expanding forest conservation areas thus remains a critical issue: How does one restore old-growth forests in Finland? In the ongoing public debate, there is the wide sentiment that at least part of modern commercial forests should be transformed into old-growth forests. This scheme sounds simple: just leave selected areas alone, and time will take care of restoration. Yet a forest is not a pendulum that simply swings back to its starting point. More importantly, is this "starting point" and pinnacle of conservation always a primeval forest? This is the central question examined in this chapter.

RESTORING A FOREST TO A NATURAL STATE

As Marcus Hall has mentioned, there are two major schools in re-establishing ecological integrity. His generalization is that North Americans *restore ahistoric* conditions, while Europeans prefer to *renature historic* milieus.[4] Yet Finns as well as Swedes and Norwegians are perhaps closer to the idea of bringing back ahistoric than historic conditions. The rationale for Nordic restoration is that in these countries one is supposed to know what changes have happened in landscapes and forests during the past centuries. In Scandinavia, the time between the present and historical forest states is not unfathomable, usually spanning centuries instead of millennia. At a result, the opinion is that re-establishing natural states requires one to set goals. One can therefore ask whether restoration should aim at a predetermined state, which is presumably closely connected to the history of each place, or whether it should be directed toward creating a timeless natural state.

AN OUTLINE OF FINNISH FOREST HISTORY

In Finland's current forest discussion, the focus is generally to restore the "natural" or "original" state of forests. However, as in North America, it is problematic to define Finland's original state of the forest. Finnish forests have gone through dramatic changes, and many of the changes before the sixteenth century took place without human interference. Prehistoric changes in forested areas resulted mainly from climatic variation. Owing to periodic cold eras, plant communities of the boreal zone are amongst the most hearty and sustainable in the world. Trees of this zone have significant resistance to climatic variation and can rapidly recolonize treeless areas.

After the Eemian epoch, which was warmer than our present climate, the latest ice age began about 115,000 years ago. Gradually a continental ice sheet spread from the North Pole toward the south, moving soil, pressing the earth, and making the entire region more flat, while inducing southward migration of biota. The ice sheet grew to two to three kilometers thick in Finland, with the southern ice edge reaching Northern Germany. The ground was permanently frozen from there to the Alps, while tundra and steppes covered major parts of Central Europe. The climate was at its coldest 20,000 to 18,000 years ago.

The European climate turned warmer approximately 14,700 years ago and melting started. Thirteen thousand years ago, ice and water still covered the main part of Finland. A thousand years later, the waning Ice Age exposed the Baltic Sea Rim by allowing the earth's crust to rebound as the ice melted. Birch *(Betula)* spread from the southeast to Fennoscandia,[5] becoming the most common tree species for centuries. Other pioneering trees were pine *(Pinus)* and grey alder *(Alnus incana)*, which grew at first in small groups.[6] Our present, Subatlantic period began approximately 3,500 years ago. In this epoch, the climate has mostly been cool, humid, and relatively maritime. Today's plants found in heaths and marshes stabilized their position. Spruce became the dominant tree species in the climax forests despite poor and arid soils, cliffs, and marshes. Pine was another widespread tree species. Meanwhile human intervention became an important factor in these forests. Human activities began increasingly to modify ecological systems, and as a result, the composition of tree species changed considerably.

A struggle between pioneer and climax species invigorates tree dynamics. In Finland, the former are all deciduous tree species, while Norway spruce is the only definite climax tree (pine can also act as a pioneer species except in nutrient poor soils). As indicated in Table 4.1, tree species can be classified according to those that tend to form forests and those that do not have this quality. Few tree species in Scandinavia and Finland are able to form forests: four deciduous and two coniferous tree species. In North America, ten deciduous and eleven coniferous tree species can build up forests. In Russia, the numbers are twelve and thirteen respectively. The relatively large number of tree species in Asiatic Russia is attributed to the fact that a large part of that region was not covered during the last glacial period.

HUMAN IMPACT ON FORESTS

After the Ice Age, the first people returned to Finland about 6,000 years ago or perhaps earlier. According to archaeological findings, Finland was inhabited even before the latest Ice Age by hunters of moose and deer. Large herds of those animals grazed grass of open heaths that had risen out of the sea following withdrawing glaciers. When forest started to occupy these rich hunting grounds, hunters began to burn forests in order to clear open areas and attract game to graze them again. As people began cultivating grains, the same burning technique was used to clear ground for sowing. Ash from trees was also excellent fertilizer in Finland, where soils tend to be too acidic for cultivation. In the course of centuries, Finns adopted various techniques for cultivating different kinds of forests. There were, however, two main swidden techniques for deciduous forests, one for pine forests and one for spruce forests.[7]

Table 4.1   The Number of Forest-forming Tree Species in the Boreal Zone
(Adapted from Simo Hannelius and Kullervo Kuusela, *Finland, The
Country of Evergreen Forest*, Forssa: Forssan kirjapaino 1995, 13)

Tree species	North America	Nordic Countries	Russia
Deciduous trees			
Birches	4	2	7
Aspen and Poplar	4	1	3
Grey alder	2	1	2
Coniferous trees			
Pines	4	1	3
Larches	1	–	2
Shade trees			
Spruces	3	1	4
Siberian silver firs	1	–	4
Thujas	1	–	–
Hemlocks	1	–	–
Total	21	6	25

The Finns also used their forest for producing tar, pitch, and potash.
Trees were also felled for lumber to build houses, ships, bridges, and fences
locally and abroad. Nevertheless, the main use of timber was heating and
cooking. With a cold climate and inefficient stoves, firewood consumption
was huge in per capita terms. These several uses of timber left their marks
in the forest. The industrialization of Finland led to cutting down most
old-growth forests and replacing them with managed commercial forests,
which are more monocultural than primeval. The main purpose of the
modern commercial forest is to produce timber for wood-processing industries. Most commercial forests grow few tree species in relatively homog-
enous age structures.

Indeed, human activities left numerous traces in the forest and such
traces remained for a long time. Besides timber cutting, there was slash-
and-burning, burning in tar-pits, building log chutes and splash dams,
grazing, constructing roads and villages, and all of these uses left indelible
marks in the forest. Slash-and-burn cultivation was still very common in
Eastern and Central Finland in the mid-nineteenth century, as shown in
Figure 4.1. However, it gradually ended in Finland between 1870 and 1940,
but remnants of this cultivation technique are still preserved in today's forests, as evidenced by the composition of plant and tree species, remnants in
soil, and traces in stones and old trees.

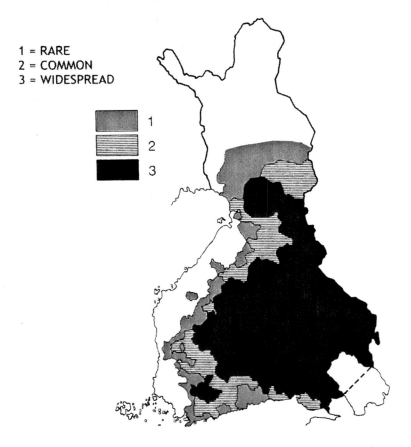

1 = RARE
2 = COMMON
3 = WIDESPREAD

1
2
3

Figure 4.1   The spread of slash-and-burn cultivation in Finland around 1860.

The several historical layers of Finnish forests may themselves become the focus of conservation efforts, reflecting as they do various economic, cultural, and religious activities. How does one conserve these human-generated layers in changing forests? And even though slash and burning belong to Finland's cultural heritage, one may ask whether these landscapes are sufficiently valuable to conserve.

CONSERVATION OF FOREST WITH CULTURAL VALUES

Conservationists therefore need to make choices about which historical environment they would like to recapture. Although one may detect certain traces of former forest use, many historical marks are hidden and covered by soil, vegetation, bushes, or trees. Observers need practice finding and evaluating those marks as well as estimating the period from which they date. For this work, foresters and allied specialists, including environmental historians and historians of technology, are needed to identify cultural and ecological layers in a forest's history, while identifying methods for restoring and achieving conservation goals.

To preserve historical traces and cultural layers, it is necessary to repeat certain human interventions. For instance, if we want to preserve the remnants of swidden cultivation in a forested landscape, it is necessary to slash and burn that area periodically at intervals varying from 30 to 250 years depending on the swidden technique used. Otherwise, the habitat will change crucially, and various ecological systems will disappear completely.

Some cultural landscapes are highly valued. A great many of them are the result of traditional agriculture: forest grazing, field clearing, firewood cutting, road and farmhouse clearing. When these activities cease, spontaneous forests and bushes can be cleared to simulate earlier activities. There is special interest in preserving woodland pastures, meadows, fields, paths, roads, and open lakesides. These open spaces with bright sunshine and sufficient rainfall have helped maintain biodiversity broad and viable. However in thousands of farmsteads, there are not enough resources or interest to preserve traditional cultural landscapes. As a result, afforestation is common, and traditional cultural landscapes are becoming especially rare in remote regions.

ECOLOGICAL RESTORATION

In the late twentieth century, foresters and forest owners in Finland developed a new kind of restoration style under the pressure of some environmental groups who criticized the rapid decline of biodiversity in monotonous commercial forests. The first challenge was to locate habitats of existing valuable species and preserve these patches within commercial forests. If such patches were insufficiently large or widespread, the second challenge was to create such areas artificially by reproducing environments characteristic of old-growth forests.

Ecological restorationists accepted that some kind of intervention in ecological processes is necessary in order to work toward certain goals or ecological states. By anthropogenic steering of natural dynamics, certain natural—or near-natural—states can be achieved. In her study on ecological restoration of forests in Fennoscandia, Saara Lilja states that the objective of restoring a commercial forest is to do so "in such a way that structures and processes typical of natural forests are re-established."[8] Like the Society of Ecological Restoration International, she considers ecological restoration to be an activity that aims to assist the recovery of an ecosystem that has been degraded, damaged, or destroyed.[9] Experimental forest restoration in Finland started in 1989, although it expanded significantly only after 2002. By the end of 2005, roughly 6,000 hectares had- been the focus of restoration activities, with special attention to conservation areas.[10]

Forest ecologists once approached restoration from the perspective of succession theory, with the climax state representing the highest goal. But observations now suggest that succession does not necessarily follow a linear development towards a climax stage, because any steady development path is irregularly shaken by disturbances from which nature tends to recover through successional phases. In reality, climax does not mean stagnation, whereby the ecosystem remains permanent; rather, succession is primarily a theoretical concept. As a result, forest ecologists now consider the goal of restoration to be one of creating "natural" or "nearly natural" forests characterized by four main features:

1. spectrum of successional phases with a multilayered age structure of the standing stock,
2. wide and sustainable biodiversity,
3. sufficient, slowly decaying tree trunks, and
4. ecological disturbances with sporadic intervals

There are now publications in Finnish detailing how to accelerate the restoration of old-growth forests and how to reproduce structures and processes of natural forests. These handbooks and Internet instructions expound on various restoration methods: damaging or killing tree stands; burning these trees; clearing small areas in the forest while restoring hydrological conditions conducive to peatlands and marshes; removing unnecessary roads while creating ecological corridors; implementing planning and monitoring.[11]

Creating Deadwood by Damaging and Killing Trees

Finnish forest ecologists prefer imitating natural disturbances as the primary method for achieving restoration goals. In boreal forests, these kinds of disturbances include damage by wind and storm, fungi, insects, and fire, which can all be lethal or harmful to trees and so can promote ecosystemic biodiversity and natural forest renovation. Dead and decomposing trees are an essential part of the habitat for thousands of endangered species. By one estimate, some 5000 or 25 percent of all species in Finnish forests depend on dead and downed wood.[12] Commercial forests contain only small amounts of rotten trees, averaging just two to three cubic meters per hectare of slash or downed wood. In conservation forests, rotten trunks account for about eight cubic meters per hectare, whereas boreal coniferous forests in a more natural state contain deadwood reaching to 60 and 120 cubic meters per hectare. Rotten trunks are an especially crucial element for Finland's endangered species and are thus a key consideration in ecological restoration.[13] Some 38 percent of Finland's 1505 endangered species are forest inhabitants.[14]

Diversity by Burning

Forest fires have been the most significant of major natural disturbances in boreal forests. Fires both renovate and diversify forests while altering succession by destroying trees and other organisms, as well as by liberating nutritive substances and creating open space and brightness. Rotten trunks and partly burnt trees are habitat for various insects adapted to a frequently burned environment. Estimates suggest that more than one hundred species of vascular plants, fungi, lichens, and invertebrates are fire-dependent in Fennoscandia.[15]

The burning of forests for restoration requires significant planning and preparation. Generally areas to be burned are only moderately fertile because the most fertile areas are too wet to burn. After the burn, some trees will fall and others are left standing. Burning requires a good deal of labor in order to prevent fires from burning out of control. Indeed, firing techniques resemble traditional slashing-and-burning.[16]

Small-scale Clear Cuts in Coniferous Forests

Small areas of some hundred square meters are cleared in monotonous coniferous forests.

The resulting gaps give living space to various plants and especially to deciduous trees, which generally lose in the competition for light and space to coniferous trees. The forest's age structure will then vary and vegetation becomes more diversified.[17]

Removing Roads and Creating Ecological Corridors

Roads tend to split forests into isolated ecological fragments. Restoration-ists therefore remove as many roads as possible, as by plowing roadways into nearby ditches or levelling them so that water can flow freely. In addition, stands of deadwood and pockets of endangered species are often small in a commercial forest. It is therefore necessary to connect these separate pockets with corridors that contain sufficient habitat for allowing species movement.[18]

From Planning to Monitoring

Conscientious ecological restoration is based on the careful study of the forested area. Information is collected on natural conditions such as soil, trees, amounts of deadwood, and deviation from a natural state. This information is saved in a systematic database so as to make future navigation easier. Restoration can be focused on "nearly natural" forests as well as on areas containing few elements of natural forests.

A system of monitoring will allow a better implementation of restoration. Fire effects and intensity will be examined and deadwood will be measured; observations will measure the rotting process, success of seedlings, and the changes in biodiversity, with an eye to insect and fungi diversity.[19] Such measuring and planning are sometimes combined with these proactive methods for better imitating natural processes so as to speed up restoration.

CONCLUSIONS

Forest restoration in Finland faces various challenges. First, the target—be it original or natural forest—should be defined more clearly. There has never really been an "original" primeval forest, as forests have never been stable in any historical epoch. Ever since the Ice Age, Finnish forests have been undergoing constant changes even before human modification. Some of the prehistoric forest types, such as pre-boreal forests, are unrealistic goals for restorationists because the colder climate once present does not provide the requisite conditions. Forest conservation must therefore be carried out within the available climatic framework. When this framework changes, forests will change as well.

Theoretically, it would be possible to restore various conservation areas to their primeval Subatlantic state. Yet the necessary restoration period would be long—probably much more than 150 years. In practice, Finnish forests contain multiple historical marks of human influence accumulating over several centuries, and removing all of them might be possible but prohibitively expensive.

Our Finnish case study has revolved around the issue that a virgin, untouched forest cannot be the only valuable forest type to restore. The many human ways of utilizing a forest have created cultural landscapes that represent significant expressions of Finnish national heritage. Such activities as slash-and-burn cultivation radically changed forests but not all of those changes were negative or detrimental to nature. Slash-and-burn cultivation opened up dim and dark coniferous forests to deciduous tree species and other biota that thrive in brighter sunshine. Natural forest fires and swidden cultivation likewise brought variability to Finnish nature. To preserve the resulting diversity of biotypes and forested cultural landscapes, it is necessary to accept the idea that conservation may require humans to intervene in certain areas.

Strict forest conservation is a complicated issue. Leaving a forest to stand on its own, without management, might lead to an old-growth forest but the result would hardly be a virgin, primeval forest. In addition, forest meadows with their versatile flower and insect communities will disappear without the grazing cattle that formed them. These meadows are important contributions to Finland's environmental diversity and cultural landscapes.

Ecological forest restoration requires a special, pragmatic approach. It does not aim at historically defined landscapes. Rather, its major goal is to create pockets in commercial forests in which certain endangered species can live and survive. Representatives of this school of forestry management think that any movement toward a more biodiverse forest is important, thereby allowing it to fulfill many ecological requirements. For a quarter of a century, multiple-use forestry has been catchword in Finland; but in the last few years, ecological restoration has been stepping in to replace the previous dogma of sustainable multiple-use forestry.[20]

Finnish forest restoration has multiple goals that depend on the particular site and its historical layers as well as on the methods of restoration. Table 4.2 describes alternative, realistic targets for forest restoration.

In 2004, the Finnish government budgeted 284,000 Euro for repairing seventeen streams, restoring 460 ha of marshes, 180 ha of forests, and removing one road in a conservation area.[21] Private landowners are currently restoring thousands more hectares of forests. Thus, Finnish forests are not destined to disappear: human activities can improve the forested condition. In 2006, fifty-three nests of white-backed woodpecker were found containing 140 to 145 young birds, and two years later somewhat more, sixty-nine nests. These and other observations over the past ten years suggest that restorative activities are bringing auspicious results—even if the 1950 level of 900 woodpecker nests is still far away.[22] Finnish restoration is still in an experimental phase but it is already providing important environmental benefits.

Table 4.2 Current Alternative Targets for Forest Restoration in Finland

- Subatlantic primeval forests
- Slash-and-burned forests and their later stages of succession
- Forests used for regular tar production
- Forests used for regular charcoal production
- Forests and meadows used for grazing and twig gathering
- Forests developing toward a set of ecological qualities as defined by restorationists
- Forest conditions that existed before the latest logging

NOTES

1. Valkoselkätikka (2006), http://www.wwf.fi/ymparisto/uhanalaiset_lajit/ kotimaiset/valkoselkatikka.html, accessed 20 Nov. 2006 [hereafter Valko-selkätikka (2006)]; Matti Koivula, *Carabid beetles (Coleoptera, Carabidae) in boreal managed forests—meso-scale ecological patterns in relation to modern forestry* (Helsinki: University of Helsinki, Department of Ecology and Systematics, Division of Population Biology 2001). Online Diss.: http://ethesis.helsinki.fi/julkaisut/mat/ekolo/vk/koivula/ accessed 24 March 2009.
2. Statistical Yearbook of Finland 2004, v. 92 n.s., (Helsinki: Statistics Finland, 2005).
3. Nature and Biodiversity Homepage (2006), http://ec.europa.eu/environ-ment/nature/home.htm accessed March 24 2009; Le portail du résau Natura 2000, http://natura2000.environnement.gouv.fr/ accessed March 24, 2009.
4. Marcus Hall, *Earth Repair, A Transatlantic History of Environmental Res-toration* (Charlottesville: University of Virginia Press, 2005), 6.
5. Fennoskandia is a geologically contiguous area that includes Finland, Nor-way, Sweden, Eastern Karelia, and the Kola Peninsula in Russia. The bed-rock of this area is composed of precambric stone species. In Finland, granite and gneiss are the most common types of rock.
6. Simo Hannelius and Kullervo Kuusela, *Finland: The Country of Evergreen Forest* (Forssa: Forssan kirjapaino, 1995), 43–45.
7. Olli Heikinheimo, *Kaskiviljelyksen vaikutus Suomen metsiin* (Helsinki, 1915); Timo Myllyntaus, Minna Hares, and Jan Kunnas "Sustainability in Danger? Slash-and-Burn Cultivation in Nineteenth-Century Finland and Twentieth-Century Southeast Asia," *Environmental History* 7:2 (2002): 267–302.
8. Saara Lilja, *Ecological Restoration of Forests in Fennoscandia: Defining Reference Stand Structures and Immediate Effects of Restoration,* Dissertationes Forestales 18 (Helsinki: Department of Forest Ecology, 2006), 4.
9. A. Clewell, J. Ricger, and J. Munro, eds., *Guidelines for Developing and Managing Ecological Restoration Projects* (2nd edition, Tucson: Society for Ecological Restoration International 2005).
10. Lilja, *Ecological Restoration of Forests in Fennoscandia*, 4.
11. Tukia, Harri, Marja Hokkanen, Sari Jaakkola, Seppo Kallonen, Tuula Kurikka, Anneli Leivo, Tapio Lindholm, Anneli Suikki, and Erkki Viro-lainen, *Metsien ennallistamisopas*, Metsähallituksen luonnonsuojelujulkai-suja, Sarja B:58, (2nd revised edition, Vantaa: Metsähallitus and Suomen ympäristökeskus 2003), English summary: The Handbook of Ecological Forest Restoration; H. Heikkilä and T. Lindholm, *Metsäojitettujen soiden ennallistamisopas* [Handbook for restoring peatlands, which were earlier ditched forestry purposes], Metsähallituksen luonnonsuojelujulkaisuja. Sarja B:25 (Vantaa: Metsähallitus, 1995, 2002).
12. J. Kouki, "Biodiversity in the Fennoscandian Boreal Forests: Natural Varia-tion and its Managements," *Annales Zoologi Fennici* 31 (1994): 35–51; J. Kouki, S. Löfman, P. Martikainen, P. Rouvinen, and A. Uotila, "Forest Frag-mentation in Fennoscandia: Linking Habitat Requirements of Wood-associ-ated Threatened Species to Landscape and Habitat Changes," *Scandinavian Journal of Forest Research*, suppl. 3 (2001): 27–37; E. Hyvärinen, J. Kouki, P. Martikainen, and H. Lappalainen, "Short-term Effects of Controlled Burning and Green-tree Retention on beetle (Celeopteera) Assemblages in Managed Boreal Forests," *Forest Ecology and Management* 212:1–3 (2005): 315–32.

13. See Web pages supplied by the Finnish National Board of Forestry (2006), http://www.metsa.fi/page.asp?Section=1979 accessed 4 Dec. 2006.
14. Experts estimate that historically there have been about 43,000 species of flora and fauna in Finland. 186 species have become extinct. See Rassi, Pertti, Aulikki Alanen, Tiina Kanerva, and Ilpo Mannerkoski, eds., *Suomen lajien uhanalaisuus 2000. Uhanalaisten lajien II seurantaryhmä*, Helsinki: The Ministry of the Environment (In Finnish with an English abstract: The 2000 Red List of Finnish Species. The 2nd Committee for the Monitoring of Endangered Species in Finland, 2001), 335, 339.
15. L-O. Wikars, "Brand beroende insekter—respons på tio års naturvårdsbränningar," *Fauna & Flora* 99:2 (2004): 28–34 (in Swedish with a summary in English).
16. Finnish National Board of Forestry (2006).
17. Web pages supplied by the Finnish National Board of Forestry (2006).
18. Ympäristöministeriö, *Metsien ennallistaminen* [Ministry of the environment, Restoration of forests] (Helsinki 2006), http://www.ymparisto.fi/default.asp?node=2730&lan=fi accessed 24 March 2009 [hereafter *Metsien ennallistaminen* (2006)], http://www.koitjarv.pri.ee/restoration/fi/metsat.htm accessed 24 March 2009.
19. Finnish National Board of Forestry (2006).
20. Marjatta Hytönen, "History, evolution and significance of the multiple-use concept," *Multiple-use Forestry in the Nordic Countries*, Ed. Marjatta Hytönen (Helsinki: Metla, The Finnish Forest Research Institute 1995), 59.
21. *Metsien ennallistaminen* (2006).
22. At the present it is estimated that 75–80 couples of white-backed woodpeckers are nesting in Finland. Valkoselkätikka (2006); Valkoselkätikka (2009), http://www.wwf.fi/ymparisto/uhanalaiset_lajit/kotimaiset/valkoselkatikka.html accessed 24 March 2009.

5 Sidebar
Clementsian Restoration In Yosemite

William D. Rowley

Tourist growth and road building in Yosemite National Park prompted park officials to seek restoration of disturbed landscapes as early as the 1930s. These restoration efforts brought famed American ecologist Frederic E. Clements and spouse Edith Schwartz Clements to the park. They had recently taken up residency in Santa Barbara, putting them in close proximity to Yosemite, located in California's central Sierra. Not only was the restoration of native vegetation along roadsides among their concerns, but they also focused on how the park's Museum Gardens might be designed for representing the local native wildflowers, especially the flowers of the Museum's site in Yosemite Valley.

As a leading and controversial figure in American ecology, Frederic believed that vegetation and its development through stages toward a stasis or climax stage formed the basis of a utilitarian science that could address agricultural problems and now restoration of a national park. In the spirit of a scientist serving the public, Clements had pursued a career in academia and as a consultant with the U.S. Department of Agriculture on drought and livestock forage production. At Yosemite Park, Frederic saw his work as offering "educational opportunities" to demonstrate the value of ecological science in directing restoration projects.

Clements's "dynamic ecology" postulated an image of vegetation growth that passed almost mechanically through stages from the primitive to the complex leading to stable vegetation complexes or climax communities. Clements believed that his ecological theories offered "great possibilities" for "restorations as are proposed for the floor of Yosemite Valley, where the original flowery meadows have been entirely dispossessed by seedy grasses, as it does likewise for refuges and reserves of various kinds."

One of his park projects, which involved "Mrs. Clements," included the reorganization of the Museum Gardens whose current state resembled a mishmash of wildflowers and shrubs brought together from all parts of the park for public admiration and display. By 1928, Edith had built a reputation in wildflower identification, having published books and articles on the subject. Their project explored how considerations of aesthetics and ecological succession could guide the design of public gardens and the display of native flowers.

The park's two-acre garden had become controversial because there were concerns over maintaining an artificial "garden" within a park, especially because its flowers derived from points across the park and did not grow there naturally. Frederic held rigid ideas on what manner and place the wildflowers should be planted, i.e., according to families or according to stages of successional transition that would lead, he believed, to a rich variety and diversity of species that represented a phylogeny or climax of the organic community. He felt that reorganizing the garden along these lines would also serve the educational goal of displaying ecological succession to the public. Yet park officials objected to his proposal because they feared its projected diversity in the succession process would fail to produce a color display for impressing visitors. Assistant Park Service Director H. C. Bryant felt that it was unimportant that the garden represent phylogenies, noting that "very few other botanists agree with him."

But Frederic Clements rarely conceded to his critics, accusing those who resisted his plans for the park's garden as sacrificing the truths of science to the "thrills" of color. "A natural landscape demands a natural treatment by means of an intelligent application of nature's processes," asserted Clements. "This is not merely the method of economy, but it produces as well the highest values in ornamentation, recreation and education." According to Clements, Nature was the perfect landscape architect who offered every desired effect. He and his collaborator, Edith, refused to abandon the

Figure 5.1 Yosemite Museum Wildflower Garden, July 7, 1933. (Courtesy of the Yosemite Research Library, National Park Service)

phylogenous and static climax model, which even today continues to serve as a conceptual aid in some restoration work.[1]

NOTES

1. All quotes located in the Central Files Yosemite Roads, 1931–1936, Box 51, Yosemite Museum Library, Yosemite, California, and the Fredric E. Clements Collection, American Heritage Center, University of Wyoming, Laramie Wyoming. On the continued use of the climax model, see Mark W. Brunson, "Managing Naturalness as a Continuum: Setting Limits of Acceptable Change," *Restoring Nature: Perspectives from the Social Sciences and Humanities*, eds., Paul H. Gobster and R. Bruce Hull (Washington, D.C.: Island Press, 2000), 236 and Hagen, *An Entangled Bank*, 32.

Part II
History in Restoration

6 Does the Past Matter in Scottish Woodland Restoration?

Mairi J. Stewart

In short, the Highlands are a devastated countryside. . . . It is quite certain that present trends in land use will lead to it, and then the country will then be rather less productive than Baffin Island.

—Frank Fraser Darling, 1955

Scotland is one of the least wooded countries in Europe. Most of its forest cover had already been lost by the Bronze Age through a long process of deforestation, the result of a combination of climate change and human intervention. This trend was essentially reversed during the twentieth century, particularly since the Second World War, largely through establishment of new commercial conifer plantations. Native woodland decline, however, continued until the closing decades of the twentieth century, when a significant "reforesting Scotland" movement developed, intent on halting this. The movement embraced the ideas of prominent ecologist, Frank Fraser Darling, who in the 1950s had described the Scottish Highland landscape as a "wet desert," at the same time lamenting the destruction of its "natural" climax vegetation, the Caledonian pine forest.[1]

There are, however, important questions to be asked about the motives behind this movement and what is meant by restoration in this context. Is the intention to try to recreate a wooded landscape analogous to its "natural" pre-human state, or to re-establish natural processes? Central to this discussion is the question of the way in which these initiatives considered the past as part of the management planning process. I will explore these issues and question the basis for the assumptions upon which woodland restoration in Scotland is currently based. By drawing on case studies from across Scotland, I aim to demonstrate firstly that motivation is varied and ranges from being blatantly financially driven, to an almost evangelical conviction that a wrong is being righted. Secondly, while history has undoubtedly become popular, it is often simply a token used to tart up promotional literature and indeed sometimes discounted altogether as a key element in management planning or simply used to rationalize pre-conceived goals. Nonetheless, I suggest that an awareness of how land, and more specifically woods, were used in the past is increasingly regarded as an important pre-requisite for future land management strategies, giving specific examples of how history can help.

This chapter concludes that it is not the role of historical studies to pre-scribe future management strategies. Nevertheless, they can inform and stimulate debate about management directions and perhaps even challenge current assumptions through a proper understanding of how landscapes have evolved and what they have meant to people in the past.

SCOTTISH WOODLAND—PAST AND PRESENT

When the ice finally retreated from Scotland over 10,000 years ago, the exposed landscape was soon recolonized by trees—Scots pine, birch, hazel, oak—that are immediately recognizable as native species today. When and exactly how this widespread "original-natural" forest was reduced to a tiny fraction of its post-Ice Age extent is a matter of debate. But it is clear that it was already dramatically reduced by the Bronze Age, a result of the com-bined impacts of deliberate clearance and other human activity, as well as long-term climate change over the previous millennia.[2]

Nor is this history a tale of abject linear decline: attempts to protect and expand the woodland resource have been encouraged since at least the fif-teenth century (and, for all we know, long before). Nevertheless, the trend was essentially downward, so that by the mid-eighteenth century it has been estimated that only about 4 percent of Scotland's land area supported woodland. Even with the brief phase of industrial coppice associated with the Napoleonic wars of the late eighteenth and early nineteenth century, which saw woodland reach a level of commercial value that seems to have checked, for a time at least, deterioration and fragmentation, it was only a slowing down of a process that was to continue through the nineteenth and twentieth centuries. By the mid-nineteenth century, when the vogue for planting introduced trees was well established and estates were producing timber to fuel the industrial revolution, the overall area under woodland was still declining and Scotland and the rest of the British Isles continued to be a net importer of timber.[3]

About 17 percent of Scotland is currently covered in woodland. This ranges from the vast area of planted conifers, about 65 percent of all wood-land, which can be found mainly in the southwest and the Highlands, to the smaller fragmented areas of semi-natural woodland accounting for about 20 percent of the cover. In European terms this is a proportion similar to Spain (17%), greater than Ireland (8%) and England (4%), but still well below Sweden (60%), Germany (31%), and the EU average of 33%. How-ever with only 4 percent of Scotland still supporting native, semi-natural woodland, Scotland contains a smaller proportion of this woodland type than most other European countries.[4]

REFORESTING SCOTLAND—ENVIRONMENTAL BACKLASH

This leads us to ponder why a small country, renowned for its inspirational landscapes of loch and glen, the stark grandeur of an essentially treeless land, should spawn such a vibrant woodland restoration movement. These statistics provide a hint of the basis for much of the motivation for Scotland's reforesting movement. From at least the second half of the twentieth century, these remnants—the last fragments of Scotland's ancient forest, which by the twentieth century was largely neglected, frequently overgrazed and diminished by felling, burning, and underplanting—were increasingly recognized as being "on their last legs" and in danger of being lost if nothing was done to save them. This, at least was the rhetoric of late-twentieth century conservationists. That these native woods were under pressure is not in dispute. For example, since the Second World War, some 40 percent of Scottish coppice woodland and 40 percent of Highland birchwoods had been converted into conifer plantation. This was a period which witnessed the UK Government *via* the Forestry Commission (FC) acting as both regulator (private and public) and manager of the state's forest holding, not only ignoring the continuing decline of native woodland, but also playing a leading role in bringing about their wholesale conversion to plantation. Following pressure from conservation groups including the Government's own nature conservation agency, policy was reviewed and new guidelines were introduced in 1985. This more or less halted the further financing of conifer planting in native woodland.[5]

By the late 1980s public opinion was shifting, a consequence perhaps of a change in perception of the value of woods and trees, which were being increasingly recognized for their recreational and ecological value. The primacy of timber production started to be challenged and notions of sustainability and biodiversity began to seep into the public psyche. The catalyst for this change in attitude included so-called "tax dodge forestry," which from the late 1970s involved a system of tax allowances for encouraging large-scale afforestation on bare ground. This resulted in a rapid expansion of commercial conifer plantations and significant environmental damage in parts of Scotland, notably in the Flow Country, an extensive area of peatland in northern Scotland. Government grants were handed out for planting this land, with little public benefit since many of these forests were unlikely ever to show good returns on maturity. Only with fiscal reform in the late 1980s, which removed these tax incentives and pressure from a disenchanted public and conservation movement, was a reversal of this government policy achieved.[6]

It was against this background that groups and initiatives emerged, championing the restoration of Scotland's native woods. Many of them purported to be dedicated to recreating forests of the past, but often without being explicit about when in the past—the so-called climax forests of the Mesolithic or perhaps the Bronze Age, or even the industrial coppice phase of the eighteenth and nineteenth centuries that had effected significant alteration in species composition and structure. Furthermore, for each area of the Scottish landscape chosen for restoration, did these *reforesters* actually know what their chosen site had been like in the past? Or were they inferring from what the likes of Fraser Darling had written fifty years earlier about the *Great Wood of Caledon*: a wood it was said, dominated by the native magisterial Scots pine, which once clothed much of Scotland and which was decimated by outsiders—Roman legions, Vikings, English ironmasters, and the like—leaving by the mid-twentieth century a fragmented remnant of the once bountiful forest, a shadow of its former glory. Such notions were grasped by Scottish environmentalists in the 1980s, captivated by the idea that they could and should be redressing the wrongs wreaked by man on nature in the past. It was a powerful image propagated by increasingly high profile individuals and activists, which compounded by the backlash created by the so-called environmental vandalism inflicted on Scottish moors and bogs by afforestation, together captured the wider pubic imagination and legitimized politically and socially a level of public and private spending on native woodland restoration hitherto unprecedented in the Scottish conservation scene.[7]

The Scottish reforesting ethos, based on a restitution approach to the environment, may have been more recently questioned by palaeoecologists and historians, but its influence has permeated deep into the nature conservation mindset in Scotland. Notwithstanding there are a diversity of approaches to woodland restoration, there remains at the core of many projects the vision of restoring the woods and the land to authentic wildwood or what Dizard (Chapter 14, this volume) refers to as the "real thing." But how do they know what the "authentic" is and if they are trying to re-create it, then how much do they consider what is known about past processes and the histories of the sites chosen for restoration? It is certainly possible to see some of the same ideas prevalent in North American restorationist approaches. Visions of landscape-scale rewilding, a focus on keystone species, and habitat corridors are all evident in Scottish projects, perhaps the difference with North America being in terms of scale and focus. Scotland certainly has most scope within the UK setting for such initiatives because of the relatively large tracts of upland semi-natural habitat with wilderness qualities. In comparison with North America and other countries with vast areas of scantily populated land however, opportunities for reforesting of the rewilding type are severely limited and arguably also inappropriate, not least because Scotland's landscape, like much of Europe is cultural, whereby "truly" wild land has not existed for thousands of years.[8]

SCOTTISH WOODLAND RESTORATION—CASE STUDIES

Foremost among groups endeavouring to rewild is Trees for Life, a charity set up by a member of the Scottish-based New-Age, Findhorn Foundation. According to their literature (strongly resonant of Fraser Darling),

Scotland is a prime candidate for ecological restoration work, as it is one of the countries which has suffered most from environmental degradation in the past. The Highlands in particular have been described as a 'wet desert' as a result of the centuries of exploitation which have reduced them to their present impoverished and barren condition.

Trees for Life's approach is based on recreating a wild forest, there for its own sake, which will be restored in a 600-square-mile target area in the Highlands, based on ecological principles such as "nature knows best" and mimicking nature, focussing on keystone species such as beaver (extinct since early sixteenth century) and working out from refugia of "intact" ecosystems. Notwithstanding that there is an inherent contradiction in their policy of non-native removal and volunteer-led tree planting program (the latter flying somewhat in the face of the nature-led emphasis), there is little recognition of the cultural origins of the landscape or of its value as such, even though there is evidence that some of the areas that they have been "restoring" have not supported trees for 4,000 years. The past only seems to matter as a place where an ideal state, the "real thing"—to their way of thinking at least—existed.[9]

The Carrifran Wildwood project, based in the Scottish borders, has taken a similar approach. A charity set up by a group of keen local activists in the late 1990s, their vision is to recreate native forest in a 1,600 acres, largely treeless, valley. This group differs from Trees for Life in that they own all of the land that they are restoring. They are more explicit about their intentions, i.e., re-introducing trees to a landscape that has been largely without them for 6,000 years. Palaeoecological research is also informing the kind of forest that they are intending to create and although they might be criticized for proceeding rather hastily—almost all the land designated for planting was planted within eight years—it is an approach driven partly by pragmatism since the government and other grants that are paying for the tree planting are time-limited.[10]

Funding for such projects is an important consideration given that any attempt to restore, regenerate, or rehabilitate usually requires considerable financial support for fencing, herbivore control, planting, etc. Much of the reforesting Scotland movement as it developed in the last decade of the twentieth century was fuelled by the millennium lottery windfall, so that many of these programs by their very nature required to be complete, or nearly so, by the turn of the century. No matter how much an ecological restoration project desires to mimic nature or adopt a minimum intervention approach, all too often the short timescale of funding sources undermines intentions to let nature take its course. Financial constraints are not the only cause for pushing nature along, of course, for while some projects appear to embrace a nature-led approach, this is only so long as nature does it in a way that is approved of; otherwise intervention takes place. There are many examples of this "gardening" approach in Scottish woodland restoration projects. For example, having invested considerable sums in deer control in the FC owned Glengarry, the prolific birch regeneration has outstripped the intended native Scots pine regeneration (a tree always favored by Scottish foresters), so much so that the birch is being controlled in order to "create a more natural woodland." Even though palaeoecological research from other Scottish pinewoods suggests that in the past there was a greater broadleaf component in such woods.[11]

An important successor to lottery funding has been the £10 million BP funded Scottish Forestry Alliance (SFA), which aims to create 10,000 hectares (ha) of new woodland in Scotland. One SFA project, on the island of Skye, provides an example of history being used as a means to justify preconceived goals. Kinloch Hills, owned by the FC, is a 7,400 ha landholding rising from sea level to over 700 meters. Most of the area is open ground and over 5,000 ha are designated as a candidate Special Area of Conservation (SAC) under European Union legislation, mainly for heath and blanket bog communities, but also for relatively limited areas of oak and birch woodland (8%). Much of the lower ground was planted with exotic conifers in the 1970s, making up 31 percent of the site. The five-year SFA funded project aims to "to recreate the past landscape," but without stating which time period is the desired outcome, or what is the basis for the clear assumption that this area was an extensively "forested" landscape in the past.

There exists for Kinloch Hills some knowledge of past occupancy, management, and land uses through archaeological survey and historical research, but other than interpreting a nineteenth-century ruined settlement, no proper use of historical information has been made in management planning. This brings us to the thorny question of whether Scots pine is native to this Scottish island. We know from palynological research at a site adjacent to Kinloch Hills that it was present from 4,600BP until 3,900BP, disappearing around the time of the big pine decline in north west Scotland. So, should it be regarded as a tree suitable and appropriate to plant today? In Kylerhea glen, which dissects the project area, Scots pine sourced from the nearest extant native pinewood on the mainland was planted in the early 1990s, but under the current project, this area was earmarked for deconiferization. This decision was not taken solely on the basis that historically Scots pine was only present for a short period several thousand years ago, but because restoring some open ground habitat was deemed desirable and this area, because it was not previously ploughed, could be easily restored. Meanwhile, on the other side of the glen, which is open ground and scenically striking, but vegetationally somewhat uninteresting, plans involve creating a significant area of new planted broadleaved woodland.[12]

The project has paid lip service to the past, using it simply to provide the language of "spin" for public consumption. A more honest approach might have been to admit that this project is mainly about planting trees, albeit mainly native ones which could well enhance future biodiversity, rather than to suggest it was about "recreating the landscape of our ancestors." Surely an understanding of the landscape and human history of this area might actually have helped evolve management policy for Kinloch Hills. For example, regarding the Scots pine issue, if we accept that it died out naturally when Scotland became wetter some 4,000 years ago, then surely it is worth considering whether it is suited to the ecological conditions prevailing on Skye today.[13]

Furthermore, it is worth remembering that humans have lived for thousands of years in this area and therefore presumably used their surrounding woods—sustainably or not—for the duration of their occupancy, even if we do not always know exactly how the land was used. It could therefore be argued that restorationists have a range of management choices to consider from the past. Grazing, for example, was an important use of most woods in Scotland, which has considerably affected structure and ground flora, and this was clearly the case for Kinloch Hills. Rather than simply focusing on reducing red deer (necessary though this strategy may be), should the Kinloch managers not be considering other forms of grazing, such as cattle? In UK terms, at 7,500 ha this area might be a candidate for naturalistic grazing along the lines advocated by Vera.[14]

The approach taken at Kinloch Hills contrasts with management of the Clyde valley woods, which lie in one of the most heavily populated parts of Scotland, and which are considered to be one of the best areas of riverine and gorge woodland in the United Kingdom. As such, around 500 ha are designated a SAC, which led to their inclusion in a recent EU-funded restoration project, based on the increasingly popular notion of habitat networks. Ecological networks, as defined by the Council of Europe (1998), are: "Systems of landscape elements that are designed and managed in a way that restores ecological functions and conserves and enhances biodiversity." This approach is essentially about habitat management and enhancement, rather than grand-scale ecosystem restoration so prevalent it seems in North America. [15]

Current management strategies for the Clyde valley woods are being exclusively determined by their perceived ecological value; however, the need to understand the history of past management and human interaction prompted the commissioning of an historical assessment of some of the woods. The subsequent report explored the potential of woodland history to inform the present and the future. The study demonstrated that the work of historians, which can reveal a great deal about woods and landscapes over the past few centuries, can also contribute to future management approaches. This is now beginning to be understood by those charged with their care, for the very good reason that their current form is a fundamental product of that past relationship. Nevertheless, the debate about how exactly the information provided by woodland histories should be integrated with specific management strategies and forestry policy, more generally, is still in its infancy in Scotland.

WHAT ROLE FOR HISTORY?

The Clyde valley has probably supported woodland for millennia; however, the current structure and composition of these woods has more to do with their more recent past. Historians are well placed to illuminate this period of their history and therefore help inform decisions about future management. This might include identifying non-wooded areas, where woodland existed in the past (say up to 300 years ago), which could form the basis of re-establishing links with existing woods as part of a Forest Habitat Network. With large upland candidate reforesting sites such as Carrifran, palaeoecology, which sheds light on deep history, can also be beneficial in developing proposals. Together these disciplines can offer a powerful tool for ecological restoration.

Some other potential applications for historical studies include the reconstruction of past management events (e.g., phases of planting) and relating these to their current condition, which would help ecologists and managers understand the woods and provide context for future management. We can also help identify features of value, so that managers can focus management in a way that will enhance or maintain the feature, such as wood pasture or industrial archaeology artefacts. Provision of plausible explanations for variation in stand structure and woodland composition can be offered, which might not be solely attributable to ecological factors. For example, identifying the extent to which oak now occurs on sites not suited naturally to it.

This list is not exhaustive and, as more historical research is undertaken as part of the management planning process, further applications will emerge. History can also provide descriptions of how people used and related to woods in the past, information that can be a valuable tool in raising awareness about a habitat that is not as relevant to people today as it used to be in the past.[16]

CONCLUSION

Many Scottish reforesters have been profoundly influenced by the deterioration and extirpation of woods and trees in living memory. Added to this the rapid expansion of conifer plantations at the expense of moor, bog, and native wood, the emergence of green politics and a liberal sprinkling of eco-romanticism, and the roots of the reforesting Scotland movement become discernible. It would be problematic to suggest that this is a rewilding movement, or that there is a uniformity of approach, but there are some common factors. Many initiatives are influenced by a rationale which had its origins in the genesis of the Forestry Commission after the First World War; that is restitution of the Scottish forests by planting. Scotland is a nation of planters.

It is perhaps now time to look back, before the twentieth century and see the history of the Scottish landscape in all its complexity and flux. Collaboration between historians on the one hand, and ecologists and managers on the other, is imperative, if we are to gain insight into the fundamental nature of our environment. Recognition by ecologists and managers of a role for history is fairly recent, as demonstrated by some of the case studies presented in this chapter. Other examples given here clearly demonstrate that it is easy to misuse and misunderstand the past. Much of the focus for Scottish woodland restoration is founded on the interpretation of the past by the likes of Sir Frank Fraser Darling, an interpretation that environmental historians would today seriously question. Such influences run deep within the Scottish restorationist's psyche, but as more collaboration develops, as it surely will, ecological restoration in Scotland will increasingly benefit from an understanding of the past.

NOTES

1. F. Fraser Darling, *West Highland Survey: An Essay in Human Ecology* (Oxford: Oxford University Press, 1955), 192; See A. L. Davies, this volume, for account of Scottish deforestation.
2. George F. Peterken, *Natural Woodland: Ecology and Conservation in Northern Temperate Regions* (London: Chapman & Hall, 1994); Also, H. J. B Birks, "Scottish Biodiversity in a Historical Context," in L. V. Fleming, A. C. Newton, J. A. Vickery, and M. B. Usher (eds.), *Biodiversity in Scotland, Status, Trends & Initiatives* (Edinburgh: Scottish Natural Heritage, 1997).
3. For estimates of eighteenth-century forest cover, see T. C. Smout, *Nature Contested, Environmental History in Scotland and Northern England since 1600* (Edinburgh: Edinburgh University Press, 2000), 46; Also T. C. Smout, A. R. MacDonald, and F. Watson, *A History of the Native Woodlands of Scotland 1500–1920* (Edinburgh: EUP, 2005).
4. N. MacKenzie, *The Native Woodland Resource of Scotland: a Review 1993–1998, Technical Paper 30.* (Edinburgh: Forestry Commission, 1999); See also C. Inskipp and M. Mathers, *Forests and Woodlands in Scotland* (Aberfeldy: WWF, 2004); and the government's *Scottish Forestry Strategy* (Edinburgh: Forestry Commission Scotland, 2006).
5. For statistics see WWF report (2004). While these guidelines were introduced in 1985 with a presumption against planting conifers in native woodland, this practice continued until the end of the decade, demonstrating the deeply engrained attitude of the forestry industry towards native and semi-natural broadleaved woodland. For a brief history of Scottish forestry in the twentieth century, see D. Foot, "The 20th Century, Forestry Takes Off," in T. C. Smout (ed.), *People and Woods in Scotland: A History* (Edinburgh: Edinburgh University Press, 2003), ch. 7.
6. Studies carried out by the Royal Society for the Protection of Birds confirmed the low rate of return to the nation from forestry in the Flows. The wetness and depth of the peat and their high exposure to wind all reduce growth rates and attacks by the pine beauty moth have already caused severe damage to some plantations. The total area of blanket bog in the Flows is estimated to have covered 401,375 ha before afforestation. By 1987 at least 79,350 ha were either planted or programmed for planting, and only eight

of 41 hydrological systems in Caithness and eastern Sutherland had been left free from afforestation. The area lost to forestry was considered by the Nature Conservancy Council (now replaced by Scottish Natural Heritage in Scotland) to represent perhaps the most massive single loss of important wildlife habitat in Britain since the Second World War, Nature Conservancy Council, *Nature Conservation and Afforestation in Britain* (Peterborough: NCC, 1986).

7. For a full critique of the "myth" of the Great Wood of Caledon, see T. C. Smout, *Nature Contested, Environmental History in Scotland and Northern England since 1600* (Edinburgh: Edinburgh University Press, 2000): ch. 2; The Scottish Green Party, *A Rural Manifesto for the Highlands* (1989) was written by Bernard Planterose, co-founder of Reforesting Scotland.

8. There is a clear problem of the definition of restoration both from a transatlantic perspective, but also between academics and practitioners. Some of the larger Scottish restoration projects could be described as rewilding and although this may be their intention, this is not their terminology, using instead rehabilitation, regeneration, and restoration. Others clearly fall into the categories that Smout (Chapter 19, this volume) identifies as "gardening" and "landscape-scale change" although again these are not terms that the practitioner would apply to their work. See Davies (Chapter 7) for a critique of current restoration approaches and Dizard (Chapter 14) for a discussion of North American approaches, both this volume.

9. For details of the work of Trees for Life see http://www.treesforlife.org.uk/tfl.eco.html accessed 25 March 2009. This includes a stated "Nature knows best" and "Let Nature do most of the work" approach. TFL now own the 10,000-acre Dundreggan estate, lying at the southern margin of their target area, which includes Glen Affric.

10. R. Tipping, "The Application of Palaeoecology to Native Woodland Restoration: Carrifran as a case study" in A. C. Newton and P. Ashmole (eds.), *Native Woodland Restoration in southern Scotland: principles and practice* (Ancrum: Borders Forest Trust, 1998): 9–21.

11. Funds for much of this planting came from the Millennium Forest for Scotland, an organization set up using money from the UK national lottery. G. Donaldson, "Native woodland management and Forest Enterprise Scotland," in *SWHDG Notes* X (Perth: Scottish Woodland History Discussion Group 10th annual conference proceedings, 2005): 14; See Davies (Chapter 7) this volume for work on Glen Affric pinewoods and S. Pratt, "Reconstructing Past Landscapes at Abernethy Forest: Some New Insights from Palaeoecology," *SWHDG Notes XI* ((Perth: Scottish Woodland History Discussion Group 10th annual conference proceedings, 2007).

12. For pine decline in Scotland see Birks, *Scottish Biodiversity*. For Skye, per. com. Althea Davies.

13. It is worth bearing in mind that the multi-national oil company BP is the principal funders and have an interest in carbon offset value of tree planting.

14. Frans W. M. Vera, *Grazing Ecology and Forest History* (Oxford: CAB International, 2000); See also Hodder and Bullock (Chapter 21) this volume for a discussion of appropriateness of naturalistic grazing in the United Kingdom.

15. Funding for the "Core Forest Sites for a Forest Habitat Network" project comprises nearly £1.9 million of which £0.76 million is provided by European Union funding. LIFE funding is made available from Europe for the conservation of habitats considered of international importance to the Union. For further information on the ecological restoration approach taken for this project, see G. F. Peterken, *Clyde Valley Forest Habitat Network, Scottish Natural Heritage Commissioned Report F99L109* (Edinburgh: SNH,1999).

16. P. Sansum, M. J. Stewart, and F. J. Watson, *A Preliminary History of the Clyde Valley Woodlands.* (Munlochy: Highland Birchwoods, 2005); R. Thompson and A. Peace, *Stand Dynamics in Tilio–Acerion Woodlands of the Clyde Valley* (Munlochy: Highland Birchwoods, 2005); Smout et al., *Native Woodlands* (2004).

7 Palaeoecology, Management, and Restoration in the Scottish Highlands

Althea Davies

It is impossible not to form opinions about a landscape or habitat from viewing or working in it. These preconceptions come to the fore in the field of ecological restoration, particularly the concept that a target habitat is damaged or somehow inadequate, and that these faults will be repaired, as there is some "better" or "more correct" state to which it ought to be returned or transformed. While some measures of degradation, such as soil erosion or biodiversity loss, and the practical implementation of restoration are based on ecological *science*, defining "degradation" and establishing restoration goals may be affected by *value judgments*, which are firmly based in current social, political, and economic values. Furthermore, there is a lack of ecological evidence to support decision-making in many instances, including the long-term impacts of reduced grazing pressures, while most of the ecological benefits of 'rewilding' are, by their very nature, uncertain and thus incompatible with current target-driven management strategies. Regardless of how it is defined, restoration calls explicitly for certain aspects of a landscape to be valued above others. In this context, history can be used to demonstrate degradation and so justify conservation and reparation, particularly in restoration schemes that seek to establish more natural systems which limit human pressures. While a long-term perspective can provide an opportunity for critical debate about management and restoration perceptions, baselines and goals, history is also at times used selectively to justify preconceptions or predetermined goals.[1]

This issue is examined from a Scottish perspective by presenting two enduring and iconic facets of the Highland landscape, both of which are frequently used to promote the area as a "wild land": Caledonian pinewoods and open upland moors or heaths. Case studies based on pollen analytical work will be drawn from a range of sites across northern Scotland to illustrate how long-term vegetation and land-use history have been applied in restoration (Figure 7.1). While there are many cases where humans have caused considerable ecosystem degradation across the globe, these examples have been selected to illustrate the complexity of human–habitat interactions since the damage caused by humans may at times be overplayed or in other instances go unrecognized, while biodiversity benefits associated with maintaining or reinstating management are discussed less frequently. This long-term perspective of the interplay between people and ecological processes will form the basis for discussing the restoration issues faced by current managers. I propose that landscape history and environmental history can provide a valuable cross-disciplinary forum for challenging preconceived ideas about landscapes and questioning the validity of current targets. I conclude by calling for greater clarity of definition and honesty with regard to the restoration debate, particularly the subjective value judgements that this involves. While particular organizations are mentioned in these case studies, no criticism of their work is intended or implied since conservation and restoration aims are, it is argued, a matter of cultural choice, regardless of whether they are labelled as rewilding, renaturing, or something else.

A TRIP TO THE "WILDWOOD": THE ANCIENT AND VENERABLE PINEWOODS

The case studies are taken from my work on environmental archaeology and environmental history, especially the analysis of pollen sequences that contain a record of long-term vegetation change and land-use history in the abundant peat and heather moorland, which make up around 17 percent of the semi-natural land cover in Scotland. With "ancient" woods (*i.e.,* woods continuously present on maps since 1750) forming only 2 percent of Scotland's semi-natural woodland cover, it is easy to understand why native woods have assumed a powerful role in Scottish conservation. This applies to Scots pinewoods in particular. Ensuring the regeneration and encouraging the spread of fragmented pinewoods are central to current management strategies to restore the pinewoods. However, the ecologist Henry Steven's statement "to stand in them is to feel the past" illustrates how these woods are often thought of as embodying more than just a currently rare habitat. Indeed one of the most enduring myths regarding the Highlands is that of the "Great Wood of Caledon," which was said to have cloaked the uplands, preventing the Romans from penetrating further north, and which was thought to have been finally destroyed over the last three centuries, first by the demand for charcoal for iron smelting and then by sheep grazing. These myths have been effectively refuted by palaeoecologists, archaeologists, and historians alike, although that does not stop them from recurring in conservation debates. While there has been a move away from seeing historical snapshots as utopian models or "Golden Ages" to try and recreate, there is continued debate over the role of humans in their loss and fragmentation, and the extent to which tree cover "should" exist in the predominantly treeless Highlands. Case studies are presented from two areas to consider how the past can inform and challenge restoration and management debates.[2]

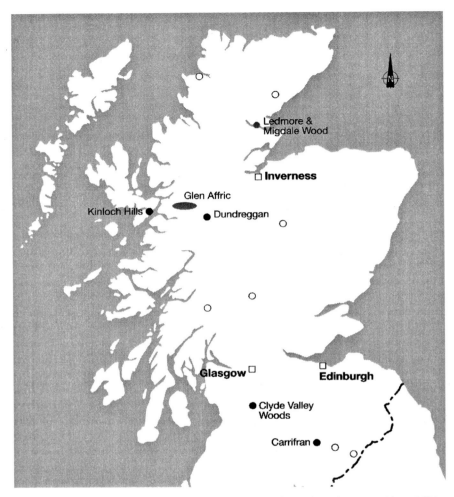

Figure 7.1 Location of Scottish sites mentioned in this chapter (Glen Affric, Migdale; open circles denote long-term biodiversity study sites) and in the previous chapter by Stewart (Kinloch Hills, Dundreggan, Clyde Valley, Carrifran).

The first relates to Glen Affric National Nature Reserve (Figure 7.2), where the pine- and birchwoods of East Affric stand in contrast to the open blanket peat of West Affric: here pine stumps eroding from the peat provide a stark indication that this may not always have been a treeless landscape. This is the view taken by the charity Trees for Life, who view the current pinewoods as a link with the past and whose goal is to establish a 600-square-mile woodland restoration area, focussed around Glen Affric. Towards this end, they are engaged in tree planting and monitoring the regeneration of the trees, particularly those in enclosed areas which aim to limit deer browsing.

The National Trust for Scotland, owners of the West Affric Estate, initiated research into past climate, vegetation cover, and land-use in order to take an informed approach to the management of the area, particularly in relation to the extent, composition, and fate of past tree cover. Much of this work has been published elsewhere, so a summary is presented here as a basis for considering the emphasis placed on woods, particularly pinewoods, and the role of climate change and people in the loss of woods.[3]

Figure 7.2 Scots pine tree. There is little doubt that gnarled "granny" pines do give a strong impression of ancientness, in addition to the rich biodiversity and threatened species which pinewoods support, including red squirrels and crossbills. (Photograph by Althea Davies)

In many ways, the story of woodland cover in Glen Affric is a classic representation of the history of much of northwestern Scotland since the disappearance of the last ice cover, some 11,000 years ago. Between c.9300 and 8000 years ago Scots pine expanded from glacial refugia to invade the birchwoods, which developed from the heath and tundra that replaced the ice sheets. However, closed-canopy woodland never blanketed the landscape and neither have the woods been stable for long periods. Instead, prehistoric Highland woods consisted predominantly of a dynamic mosaic of pine and birch, with rowan (mountain ash), some alder and hazel, all of which formed an open or scattered distribution amongst heather and expanding peat communities. Vegetation patterning was continually shifting, as each generation or cohort of trees became moribund and decayed, followed by the regeneration of a different configuration of these species. This was in addition to disturbances such as windthrow and possibly fire, although burning was more common in heaths than in woods. These wood-heath mosaics existed on the extensive acidic and peaty soils that dominate Highland Scotland throughout the former distribution range for pine.[4]

Despite the impression conveyed by the huge interest (both ecological and palaeoecological) in Scots pine, past Highland woods were not dominated solely by pine. High resolution pollen sequences from West Affric, which allow past vegetation to be understood on the scale of woodland stands, show that open deciduous woods with birch, rowan, alder, and willow were more prevalent on better-drained alluvial soils beside rivers and streams. Here pine could not compete with deciduous species, and grass and fen communities, rather than heather, were more common in openings and beneath the trees. In addition, contrasting with the long-term persistence of pine in East Affric, the abundance of pine declined markedly further west, to the extent that it was probably not present in the lower altitude coastal valleys adjoining Glen Affric. This suggests that current pine-dominated stands, such as those in East Affric, may provide an inadequate baseline or model for long-term woodland expansion or the establishment of new woods since they are not representative of past competition and climatic limits on pine growth, and underestimate the abundance and variety of arboreal species and open habitats that co-existed with pine.[5]

The fate of these woods and the manner in which they were reduced to their present, limited distribution pattern are also relevant to current restoration plans since climatic change is central to woodland loss and to our current and future global situation. The main woodland decline across northwest Scotland occurred around 4400 years ago: this affected all trees but is most spectacularly demonstrated by pine. The primary reason for the extensive failure of regeneration was probably a shift to more oceanic climatic conditions. In some cases, this may have been compounded by the expansion of people into the uplands, although there is no evidence that human activity was sufficiently extensive or intensive to be the primary cause of the "mid-Holocene pine decline," as this event is known. Further instances of woodland loss are documented during the medieval and historic periods, but the most extensive woodland contraction across northern Scotland occurred over 4000 years ago, after which their distribution pattern was rather similar to that recorded on maps dating from the sixteenth century AD onwards. In part, this is because native timber was recognized as a scarce resource by the late fifteenth century and many (although certainly not all) woods were managed to maintain this resource. There was definitely no "great wood of Caledon" to halt the Romans when they arrived around 2000 years ago. The past, therefore, does not provide justification for the "reparation" of past human devastation through woodland restoration, but it does provide a cautionary tale regarding the vulnerability and uncertain future of woods at times of climatic change. The evocative pine stumps preserved in peat across northwestern Scotland are an artefact of the preservational powers of peat; they reflect the cumulative product of shifting patterns of open pine-birch wood-heath between c.8000 and 4000 years ago, and are not a secure indicator of past woodland density, continuity, or longevity.[6]

How, then, does this relate to the extant pinewoods and their central place in Scottish woodland ecology? Some pine populations, such as those in East Affric appear to have continued regenerating uninterrupted, although they too remained dynamic rather than stable pine monocultures. However, in other woods where pine is currently of conservation value, long-term histories again suggest that current composition may not provide a good model for setting conservation goals, posing a potential dilemma for managers. This was recognized by the Woodland Trust Scotland at Ledmore and Migdale wood, in southeastern Sutherland (Figure 1). At present, the wood includes a diverse range of habitats, including open heather moorland, valley mire, Scots pine, oak and birch woodland, juniper scrub, stands of aspen, and coniferous plantation of varying ages. The broadleaf woodland and much of the pinewood on Migdale Rock are described by ecologists as semi-natural ancient woodland. However, the history of a mature birch stand at the heart of Migdale Wood indicates firstly that pine never achieved local dominance and secondly that the current pines may have been planted. Pine was in fact a minor component of the local woodland for millennia because it was out-competed by alder and oak until c.4700 years ago. Then, following a short-lived recovery in pine during a drier climatic phase, it was ousted by birch around 4400 years ago as part of the "pine decline" discussed previously. The subsequent history and current composition of these woods is firmly entwined with the history of local land-use. The period with the greatest diversity, in terms of species composition and heterogeneity, corresponds with the most prolonged period of agricultural use, which lasted from about 2900 years ago until at least the fifteenth century AD. Rather than representing continuity or longevity, the current abundance of pine dates back only two to three hundred years in this part of the woods, at least, and these may have been planted or deliberately nurtured when the woods came under silvicultural management. This brought to an end c.2500 years of combined farming and woodland management. Over the last c.100–150 years, under this silviculture system, the stand has been both less stable and less diverse than its predecessors. The ancient look of the pinewood thus belies its recent origin and current stand ecology provides a poor model for sustainable planning.[7]

Given the current emphasis on woodlands and on the expansion of pine in particular, the open structure, diverse composition, and dynamism of past woods are highly relevant to management and restoration plans. While pine continues to take center stage, there is some recognition that the wider tree and ground flora diversity associated with current pinewoods merits more attention. However, some communities have been so depleted that their current ecology is very much a partial or distorted echo of past diversity. This is the case for riparian woods such as those recorded in West Affric, since these are now rare across Scotland, in part due to their location on relatively flat and fertile ground which has been drained and developed for agriculture, industry, and housing. The value of low-lying woods and mires is now recognized in biodiversity enhancement and flood mitigation strategies. With such a small and depauperate range of extant examples to draw on, the past can provide insights that are highly relevant to present management issues.[8]

Long-term vegetation history thus provides a reminder that plant communities are not fixed entities and that current patterns need to be understood in a longer context if they are to provide a sustainable guide or goal for future conservation or restoration choices. These pinewood examples indicate how modern patterns and perceptions can skew our views of the composition and stability of long-lived woodland communities. In areas like Glen Affric, woodland restorationists need to recognize that woodland loss was primarily the result of environmental change, not anthropogenic pressures, some 4000 years ago, and has not regained this ground through natural regeneration since that time, although natural factors curtailing regeneration have certainly been compounded by increased grazing pressures from sheep and then deer over the last c.250 years. Furthermore, in view of predicted future climatic change, evidence that these woods were unstable during periods of increased climatic instability must be taken into account in any woodland restoration initiative. A long-term perspective stresses that change is the norm and should be expected, making it difficult to establish definitive, static "baselines" or goals to define "success" in conservation or restoration initiatives. Are managers willing to accept that current stands, such as the study site in Migdale, are not representative of stable or long-lived communities and that a desire to promote natural ecological processes may mean unpredictability or damage resilience? If ecosystem stability and high diversity are central to management aims, then human intervention may need to be reinstated at such sites rather than withdrawn or minimized. This may amount to conservationists taking on the role of gardeners if they wish to meet definite targets and reduce uncertainty.[9]

"SO WONDROUS WILD": THE NATURE/CULTURE DIVIDE AND THE ROLE OF PEOPLE[10]

As the foregoing discussion intimates, the predominantly open heaths and peatlands of northern Scotland have existed for at least 4000 years. To some these are bleak and empty landscapes, but to others the lack of obvious human impacts or development gives the impression of a wild land, threatened by and vulnerable to human pressures. There is little doubt that this view arises, in part, from increasing urbanisation, development, and growing consciousness of our impacts on the environment, on a global scale. In Scotland this perspective has been bolstered by allegations that human intervention has, historically, caused the degradation of the uplands (although, ironically, it is rare to hear Scottish native woods viewed as depauperate, as illustrated earlier). Even though moorlands are recognized as a cultural landscape, preserving natural heritage remains a priority. While this may help create a more sustainable balance between people and nature it is important that the archaeology of settlement, land-use, and other monuments does not come to resemble a museum exhibit, dissociated from the landscape or treated as a reminder of a more "harmonious" time. To what extent is a historical perception of human degradation of this environment a reality that can support conservation or restoration strategies?

As shown by Mairi Stewart in the previous chapter, influential historical and ecological studies have labelled the Scottish uplands as a degraded or "devastated countryside." These can be traced back to reports (some anecdotal) of the deterioration of Highland pastures and woods during the mid- to late-nineteenth century, and have been blamed on the large "sheep walks" which were introduced during the Agricultural Improvements of the later eighteenth and nineteenth centuries. This process was part of the infamous Highland Clearances, during which tenants were evicted from the many small farming townships that had previously been distributed throughout the now empty upland valleys (Figure 7.3). Extensive sheep grazing has thus been labelled as the destroyer of upland Scotland's rich ecological diversity. Indeed, there is growing palaeoecological evidence from across the Scottish uplands that extensive hill sheep farming coupled with the abandonment of small-scale, labor-intensive farming systems caused the loss of biodiversity and formation of more homogeneous vegetation. The twentieth century is characterized by continued or further loss of floristic diversity at some sites. Similar long-term patterns are also evident in other parts of the United Kingdom and from across Europe.[11]

While these damaging legacies sound a warning about the current state of the uplands, degradation is not the whole story. First, these changes were not uniform or universal: instances of heather expansion can be found to set against the many examples of heather loss, for example. Second, focussing on recent centuries, during a period of marked agricultural intensification, overlooks the ways in which people have, at other times, made beneficial

Figure 7.3   An abandoned (cleared) settlement. The crumbling remains of farms provide an evocative reminder of past human inhabitants, but how did they alter the landscape? (Photograph by Althea Davies)

contributions to the diversity of the landscape. Under lower-intensity systems, the multitude and heterogeneity of disturbance and soil enrichment processes in fields, routeways, and around settlements often increased and maintained diversity, especially on small-scale farms. In a landscape such as northern Scotland, where acidic soils and plant communities dominate, the labor-intensive tillage and manuring of small arable and pastoral fields around settlements created "islands" of diversity and fertility within the extensive, acidic heath- and peatlands. This is evident around settlement sites from the Bronze Age to the eighteenth and nineteenth centuries AD, and around the crofts or small-holdings created after the Clearances. Since prehistory people may also have managed and perhaps expanded species-rich floodplain meadows, as these provided an essential "early bite" for over-wintered livestock. Prior to the Clearances, relict woodlands were managed as wood pasture (a source of sheltered grazing) and as coppice. Both of these practices are now viewed as beneficial for biodiversity, but such views have been slow to receive recognition in Scotland's "ancient, semi-natural woods." Palaeoecological and ecological studies outside this region also show that higher floristic diversity is closely linked with human activities, particularly "traditional" systems which were either lower intensity and/

or smaller-scale than modern agriculture. However, the beneficial effects of small-scale farming for biodiversity are acknowledged far less readily in the Scottish uplands than in lowland zones, where a high value is placed on careful management of some arable land as a means of maintaining long-established field floras (anthropocores). In part, no doubt, this reflects the wilder appearance of the uplands, far lower density of settlements and enclosed fields, and an understandable desire to prevent any further damage, particularly when compared with lowland areas which are more clearly "cultural" in origin. Knowing which practices or combinations of factors are harmful and which are beneficial to biodiversity can broaden the ecological evidence-base for management.[12]

CONCLUSIONS

This chapter has examined the role of the past in current upland management and questioned some of the preconceptions and motivations within current conservation and restoration strategies. Although the uplands may—in relative terms—be *wilder* than many landscapes, they are the product of at least four millennia of interactions with humans. Human involvement in nature conservation has never been so intense, yet "naturalness" remains a common goal. How different is designed naturalness from wilderness gardening, particularly when "traditional" agricultural techniques and older breeds of livestock are often used to help maintain or regain the desired "natural" processes, such as woodland or heathland regeneration?

Restoration goals are often influenced by perceptions of the past and our own culturally determined choices and modern world outlook rather than by any fixed reality. If this is the case, does the past actually have a role to play in management, conservation, or restoration? Certainly studies of the past (encompassing palaeoecology, archaeology, and history, to name but a few) cannot claim to have a prescriptive role, but care needs to be taken that misconceptions about the past, especially long-term drivers of habitat change, are not used to justify preconceived management ideals. Since ecological datasets often span only a short part of an ecosystem's history and variability, especially for woodlands and peatlands which have long turnover times, it can be difficult to assess whether current knowledge is representative or robust. Long-term studies thus provide a valuable reality check. By providing a critical evidence-base for disentangling natural and cultural drivers, establishing where ecological thresholds lie and what underpins current landscape values, a historical perspective can be used to test the applicability and sensitivity of baselines and targets derived from shorter-term ecological knowledge. This provides a more robust, scientifically, and socially defensible basis for making sustainable policy and management decisions. No less usefully, stories from the past may also serve as a source of inspiration, strengthening our links with the landscape and celebrating its rich diversity. All of this requires more open-minded debate and collaboration between ecologists, managers, the public, and those who study the past.[13]

NOTES

1. For restoration science and choices, see M. B. Davis and L. B. Slobodkin, "The science and values of restoration ecology," *Restoration Ecology* 12 (2004): 1–3. For the shortage of data on the impacts of reduced grazing, see F. J. G. Mitchell and K. J. Kirby, "The impact of large herbivores on the conservation of semi-natural woods in the British uplands," *Forestry* 63 (1990): 333–53; J. A. Milne and S. E. Hartley, "Upland plant communities— sensitivity to change," *Catena* 42 (2001): 333–43. The uncertain benefits of 'naturalistic grazing' are discussed by K. H. Hodder, J. M. Bullock, P. C. Buckland, and K. J. Kirby, *Large herbivores in the wildwood and modern naturalistic grazing systems* (Peterborough: English Nature Research Report Number 648, 2005).
2. For the management and place of pinewoods in Scottish ecology, past, present, and future, see J. R. Aldhous, *Our Pinewood Heritage* (Farnham, Surrey: Forestry Commission/Royal Society for the Protection of Birds/Scottish Natural Heritage, 1995); special issue of *Forestry* 79 (2006). For the mythical great wood of Caledon, see T. C. Smout, Highland land-use before 1800: misconceptions, evidence, and realities, in *Scottish Woodland History*, ed. T. C. Smout (Aberdeen: Scottish Cultural Press, 1997), 5–23. For an alternative argument for the 'naturalness' of treeless uplands, see J. H. C. Fenton, "A postulated natural origin for the open landscape of upland Scotland," *Plant Ecology and Diversity* 1 (2008): 115–27.
3. The woodland restoration scheme in Glen Affric is one of numerous such proposals across Scotland.
4. For more detailed discussions of former woodland diversity and dynamics in Glen Affric, see C. A. Froyd and K. D. Bennett, "Long-term ecology of native pinewood communities in East Glen Affric, Scotland," *Forestry* 79 (2006): 279–91; H. Shaw and R. Tipping, "Recent pine woodland dynamics in East Glen Affric, northern Scotland, from highly resolved palaeoecological analyses," *Forestry* 79 (2006): 331–40; R. Tipping, A. Davies, and E. Tisdall, "Long-term woodland dynamics in West Glen Affric, northern Scotland," *Forestry* 79 (2006): 351–59.
5. For fine resolution pollen theory and method, see R. H. W. Bradshaw, Spatially precise studies of forest dynamics, in *Vegetation History*, eds. B. Huntley and T. Webb III (Dordrecht: Kluwer Academic Publishers, 1988), 725–51. For more information on past pine limits, see K. D. Bennett, Postglacial dynamics of pine (*Pinus sylvestris* L.) and pinewoods in Scotland, in J. R. Aldhous, 23–39; B. Huntley, R. G. Daniell, and J. R. M. Allen, "Scottish vegetation history: the Highlands," *Botanical Journal of Scotland* 49 (1997): 163–75. For past alluvial woodland diversity, see A. Davies, Carnach Mor, and Camban: woodland history and land-use in alluvial settings, in *The Quaternary of Glen Affric and Kintail, Field Guide*, ed. R. M. Tipping (London: Quaternary Research Association, 2003), 75–83.
6. For the relationship between pine, climate change, and human activity, see D. E. Anderson, H. A. Binney, and M. A. Smith, "Evidence for abrupt climatic change in northern Scotland between 3900 and 3500 calendar years BP,"

The Holocene 8 (1998): 97–103; T. C. Smout, "The pinewoods and human use, 1600–1900," *Forestry* 79 (2006): 341–49; R. Tipping, A. Davies, and E. Tisdall (2006). For historic woodland management in Scotland, see A. Crone and F. Watson, Sufficiency to scarcity: medieval Scotland, 500–1600, in *People and woods in Scotland*, ed. T. C. Smout (Edinburgh: Edinburgh University Press, 2003), 60–81; T. C. Smout, A. R. MacDonald, and F. Watson, *A history of the native woodlands of Scotland, 1500–1920* (Edinburgh: Edinburgh University Press, 2005).

7. For other examples where current stand ecology provides a poor model for future planning, see R. Bradshaw and G. Hannon, The Holocene structure of NW European forest deduced from palaeoecological data, in *Forest Biodiversity: lessons from history for conservation*, eds. O. Honnay, K. Verheyen, B. Bossuyt, and M. Hermy (Oxford: CAB International, 2004), 11–25; M. Grant and M. Edwards, "Conserving idealized landscapes: past history, public perception, and future management in the New Forest (UK)," *Vegetation History and Archaeobotany* 17 (2008): 551–62.

8. For alluvial woodland loss and current value, see G. F. Peterken and F. M. R. Hughes, "Restoration of floodplain forests in Britain," *Forestry* 68 (1995): 187–202; A. Davies, Carnach Mor, and Camban: woodland history and land-use in alluvial settings, in *The Quaternary of Glen Affric and Kintail, Field Guide*, ed. R.M. Tipping (London: Quaternary Research Association, 2003), 75–83. For past insights into current floristic diversity, see R. Tipping, J. Buchanan, A. Davies, and E. Tisdall, "Woodland biodiversity, palaeo-human ecology and some implications for conservation management," *Journal of Biogeography* 26 (1999): 33–43.

9. For recent insights into changing ideologies behind woodland management in Scotland, see A. C. Midgley, "The social negotiation of nature conservation policy: conserving pinewoods in the Scottish Highlands," *Biodiversity and Conservation* 16 (2007): 3317–3352.

10. Extract from "The Lady of the Lake" (1810), by Sir Walter Scott. This poem, which is set in the Western Highlands, provides a typical example of late eighteenth- and nineteenth-century Romantic visions of the Scottish mountains.

11. The quote about the state of the uplands is from F. F. Darling *West Highland Survey: an essay in human ecology* (Oxford: Oxford University Press, 1956), 192. For long-term analyses of land management impacts on upland habitat and biodiversity see A. C. Stevenson and D. B. A. Thompson, "Long-term changes in the extent of heather moorland in upland Britain and Ireland: palaeoecological evidence for the importance of grazing," *The Holocene* 3 (1993): 70–76; R. Tipping, Palaeoecological approaches to historical problems: a comparison of sheep-grazing intensities in the Cheviot hills in the medieval and later periods, in *Townships to Farmsteads: Rural Settlement Studies in Scotland, England and Wales*, eds. J. Atkinson, I. Banks, and G. MacGregor (British Archaeological Reports (British Series) 293, 2000), 130–43; F. M. Chambers, D. Mauquoy, A. Gent, F. Pearson, J. R. G. Daniell, and P. S. Jones, "Palaeoecology of degraded blanket mire in South Wales: data to inform conservation management," *Biological Conservation* 137 (2007): 197–209; A. L. Davies and P. Dixon, "Reading the pastoral landscape: palynological and historical evidence for the impacts of long-term grazing on Wether Hill, Cheviot foothills, Northumberland," *Landscape History* 30 (2008): 35–45; N. Hanley, A. Davies, K. Angelopoulos, A. Hamilton, A. Ross, D. Tinch, and F. Watson, "Economic determinants of biodiversity change over a 400-year period in the Scottish uplands," *Journal of Applied Ecology* 45 (2008): 1557–1565. Comparable patterns of diversity loss over

the last *c.*250 years are increasingly recognized across Europe, *e.g.*, M. Lind-bladh and R. Bradshaw, "The origin of present forest composition and pattern in southern Sweden," *Journal of Biogeography* 25 (1998): 463–77; E. Gustavsson, T. Lennartsson, and M. Emanuelsson, "Land use more than 200 years ago explains current grassland plant diversity in a Swedish agricultural landscape," *Biological Conservation* 138 (2007): 47–59; B. E. Berglund, M.-J. Gaillard, L. Bjorkman, and T. Persson, "Long-term changes in floristic diversity in southern Sweden: palynological richness, vegetation dynamics and land-use," *Vegetation History and Archaeobotany* 17 (2008): 573–83.

12. For positive relationships between diversity and human activities, see M. Lindbladh and R. Bradshaw (1998); Davies (1999); V. Vandvik and H. J. B. Birks, "Mountain summer farms in Roldal, Western Norway—vegetation classification and patterns in species turnover and richness, *Plant Ecology* 170 (2004): 203–22. For the importance of management in ancient woods, see H. J. B. Birks, "Contributions of Quaternary palaeoecology to nature conservation," *Journal of Vegetation Science* 7 (1996): 89–98; K. Holl and M. Smith, "Scottish upland forests: History lessons for the future," *Forest Ecology and Management* 249 (2007): 45–53. For arable biodiversity and land-use see Smout, Chapter 10, this volume.

13. For other perspectives on the value and role of long-term and historical studies in restoration, see S. R. Dovers (2006). On the contribution of environmental history to current debate and policy, see *Environment and History* 6, 131–50; F. W. M. Vera (2000), *Grazing Ecology and Forest History*, CAB International, Oxford. For the past as inspiration, see W. Cronon (1992), A Place for Stories—Nature, History, and Narrative, *Journal of American History* 78, 1347–1376.

8 Conservation Lessons from the Holocene Record in "Natural" and "Cultural" Landscapes

Nicki J. Whitehouse

This chapter highlights the value of archaeological and palaeoecological studies to conservation by drawing upon research undertaken on two raised mires, Thorne and Hatfield Moors, both located in the Humberhead Levels of eastern England. The area today is typically viewed as a rather unattractive, backwater landscape, yet its archaeological ("cultural") and palaeoecological ("natural") record provides an amazing insight into a complex landscape of raised mires, floodplain wetlands, and heathlands that have all evolved and changed over spatial and temporal scales spanning thousands of years. The records emphasize the changing character of the landscape: it is never static, always evolving in ways that we sometimes fail to appreciate or anticipate. Today we are left with component parts of this ancient landscape fundamentally affected by human action; in that respect, it is a cultural product, yet at its core one that is still dominated by natural landscape features and systems. Culture and nature cannot easily be separated in this place, although much of the current policy has been towards total eradication of the former record with a view to rewilding the landscape through natural processes. It is my contention that it is impossible to rewild the landscape to what it once was, but that we should try to conserve what is left, while creating, or gardening, the landscape into something new that recognizes the legacy of both ecological and human change.

PEATLAND PALAEOECOLOGY AND THE "WILDWOOD"

Sir Harry Godwin's *Archives of the Peat Bogs* (1981) helped popularize the usefulness of using peat as a window into the past. In fact, palaeoecological and archaeological wetland archives have immense scientific value to a diverse range of subjects, especially the management and conservation of biodiversity of peatlands and other wetlands.[1] This is because peatlands lay down a continuous three-dimensional record of their own history, layer upon layer of animals and plants which used to live there, as well as that of surrounding animal and plant communities. These deposits allow us to study past biodiversity, climate and environmental change over the last 10,000 years (the Holocene). Peatlands also contain invaluable information on human impact on the environment and human activity in general, through archaeological artifacts such as trackways (logs set across the bog to facilitate human transit across wet areas), bog bodies (or corpses), and votive offerings, suggesting these sites had important cultural significance in the past. My own research has focused upon the record of faunal changes recorded from beetle (Coleoptera) fossils in peatlands. Fossil beetles are preserved in peat and other waterlogged deposits, such as archaeological deposits. This record allows the palaeoecologist to study issues as past biodiversity, changing environmental conditions, the succession of insects and their host plants, and the differing water regimes throughout a bog's development. Changes to fossil beetle fauna can provide insight on how the biota responded to environmental stimuli over time, creating a story of the life cycle of the bog and its ecological processes.[2]

Two sites that have been particularly well studied for their palaeoecological records are Thorne and Hatfield Moors SSSI's (Site of Special Scientific Interest) in the Humberhead Levels (Figure 8.1). The sites are the two largest surviving examples of ombrotrophic (acidic) lowland raised mire in eastern England. They are the remnants of the once large mosaic of wetlands and raised mires that formed in the region about 6000 years ago (see also Rotherham and Harrison, this volume). The sites remained relatively undisturbed for 4000 years, until modern large-scale peat extraction and water abstraction began in the 1960's. Extensive peat cutting has meant that large areas now consist of un-vegetated peat, although these are slowly being re-wetted. Thorne and Hatfield Moors have been the *foci* of considerable archaeological and palaeoenvironmental research over the years, with trackways, ritual artifacts, and bog bodies recovered within the peats as well as pollen, beetles, tree, and other plant macrofossils.[3]

Several sections of the peat archive at Hatfield Moors have been studied, revealing the complex evolution of the site and the mosaic-like nature of the landscape. On the western side of the Moor, a sub-fossil forest preserved in the basal peats shows what the landscape looked like before peat formation, indicating oak-birch-pine dominated woodland. A few hundred years later, increasing water levels and acidity had favored the transition to pine-dominated woodland, and peat had started accumulating. Heather-cotton grass vegetation and pine-birch forest covered much of the land surface, but increasingly ombrotrophic, wet conditions are indicated. Shortly afterwards, the last pine trees had fallen into the peat to be preserved as the mire turned acidic, no longer favouring tree growth. Fully acidic conditions are evident within c.500 years of peat initiation in this area, highlighting the amount of time it can take to go from eutrophic to ombrotrophic conditions. This contrasts to a study site on the north side of the site, where the beetles first indicate sandy, pine-heath areas interspersed with broad-leaved trees such as oak and birch in areas with better drainage, with heather

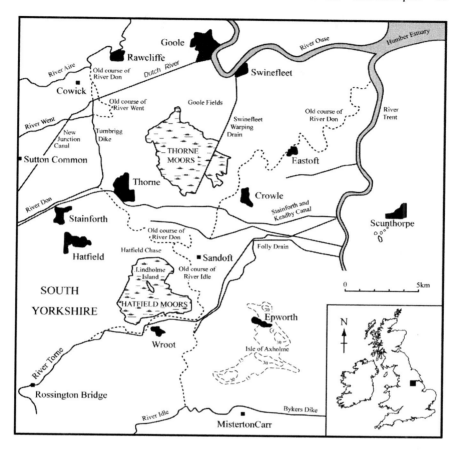

Figure 8.1   Map of Humberhead Levels of South Yorkshire. Re-drawn from an original by P. I. Buckland. (Copyright Thorne and Hatfield Moors Conservation Forum)

forming an understory. Within a few decades, fully acidic conditions are indicated, with beetles typical of acid raised bogs—a significantly faster transition than that indicated on the western side of the mire. Fifteen hundred years later, the mire was still actively growing and expanding, as shown by assemblages on the southern edges of the present site, representing a lagg fen community. Here also, the fossil record shows that it took about 500 years for fully acidic conditions to develop on the site, but with an important wet heath component.[4]

Overall, the palaeoecological story highlights the complexity and life cycle of the peatland, both in space and time, a complexity which is sadly missing from the site today as it has become decimated by peat cutting. This complexity is dictated by a host of interacting factors and processes which we are not entirely sure about, but which probably included site (edaphic) factors, and external hydrological issues connected to sea level rise, climate change as well as human activities. The basal deposits of both Thorne and Hatfield Moors include extensive deposits of burnt pinewoodland and there has long been speculation that the twinned roles of human activities and/or natural forest fires may have been important in causing these deposits. It is probable that humans continued to use the landscape long after mire formation, as discussed here. The modern insect fauna and flora is thus the result of all these complex factors and their inter-relationships, modifying conditions in often subtly different ways.[5]

Indeed, when fossil beetle lists from both Thorne and Hatfield Moors are compared with modern beetle populations, there is an astonishing similarity between fossil and modern beetle species lists. More astonishingly, the research indicates that some of Thorne and Hatfield's national endemics (e.g., *Bembidion humerale* Strum. and *Curimposis nigrita* Palm.) have been present on the site since peat inception, having been found as fossils in the basal deposits of both sites. The presence of these rarities together with the overlap between historic and modern faunas indicates that the faunal characteristics of the mires became established during the early phases of peat development. Re-creating or rewilding raised mire with the 'correct' (climax) vegetation, the goal of present management practices, may not allow the longer-term restoration of faunal communities, since these characteristics were generated 4,000 years ago—something which cannot be reproduced over a short time span.[6]

One of the characteristics of these peatlands is that during the earlier stages of mire development, trees regularly expanded upon peat surfaces at regular intervals, forming mire woodlands. The abundance of material sometimes gives the impression that many trees were growing at the same time, but in fact many mire woodlands were growing over considerable periods of time and re-expanding across bogs during periods of mire dryness, sometimes over periods of millennia. This has been shown at many peatland sites. The trees growing on the peat bog can thus be seen as part of the raised mire ecosystem and play an important biodiversity role—indeed both sites are well known for their woodland fauna. Yet today, trees which invade mire sites are regularly removed in an attempt to maintain a high water table and retain sites in an idealised state, devoid of trees.[7]

One advantage of these trees is that they have provided important insights into the nature of the Holocene "wildwood." Abundant dead wood appears to have been an important ecological characteristic of these mire-woodlands, as evidenced from the fossil beetle record associated with these phases. Evidence from this material is now providing fascinating insights into recent debates sparked by the Dutch ecologist, Frans Vera (next chapter), concerning the true nature of the wildwood. Significantly, this work draws attention to the fact that human impact has had considerable effect on the European landscape, perhaps for far longer that many, including Vera, have realized (see also Smout, Chapter 10, this volume). Perhaps the most fascinating aspect of this debate concerns the fact that Vera has not only challenged our scientific basis of understanding, but he also challenges fondly held images of the wildwood. I believe part of the reason this debate has been so fierce is because it challenges our cultural concepts of wilderness, which in Europe is close to the notion of wildwood. Indeed, the debate sparked by Vera's work represents a fascinating moment in the development of palaeoecological science when long-cherished assumptions are opened up for challenge, and culturally imbued intuitive feelings about past landscapes battle with the desire to uphold the ideals of scientific thinking.[8]

MIRE DESTRUCTION, RESTORATION, AND REWILDING

Until recently, peat extraction continued stripping the landscapes of these raised mires, along with their attendant archaeological and palaeoecological record and thus their landscape continuity and mosaic structure so essential to many of these invertebrates. Peat cutting companies had been keen to promote the concept that peat could be regenerated. Active cutting of peat continued to be advocated by English Nature for many years (now Natural England, the government's body charged with nature conservation) and DEFRA (Department of Food, Environment and Rural Affairs, and formerly DETR, Department of Transport, Local Government and the Regions), the argument being that the quicker the peat was removed the quicker restoring and rewilding schemes could be put into operation. The palaeoecological record suggests that while poor fen can be rapidly achieved, initiation and development of true raised mire may take hundreds of years—as shown from the record at Thorne and Hatfield Moors—and without regaining much of its former invertebrate species' diversity.[9]

Perhaps more importantly, the record of the development of the mire biota, once destroyed, can never be regenerated. Yet modern management practices on site are underpinned by a belief that it is possible (and desirable) to rebuild the sites; as Paul Buckland has so eloquently put it: "The philosophy of 'take one sprig of Sphagnum, and we can reconstruct a raised mire' leads to a cavalier approach to conservation, with an attitude, actually expressed by EN (English Nature) staff, that the sooner all the peat is gone, the sooner we can start again."[10] This, of course, makes the assumption that all the mechanisms for mire development are known and that the right conditions exist today as they did several thousand years ago. Factors such as pollution, continued dewatering, and drainage are not likely to help the process, while faunal groups cannot re-establish without appropriate *refugia*. In truth, the regeneration of milled areas is still at a pioneering stage; many of the expectations of full mire restoration are based upon short-term analysis of sites.

Figure 8.2   Hatfield Moors after the peat cutters had left, on the day the site was handed over to English Nature, September 2004. (Photograph by Helen Kirk)

When the peat cutters eventually departed Hatfield Moors they left a flat wilderness, with lagoons and areas of standing water, and tufts of cotton grass (Figure 8.2). One is left wandering how an SSSI, in the care of the state, has been allowed to reach this point of almost complete destruction with its associated irreplaceable loss to biodiversity.

PEATLANDS, ARCHAEOLOGY, AND THE "CULTURAL" LANDSCAPE

Only when the peat cutters left Hatfield Moors was some of the site's archaeology discovered by a local man, Mick Oliver, walking across the site in October 2004. Sitting upon the drying surface of the exposed peats, was a 50-meter pinewood "corduroy" later Neolithic trackway and platform (Figure 8.3), dated to 2900–2500 cal BC. The trackway, known as the Lindholme trackway because of its proximity to Lindholme Island, is the earliest corduroy track known in Britain, with only two earlier known examples in Europe. The trackway is the only site to have both ends (termini) identified, one end abutting the basal sand dunes which characterized the area prior to peat formation, the other end forming a platform, above pool peats. The interpretation put forward is that the trackway and platform served as an approach avenue in which control of the ritual proceedings may have meant that only certain players would have participated in proceedings on the platform, with onlookers on the dry sand dunes, separated by the wetlands in between.[11]

A Bronze Age trackway was also discovered in the 1970s on Thorne Moors and there are historical accounts which suggest that human remains such as bog bodies have been found in and around Hatfield Chase, particularly during peat cutting for the seventeenth century drainage and canalization works. The archaeological record therefore indicates that these areas have long been important places for humans, whether for hunting or more ritualised activities.[12]

Several historical accounts provide a fascinating picture of what the Humberhead Levels and mires prior to large-scale drainage and peat extraction must have looked like. The earliest description is by John Leland from 1538; he provides a window onto what must have been a truly fabulous "Everglades-like" landscape. As recently as 1966, an eyewitness account by Peter Skidmore provides a colorful description of Hatfield Moors: "Drier

Figure 8.3 Lindholme trackway, Hatfield Moors, when it was first discovered, October 2004. (Photograph by Mick Oliver)

even then than Thorne Moor but similarly a superb, vast-wilderness-like place, it differed from Thorne Moors on its clumps of Scots Pine, its beds of Sweet Gale and the old oaks on the low moraine of Lindholme. It had a truly medieval appearance and a character of its own. . . ."[13] One of the striking aspects of many historical accounts concerns the importance of the landscape to the wider community. Much of the area during the medieval period remained a swampy wilderness in which commoners had been granted fishing, fowling, timber, and turbary rights. These areas provided important means of livelihood for the people of the Humberhead Levels and were teeming with wildlife (Rotherham and Harrison, this volume). When in 1625, Charles I granted permission to a Dutchman, Cornelius Vermuyden, to drain the area, commoners tried to stop the drainage. Both Thorne and Hatfield Moors remained unimproved by the drainage schemes, although the surrounding wetlands were substantially drained and changed forever, creating the rather flat and featureless character of the landscape today. By the early 1900s, the area was almost completely destroyed. Vermuyden remained a loathed figure within historical accounts; even today, it is common to encounter strong views against him and his successors who effectively disenfranchised local people from their landscape and removed the existence of these former extents of wetland from memory. For this very reason, Thorne and Hatfield Moors, as the last vestiges of this long-vanished wetland, are considered integral to local people's cultural identity.

So here is an effectively natural ecosystem, with copious plants and animals; yet the discovery of the trackways, other human artifacts, and the historic record serve to emphasize that "wild" places like Thorne and Hatfield Moors have long been culturally important and culturally manipulated to a greater or lesser extent. Humans have interacted within this landscape for millennia, through everyday, routine activities of collecting, gathering, grazing, burning, hunting, and ritual activities to produce a distinct cultural landscape.

MYTHS OF PRISTINE LANDSCAPES, AGAIN

The palaeoecological and archaeological record suggests that we should raise important questions concerning the nature of *what* site managers or those involved in nature conservation should be attempting to restore on such sites, especially where large-scale destruction is taking place. Modern management practices are underpinned by a belief that it is possible (and desirable) to restore sites, with goals focusing on the re-creation of raised mire with 'correct' (climax) vegetation. Perhaps it is possible to rewild sites to their former botanical appearance, but this may not allow the longer-term restoration of faunal communities, since the faunal characteristics of these sites were generated 4,000 years ago. The end result of any restoration work on these sites may be a visually satisfying, but faunally depleted, landscape. More often than not, habitat conservation is often a cosmetic exercise relying on a few species of plants, when the real diversity lies in the invertebrate fauna.

Many of these restoration activities are not supported by the historical evidence of biodiversity. In fact, the concept of restoring any site can be ingenuous. After all, it gives the impression that anything can be "repaired" or that "we can rebuild it." This is the kind of thinking we have spent the last fifteen years fighting against on Thorne and Hatfield Moors and which have led to vast areas of destroyed raised mire, leading to huge biodiversity losses. From a palaeoecological perspective, it is clear that no two sites are identical, and that each is characterized by a particular suite of species formed as a result of unique historical site characteristics.

CONCLUSIONS

If we look carefully at the layers in the peat, it is clear that we still have much to learn about ecosystems and the complex interactions between natural processes and human–animal induced changes. Ecosystems are constantly changing and never static. We need to embrace changes that the modern world produces, while facilitating biodiversity conservation. It is impossible to turn back the clock and thus one should be doubly cautious in doing so with large-scale rewilding projects. Within the current context of a warming climate, we need to concentrate on safeguarding what we already have and protecting habitats and species from further degradation rather than being diverted toward ecologically and financially risky experiments. We can enhance, regarden, and accept the inherent instability of systems. We can use the Holocene record to help us identify the options we might have, while accepting that what we will achieve will always be different than what was once there. The case study of Thorne and Hatfield Moors presented here illustrates the consequences of human actions on our landscapes and the wish to restore them. Central to such an approach needs to be an active debate that helps promote greater awareness of the historical and epistemological context of restoration.[14] What restoration leads to will always be something new, and perhaps also unpredictable.

ACKNOWLEDGMENTS

I would like especially to thank Prof. Paul Buckland who has always been a great source of encouragement and first introduced me to the Moors; he has been greatly influential in many of my ideas. I have benefited from many insightful discussions with members of the Thorne and Hatfield Moors Conservation Forum and to whom I give thanks. I would like to thank Helen Kirk and Mick Oliver for allowing me to use some of their images of the Moors. This chapter was written during a period of study leave spent at the Department of Geological Sciences, University of Canterbury, Christchurch, New Zealand. I would like to thank the department and especially my host, Prof. Jamie Shulmeister, for their hospitality. Finally, I'd like to thank Marcus Hall for the opportunity of attending the workshop which gave rise to the chapters in this volume and for his editorial input.

NOTES

1. On the use of palaeoenvironmental data to assist with management and conservation of species, see Kathy Willis and John Birks, "What is natural? The need for a long-term perspective in biodiversity conservation," *Science* 314 (2006), 1261–1265; Nicki Whitehouse, et al., "Sub-fossil Insects and ecosystem dynamics in wetlands; implications for biodiversity and conservation," *Biodiversity and Conservation* 17 (2008): 2055–2078.
2. Nicki Whitehouse, "Mire ontogeny, environmental and climate change inferred from fossil beetle successions from Hatfield Moors, eastern England," *The Holocene* 14, (2004), 79–93.
3. Reviews include Henry Chapman and Ben Gearey, "Archaeological predictive modelling in raised mires—concerns and approaches for their interpretation and management," *Journal of Wetland Archaeology* 2 (2003): 77–88; Brian Smith, *A Palaeoecological study of raised mires in the Humberhead Levels*, edited by P.C. Buckland and M. Limbert (Doncaster and Oxford: Thorne and Hatfield Moors Conservation Forum and British Archaeological Reports, 2004).
4. Nicki Whitehouse, "Forest fires and insects: palaeoentomological research from a sub-fossil burnt forest," *Palaeogeography, Palaeoclimatology, Palaeoecology* 164 (2000): 231–246; Gretel Boswijk and Nicki Whitehouse, "*Pinus* and *Prostomis*: a dendrochronological and palaeoentomological study of a mid-Holocene woodland in eastern England," *The Holocene* 12 (2002): 585–596; Nicki Whitehouse (2004) op. cit.
5. Paul Buckland and Harry Kenward, "Thorne Moor: a palaeo-ecological study of a Bronze Age site," *Nature* 241 (1973): 405–407; Paul Buckland, *Thorne Moors: a palaeoecological study of a Bronze Age site; a contribution to the history of the British insect fauna* (Birmingham: Dept. of Geography, University of Birmingham, 1979); Gretel Boswijk and Nicki Whitehouse (2002) op. cit; Nicki Whitehouse (2000, 2004) op. cit.
6. Nicki Whitehouse (2004) op. cit.; Peter Skidmore, *An inventory of the Invertebrates of Thorne and Hatfield Moors* (Doncaster: Thorne and Hatfield Moors Conservation Forum, 2006).
7. Gretel Boswijk and Nicki Whitehouse (2002) op. cit. and Althea Davies, this volume.
8. On wildwood, see Nicki Whitehouse, "The Holocene British and Irish ancient woodland fossil beetle fauna: implications for woodland history, biodiversity and faunal colonisation" *Quaternary Science Reviews* 25 (2006): 1755–1789.
9. See Frans Vera, *Grazing Ecology and Forest History* (Oxon: CABI Publishing, 2000).
10. See Paul Buckland "Peatland archaeology: a conservation resource on the edge of extinction" *Biodiversity and Conservation* 2 (1993): 513–527; Hans Joosten, "Time to Regenerate: long-term perspectives of raised bogs

regeneration with special emphasis on palaeoecological studies," in Brian Wheeler, et al., eds., *Restoration of Temperate Wetlands* (Chichester: J. Wiley & Sons, 1995), 379–4.

11. Paul Buckland, "Conservation and the Holocene Record: an invertebrate view from Yorkshire," *Bulletin of the Yorkshire Naturalist Union* 37 (Suppl., 2002), 23–40, p. 26.
12. Henry Chapman and Ben Gearey, "The Henge that went straight," *British Archaeology* 90 (2006), 43–47.
13. Paul Buckland (1979), op. cit.
14. Peter Skidmore, "Balaam's Donkey and the hairy Canary: personal reflections on the changing invertebrates of Thorne and Hatfield Moors," *Thorne and Hatfield Moors Papers* 3 (1993): 66–70.

9 The Shifting Baseline Syndrome in Restoration Ecology

Frans Vera

The Shifting Baseline Syndrome is a concept formulated by Daniel Pauly in 1995.[1] It results in a drift away from true natural conditions, and as a consequence a change in perception of ecological change varying from generation to generation. It eventually causes a continuous lowering of standards of nature and the acceptance of degraded natural ecosystems to be the normal state of nature. The Shifting Baseline Syndrome arises if scientists:

- Lack a clear unequivocal reference point of how the natural situation used to be;
- Examine an environment that is hard-to-notice and continuously changing because of man;
- Redefine what nature and natural is according to their personal experience.

If restoration ecology aims to restore natural conditions or natural processes, this syndrome will result in an erroneous starting point for restoration projects, such as a state of degradation of nature.[2] In this chapter I will show how a shift in the meaning of certain words that took place centuries ago caused an erroneous starting point in Europe for restoration projects in reserves and national parks aiming to restore natural conditions. This resulted in a loss of biodiversity that was naturally present. Because of a shifted baseline, this loss was accepted as normal, because it was in agreement with what was defined as the baseline for natural.

THE RECONSTRUCTING OF THE BASELINE FOR TRUE NATURAL CONDITIONS

In the nineteenth and the beginning of twentieth century, in Europe a baseline for natural conditions has been formulated. This was done at a time that with the exception of some raised bogs and remote high treeless elevated mountainous areas, all natural ecosystems had been cultivated. Because true natural conditions were lacking, a theory of what were the natural conditions was formulated. This theory was and still is based on three basic assumptions. First, mankind disturbed the natural conditions by cutting trees, plowing, and by introducing and grazing domestic animals like cattle. Second, when mankind withdraws, nature rebounds spontaneously to its natural state.[3] Third, because herbivores are completely dependent on plants, they follow the development of the vegetation and do not play a determining role in succession. Based on these assumptions, the spontaneous development of forest in Europe on abandoned agricultural land and pastures where domestic stock was excluded by fences, was considered to be the return of natural conditions. Therefore under natural conditions in Europe, having a temperate climate favoring trees, it was supposed to be covered with a closed canopy forest in its natural state.[4] The regeneration of the forest would have taken place in gaps in the canopy or in windblown areas, where seedlings and saplings of trees were able to grow up successfully.[5] This theory is still used extensively across Europe as a baseline for natural conditions and restoration projects in reserves and national parks.

Certain words in historic texts from continental Europe, dating from the time that the natural conditions were supposed to have been present were read as support for this theory: such words include "Forst" and "Wald" in Old-German, "forest" in Old-French, and "foreest" and "woud" in Old-Dutch. The modern meaning of these words is unequivocably, closed canopy forest. Support for former dense forests was based on the extrapolation of the modern meaning of these words back into the past.[6]

Similar to this backward extrapolation of word meanings, regulations temporarily prohibiting livestock grazing—arising the thirteenth century onwards—have been interpreted as regulations for protecting seedlings and saplings of trees in forests in order to ensure the survival of the forest.[7] This interpreation was based on the experience with livestock grazing and wild ungulates like deer from the nineteenth century onwards in forests as we know them today. These herbivores kill seedlings and saplings in the forest by trampling and browsing, and were therefore labeled as the greatest enemy of the forest.[8] Livestock, especially cattle, was said to degenerate forests by way of a retrogressive succession to a park-like landscape (or so-called wood-pasture) and then to open grassland or heathland. With the exception of perhaps small glades in the forest, open grassland was considered an anthropogenic phenomenon caused by cattle and horses that were considered alien species introduced by humans. The prolific regeneration of trees in fenced parts of pastures and wood-pastures were offered as proof of how destructive these animals were for the forest.[9]

Finally, in the first half of the twentieth century, palynologists claimed to have reconstructed by pollen analysis the history of the forest back to pre-historic times.[10] Their argument was that up to 90 percent of the pollen derived from trees. They adopted the view of the destructive role of large ungulates in the forest, concluding that under natural conditions

large, indigenous ungulates must have lived in very low densities, otherwise the former natural vegetation would not have been a closed canopy
forest.[11]

THE INTERPRETATION OF HISTORICAL TEXTS

There must be a number of cautions placed on these interpretations of former texts. The landscapes referred to proved to contained not only trees
and shrubs, but also open treeless areas, like open grasslands and raised
bogs. Therefore, the meaning of these words shifted over the centuries.
How did this shift occur?

From the seventh century onwards a new word "forestis" appeared in
deeds of donation of Merovingian and Frankish kings, written in Latin.
During the following centuries this word evolved in "Forst" and "Vorst"
in Old-German, "forest" and "fôret" in Old-French, and "forest," "foreest," and "voorst" in Old-Dutch. The word "Forst" still exists in modern
German, "forêt" in modern French, and "forest" in modern English all
currently mean a closed canopy forest.

The Merovingian and Frankish kings declared the uncultivated wilderness as "forestis nostra" (our "forestis"). They did so on the basis of
Roman law that stated that everything without a clear owner (such as wilderness) belonged to the "authority." "Forestis" would have been derived
from the Latin "foris" or "foras," which means "outside," "outside it,"
and "outside the settlement."[12] The "forestis" was the uncultivated outside
settlements, arable land (fields), and hay-fields, that all had clear owners,
namely person who cultivated that particular piece of wilderness. "Forestis" was a legal concept that described or confirmed the royal rights concerning ownership and user rights to the uncultivated (= wilderness). To it
applied the "ius forestis" (or "forestis" law).[13] The law applied to an area
in general and to every individual grass, herb, shrub, tree, and animal that
lived there on land or in water. Only the king had the right to make use
of these. Others needed express consent of the king, which was given by
officials appointed by the king, so-called "forestarii." They issued regulations as an implementation of the "ius forestis" for local communities for
pasturing cattle and pigs, collecting leaf-fodder for their livestock, and
cutting firewood and getting timber in the "forestis," in order to fulfil the
needs of their household .[14]

What was claimed as "forestis," was termed in common Germanic
languages including Old-German, Old-Dutch, Old-Frisian, and Old English as "wold," "weld," "wald(e)," "weald," "woulds," and "woud." The
word "Wald" survived up to modern German, as did the word "woud"
in modern Dutch; both of which now mean: closed canopy forest. The
words "wold," "wald," "weald," and "woulds" only remained in Dutch
and English as place names.[15]

The grazing regulations make clear that these words referred to areas containing trees and shrubs, as well as to areas that were treeless, like open grassland where livestock was grazed and raised bogs where peat was cut. These words also included the meaning of places where the food for animals was, like grasses as well as the foliage of trees and shrubs that was cut and dried to serve as winter food were food for livestock.[16]

The wilderness that was declared "forestis" also contained light-demanding tree species such as oak (*Quercus spp.*), wild apple (*Malus sylvestris*), wild pear (*Pyrus pyraster*), and wild cherry (*Prunus avium*). These trees bore fruits (acorns, apple, pears, and cherries) called the "mast," on which pigs were fattened, while oak also delivered timber to construct buildings and ships.[17] There were also light-demanding shrub species like hazel and hawthorn (*Crataegus monogyna*) and sloe (*Prunus spinosa*) that delivered firewood.

Areas containing light-demanding species cannot have been closed canopy forests, because spontaneous developing forests in National Parks and forest reserves all over Europe show that these species do not regenerate successfully in closed canopy forests. They become ousted by shade-tolerant tree species like broad-leaved lime (*Tilia cordata*) and small-leaved lime (*T. platyphyllos*), elm species (*Ulmus spp.*), ash (*Fraxinus excelsior*), beech (*Fagus sylvatica*), sycamore (*Acer pseudoplatanus*), field maple (*A. campestre*), and hornbeam (*Carpinus betulus*). All the light-demanding tree species and shrub species do however regenerate successfully in wood-pastures. They do so in the presence of shade-tolerant tree species that also regenerate successfully in a wood pasture.[18]

A wood pasture consists of a mosaic of grassland, thorny scrub thickets with and without trees, and dispersed forests (groves) surrounded by thorny shrubs called mantle and fringe vegetation. This mantle and fringe vegetation marks the transition between grassland and grove. The regeneration takes place under densities of cattle, deer, and horses that would prevent regeneration of closed canopy forests.[19]

LARGE UNGULATES AND THE REGENERATION OF TREES IN WOOD PASTURES

A characteristic of a wood-pasture is the grazing of livestock like cattle, horses, and pigs. As mentioned before, the theory that a closed canopy forest is the natural vegetation assumes that a wood-pasture is in a state of degradation of a closed canopy forest, made that way by grazing livestock that destroy seedlings and saplings in the forest. However, in a wood-pasture trees regenerate successfully. Nonetheless, it does not take place in the forest, but outside the forest in open grassland. Seedlings and saplings grow up there close to thorny and spiny shrubs like Blackthorn (*Prunus spinosa*), Hawthorn (*Crataegus monogyna*), Juniper (*Juniperus communis*), and

Brambles (*Rubus spp.*), and plant species containing chemical substances that make them unpalatable for large ungulates, such as Bracken (*Pteridium aquilinum*) and Heather (*Calluna vulgaris*). They protect seedlings, saplings, and young trees against grazing and browsing by large ungulates. They are called nurse-species.[20] These nurse species establish themselves in open grazed grassland. Nurse species that spread clonally by root suckers into open grassland like blackthorn form a convex shaped scrub in which tree seedlings establish themselves on the fringe of this advancing scrub as this thorny scrub spreads.[21] In this way, a characteristic convex-shaped assemblage of trees develops, a so-called grove, in Old English called "graf," "grave," or "grove." This grove may cover many hundreds of hectares. The trees expand their crowns, shading out the light-demanding nursing scrub beneath them. Because of this, the grove becomes surrounded by a scrub called (in Old-Germanic) "hage," "haga," or "haye," but lacking a shrub layer to the interior. From the inside the grove looks like a closed canopy forest. Nurse species that do not spread clonally like hawthorn, will promote the development of an open-grown tree, that is a tree with a short trunk and a huge crown. Scattered hawthorns will promote scatted trees, forming a kind of savannah landscape.[22]

As is known from present-day wood-pastures, large ungulates enter a grove by small corridors through the scrub, and prevent inside the grove the regeneration of trees. In this way shade-tolerant tree species that can grow up under the canopy of oaks are prevented from doing so. This mechanism causes oak and other light-demanding tree species to remain part of the canopy of the grove in the presence of shade-tolerant tree species. This is contrary to what happens in forest reserves and National Parks where there is no livestock grazing; here, shade-tolerant tree species grow up under oaks species, then overgrow and kill them.

When trees became senescent and die, a gap in the canopy of the grove is formed. In the gap regeneration of trees is prevented by the large herbivores because they kill the seedlings that emerge in the gap by trampling and browsing. Fungi facilitate the process of the demise of trees, as do drought and storms.[23] As more trees die, the area of the gap grows bigger. Large ungulates bring in seeds of grasses and herbs with their dung and fur, thereby forming a grazed lawn in the centre of the grove. As more trees die, the grove changes over years from the center outwards into an ever-increasing surface of open grassland.[24] When large tracks of open grassland have developed, a micro-pattern of intensive and less intensive grazed patches develops that give light-demanding thorny shrubs the possibility to establish themselves in the less intensive, periodically used patches. There their spines get the change to harden, which takes one growing season. These shrubs then act as nurse species for young trees, and new groves will emerge from the grassland. This process is a non-linear, cyclical succession of grassland → shrubs → grove → grassland → shrub → grove → grassland, etc. The result is a shifting mosaic of open grassland, with or without scattered trees and groves.[25]

The regeneration of light-demanding trees and shrubs in a wood-pasture explains the presence of these species in a wilderness that contains shade-tolerant tree species as well as light-demanding shrub and tree species in the presence of large indigenous grazers as Aurochs (*Bos primigenius*) and Tarpan (*Equus przewalski gmelini*). These wild ungulates were still part of the European wilderness when it was declared a "forestis," but became extinct in 1627 and 1887, respectively.[26] The natural processes and landscape connected with these large ungulates, however, persisted because their domesticated cattle and horse descendants acted as proxies. Indeed, the wood-pasture system of tree regeneration in the presence of high densities of large ungulates is a proxy of the natural conditions that in the Middle Ages were called "forestis," "wald," "wold," "weld," and "woud."

CUTTING FIREWOOD AND THE REGULATION OF GRAZING DOMESTIC STOCK

The wood-pasture system also explains the regulations that were established for temporarily prohibiting the grazing of livestock, which, according to the classic theory of the high forest, are interpreted as to allow the regeneration of trees in the forest. The earliest regulations that temporarily prohibited livestock grazing date from the thirteenth century and were connected with cutting "thorns and hazel" as firewood. During this cutting, the regulations mentioned a certain number of saplings and young trees per unit area needed to be saved, namely oak and wild fruit that provided food for pigs, and oak provided timber. After the cutting (or coppicing) grazing livestock was forbidden to enter an area for three to six, and sometimes nine years.[27]

The presence of thorny species and hazel interspersed saplings of trees along with the presence of livestock answers the description of the mantle and fringe vegetation bordering the grove in a wood-pasture. The temporary prohibition of grazing livestock can be explained by the demand to protect the young sprouts growing up from the stools of the spiny shrubs and hazel after the cutting. The young sprouts of blackthorn and hawthorn will have been browsed immediately by livestock, as the spines do not harden until the end of the first growing season. The spared saplings and young trees could also be browsed because they were stripped from their spiny protectors. Regeneration by sprouting as well as the spared seedlings and saplings therefore needed protection from the animals. After only one growing season, blackthorn, hawthorn, and hazel sprouts can reach two meters high. Blackthorn and hawthorn have developed hard sharp spines then, and can nurse the spared saplings of the trees again. The sprouts of hazel are after a few years so thick and form such a shrub that it is impossible for the animals to bend a sprout over to browse its top. This makes it clear why forest regulations also mentioned that regenerating plots could again be grazed once the shoots had grown above the reach of the cattle.[28]

The regulations of grazing livestock make sense if they are read within the context of the wood-pasture system. They refer then to the thorny scrub that as the mantle and fringe vegetation borders the groves and nurses the saplings of the trees. The regulations aimed to protect the sprouting stools of the thorny scrub and hazel that were cut as firewood and the saplings that temporary were deprived from the protection of the spiny scrub, because it was cut as firewood. As the regulations show, some thinning among saplings was done to promote the forming of trees with big crowns that produced much mast (acorns, pears, apples, and cherries) for pigs.[29] Stools of young oaks that were cut because of the thinning also sprouted. So if one wanted a tree, just one sprout on a stool had to be spared. In this way trees of different ages could be grown to deliver mast as well as timber for ships and buildings, resulting in standards being developed from the thinned saplings. If the temporary prohibition was meant to protect seedlings and saplings against livestock in a high forest—as foresters and scientists explained these regulations in the nineteenth and twentieth centuries—instead of the three to six years written in the regulations, some fifteen to twenty years would have been necessary for the stems of seedlings to grow sufficiently thick to withstand the animals bending them down to browse the top.[30]

In conclusion: For protecting seedlings in the forest, the regulations make no sense. For protecting vegetative regeneration of sprouting stools and saplings or young trees in the mantle and fringe vegetation of groves in a wood pasture, they make excellent sense.

THE DEVELOPMENT OF NATURAL REGENERATION IN THE FOREST

In the eighteenth century people wanted firewood in blocks instead of bundles of sprouts that were delivered by the coppice. For blocks the sprouts had to grow thicker. To achieve this the time between two successive cuts was extended from three to six years in the Middle Ages by way of thirty and fifty, even eighty years. The number of sprouts on the stool was eventually diminished to one.[31] The single stem was cut after eighty years. At such age stools do not sprout again. To obtain the "regeneration" of wood (as material) a new generation of trees had to be planted. Beech was favored, because it produced good firewood for a household and the best charcoal for the industry, whose demand for charcoal rose strongly because of the industrial revolution. A beech of eighty years flowers from the age of thirty years onwards and forms seed from which seedlings emerge. Because they are shade-tolerant, seedlings can sustain the shade of the canopy for several years. Foresters in the German country Hessen discovered in the first half of the eighteenth century that if they cut a tree at the age of eighty years, seedlings grew up in the gap because they received more daylight. In this way in the nineteenth century a technique was developed whereby the canopy was thinned by harvesting single trees in order to give seedlings in the gaps the possibility of growing up, while remaining trees were left standing in order to create a micro climate that sheltered the seedlings and saplings against frost and dryness. After forty years of successive thinning, the last old trees were felled and a new generation of trees had replaced the old one. This technique is today known as shelterwood or selective cutting.[32]

The regeneration of trees from this technique took place in the forest and was called "natural" because the new generation of trees emerged from seed that has spontaneously fallen from standing trees. This was opposite to artificial regeneration that consisted of sowing seed and planting young trees.[33] However, "natural" regeneration was not the regeneration of trees in the natural situation. In Germany initially the "natural" regeneration (of *natürliche Verjungung*) was distinguished from the regeneration in the wild (*Holzwildwuchse*). During the nineteenth and twentieth centuries this distinction disappeared and "natural regeneration" became current for the forestry technique as well for the regeneration in the wild.[34] Human intervention like plowing the soil to create a good germination bed for seed and removing undesirable species of trees, shrubs, herbs, and grasses with chemicals were and still are part of "natural" regeneration. Therefore "natural regeneration" in all the forestry books does not mean what it suggests.[35]

"Natural" regeneration was first developed with the shade-tolerant beech; later it was copied for the light-demanding oak. As with beech, the canopy of standing oak trees was gradually thinned over a period of 40 years. Yet for almost a century all "natural" regenerations of oak failed. Oak seedlings died. By trial and error was found that the opening of the canopy of an oak forest should be faster and the last trees to be removed within ten years so as to give oak seedlings full daylight. During this ten-year period and afterwards, much human assistance was necessary. Shade-tolerant tree species, such as beech, lime, ash, elm, and hornbeam, had to be eradicated, along with grasses and forbs that produced shade.[36] Thus, without significant human management, oak could and still cannot regenerate "naturally." This empirical evidence supports what is observed in forest reserves all over Europe: Oak cannot regenerate spontaneously in a forest growing also shade-tolerant tree species.

HOW LIVESTOCK BECAME A THREAT FOR THE FOREST

With the development of "natural" regeneration, the regeneration of trees moved from outside a grove (forest) in a wood-pasture system in which seedlings were protected against livestock by thorny scrub, to the inside of a forest in which seedlings were not protected against large ungulates, as thorny scrubs could not thrive because of shade. Grazing livestock therefore became a problem for "natural" regeneration of trees and judged as the greatest threat to the forest. Foresters promoted removing livestock from what was still called "Forst," "forêt," "Wald," and "woud," while reserving "Forst," "Wald," and "woud" for the production of wood materials.[37] This new way of managing the forest became possible in the eighteenth century after the development of the so-called New Agriculture. The potato was introduced on a large scale as a foodstuff for mankind as well as for pigs, making oak as a source of mast for pigs useless. Grass species that were specially bred for a high production of food livestock became available as seed and made it possible to create grazing areas for livestock that were more productive than grazing areas in the "Forst," "Wald," and "woud." The total spatial separation of livestock grazing and wood production was the result in the nineteenth century. Livestock grazing was abolished in the "Forst," "Wald," and "woud."[38] The "Forst," "forêt," "Wald," and "woud" foremerly meaning a park-like landscape came to mean a closed canopy forest in which trees regenerated "naturally."

In combination with the assumption that spontaneous vegetation on abandoned agricultural land resulted in the forest as natural vegetation, the baseline for natural conditions shifted from a wood-pasture system to a closed canopy forest. This shift meant that in wood-pastures that were declared forest reserves or national parks in the nineteenth and twentieth centuries—among them the famous National Park Bialowieza—cattle and horses were removed as they were considered to be alien, introduced species, and so an "unnatural" part of the system. This forest as the baseline for natural included that remaining wild ungulates such as red deer were reduced by culling to such low densities, that they did not prevent trees to regenerate in the forest. As a consequence, the park-like wood-pastures formerly rich in species developed into closed canopy forests low in biodiversity. All light-demanding plant species disappeared, among them two oaks that are associated with more insect species than any other plant species.[39] All wild fruit species likewise disappeared together with indigenous shrub species. Animal species thriving in this mosaic landscape of open grassland, groves, solitary trees together with all the edges that are characteristic for this landscape, disappeared—to include many butterflies and song birds. The result was an enormous loss of biodiversity.

REWILDING OR GARDENING?

The shift of the meaning of words in historic texts, currently meaning closed canopy forest, deprived indigenous, large ungulates in Europe from their natural role of structuring and functioning natural ecosystems. This caused and still causes a great loss in biodiversity in the name of nature conservation. In order to prevent this, their role needs to be restored. To achieve this goal, large natural areas must be established, and the role of the large indigenous ungulates, in particular, must be reincorporated in these systems, as ungulates fulfill key roles in creating park-like landscapes

Figure 9.1 The wood-pasture Borkener Paradise in Germany. In the foreground is an oak surrounded by hawthorns that act as nurse species for the oak. Behind it lies a grove surrounded by a scrub of flowering blackthorn. Such groves advance into the grazed grassland at a rate equal to that of the advancing outer edge of the scrub. (Photograph by Frans Vera)

that harbour the indigenous biodiversity. The promotion of the role of large ungulates demands human interference, because in many places they have to be reintroduced, especially wild cattle and horses. From the perspective of plant and animal species currently deprived of their partners by human interference, reinstating their role is not gardening, but simply (re)wilding.

NOTES

1. Daniel Pauly stated in 1995 in his essay *Anecdotes and the shifting baseline syndrome of fisheries*: "Essentially, this syndrome has arisen because each generation of fisheries scientists accepts as a baseline the stock size and species composition that occurred at the beginning of their careers, and uses this to evaluate changes. When the next generation starts its career, the stocks have further declined, but it is the stocks at that time that serve as a new baseline. The result is a gradual accommodation of the creeping disappearance of resource species, and inappropriate reference points for evaluating economic losses, resulting from over fishing, or for identifying targets for rehabilitation measures."
2. D. Pauly, "Anecdotes and the shifting baseline syndrome," *Trends in Ecology & Environment* 10 (1995), 430; C. Sheppard, "Shifting baseline Syndrome. *Marine Pollution Bulletin* 30 (1995): 766–67; E. Duffy, "Biodiversity loss, trophic skew and ecosystem functioning," *Ecology Letters* 6 (2003): 680–97.

3. H. Cotta, *Anweisung zum Waldbau* (Neunte, neubearbeitete Auflage; Leipzig: Arnoldische Buchhandlung, 1865); E. Landolt, *Der Wald, seine Verjüngung, Pflege und Benutzung* (Zürich: Schweizerischen Forstverein, 1866); F. E. Clements, *Plant succession. An analysis of the development of vegetation* Publication nr. 242 (Washington, D.C.: Carnegie Institution, 1916); A. G. Tansley, *The British Islands and their Vegetation.* v. 1 & 2, 3rd ed. (Cambridge: Cambridge University Press, 1953).
4. A. G. Tansley, "The Use and Abue of Vegetational Concepts and Terms," *Ecology* 16 (1935): 284–307; A. C. Forbes, "On the regeneration and formation of woods from seed naturally of artificially sown," *Transactions of the English Arboricultural Society* 5 (1902): 239–70.
5. A. S. Watt, "Pattern and process in the plant community," *Journal of Ecology* 35 (1947): 1–22; H. Leibundgut, "Über Zweck und Methodik der Struktur und Zuwachsanalyse von Urwäldern," *Schweizerische Zeitschrift für Forstwesen* 110 (1959): 111–24; H. Leibundgut, "Über die Dynamik europäischer Urwälder," *Allgemeine Forstzeitschrift* 33 (1978): 686–90.
6. F. W. M. Vera, *Grazing Ecology and Forest History* (Wallingford: CABI Publishing, 2000).
7. A. Bühler, "*Der Waldbau nach wissenschaftlicher Forschung und praktischer Erfahrung*," II Band (Stuttgart: Eugen Ulmer, 1922); K. Mantel, *Wald und Forst in der Geschichte* (Alfeld-Hannover: M. und H. Schaper, 1990).
8. Landolt, *Der Wald*; E. H. L. Krause, "Die Heide. Beitrag zur Geschichte des Pflanzenwuchses in Nordwesteuropa," *Engleis Botanisches Jahrbuch* 14 (1892): 517–39.
9. Tansley, *The British Islands and their Vegetation*; Tansley, "The Use and Abuse of Vegetational Concepts and Terms;" Krause, "Die Heide. Beitrag zur Geschichte des Pflanzenwuchses in Nordwesteuropa."
10. F. Firbas, "Über die Bestimmung der Walddichte und der Vegetation Waldloser Gebiete mit Hilfe der Pollenanalyse," *Planta* 22 (1934): 109–46; F. Firbas, "*Die Vegetationsentwicklung des Mitteleuropäischen Spätglacials*," Bibliotheca Botanica 112 (1935): 1–68; H. Godwin (a), "Pollen analysis. An outline of the problems and potentialities of the method. Part. I. Technique and interpretation," *New Phytologist* 33 (1934): 278–05; H. Godwin (b), "Pollen analysis. An outline of the problems and potentialities of the method," Part. II. General applications of pollen analysis. *New Phytologist* 33 (1934): 325–58.
11. Vera, *Grazing Ecology and Forest History*; J. Iversen, "Problems of the Early Post-Glacial Forest Development in Denmark," *Danmarks Geologiske Undersøgelse* IV. Raekke Bd. 4, nr 3 (1960, *Geological Survey of Denmark. IV Series* Vol. 4 No. 3); K. Aaris-Sørensen, "Depauperation of the Mammalian Fauna of the Island of Zealand during the Atlantic period," *Videnskabelige Meddelelser fra Dansk Naturhistorisk Forening* 142 (1980): 131–38.
12. H. Kaspers, *Comitatus nemoris. Die Waldgrafschaft zwischen Maas und Rhein*, Beiträge zur Geschichte des Dürener Landes, Band 7 (Düren und Aachen, 1957); H. Hesmer, *Wald- und Forstwirtschaft in Nordrhein-Westfalen* (Hannover: 1958); J. Buis, *Historia Forestis: Nederlandse bosgeschiedenis*, Deel 1 en 2 (Utrecht: H & S Uitgevers, 1985).
13. Vera, *Grazing Ecology and Forest History*; This law should not be confused with the so-called Forest Law that was centuries later, in the eleventh century, introduced by William the Conqueror from the continent to England, when he became king of England. The Forest Law was in fact derived from the "ius forestis."
14. Mantel, *Wald und Forst in der Geschichte*; Kaspers, *Comitatus nemoris*.

15. K-H. Borck, Zur Bedeutung der Wörter *Holz, Wald, Forst* und *Witu* im Althochdeutschen. *Festschrift für Jost Trier* (Meisenheim: 1954): 456–76; J. Trier, *Holz. Etymologien aus dem Niederwald*, Münstersche Forschungen 6 (Münster/Köln: Böhlau Verlag, 1952); J. Trier, *Venus: Etymologien um das Futterlaub*, Münstersche Forschungen 15 (Münster/Köln: Böhlau Verlag, 1963); W. A. Ligtendag, *"De Wolden en het water. De landschaps- en waterstaatsontwikkeling in het lege land ten oosten van de stad Groningen vanaf de volle Middeleeuwen tot ca. 1870,"* Regio en Landschapsstudies n.2, Stichting Historisch Onderzoek en Beleid. REGIO-PRoject (Groningen: Uitgevers, 1995); S. Wager, *Woods, Wolds and Groves: The Woodland of Medieval Warwickshire*, British Archaeological Reports, BAR British Series 269 (1998).

16. Trier, *Holz;* Trier, *Venus.*

17. Kaspers, *Comitatus nemoris;* G. E. Hart, *Royal Forest. A History of Dean's Woods as Producers of Timber* (Oxford: Clarendon Press, 1966); N. Flower, "An Historical and Ecological Study of Inclosed and Uninclosed Woods in the New Forest," MSc thesis, King's College, University of London (Hampshire: 1977); H. Hausrath, *Geschichte des deutschen Waldbaus. Von seinen Anfängen bis 1850* (Freiburg, Breisgau: Hochschulverlag, 1988).

18. Vera, *Grazing Ecology and Forest History;* C. Smit, D. Béguin, A. Buttler, and H. Müller-Schärer, "Safe sites for tree regeneration in wooded pastures: A case of associational resistance?," *Journal of Vegetation Science* 16 (2005): 209–14; C. Smit, J. Den Ouden, and H. Müller-Schärer, "Unpalatable plants facilitate tree sapling survival in wooded pastures," *Journal of Applied Ecology* 43 (2006): 305–12.

19. Tansley, "The Use and Abuse of Vegetational Concepts and Terms;" Vera, *Grazing Ecology and Forest History;* R. Pott and J. Hüppe, *Die Hudenlandschaften Nordwestdeutschlands*, Westfälisches Museum für Naturkunde, Landschafsverband Westfalen-Lippe, Veröffentlichung der Arbeitsgemeinschaft für Biol.-ökol. Landesforschung, *ABÖL*, n. 89 (Münster: 1991).

20. E. S. Bakker, H. Olff, C. Vandenberghe, K. De Maeyer, R. Smit, J. M. Gleichman, and F. W. M. Vera, "Ecological anachronisms in the recruitment of temperate light-demanding tree species in wooded pastures," *Journal of Applied Ecology* 41 (2004): 571–82.

21. Pott and Hüppe, *Die Hudenlandschaften Nordwestdeutschlands;* A. S. Watt, "On the ecology of British beech woods with special reference to their regeneration," Part II. The Development and Structure of Beech Communities on the Sussex Downs. *Journal of Ecology* 12 (1924): 145–204.

22. Vera, *Grazing Ecology and Forest History;* Ligtendag, *De Wolden en het water.*

23. T. Green, "The forgotten army—woodland fungi," British Wildlife 2 (1992) 85–86; A. Dobson and M. Crawley, "Pathogens and the structure of plant communities," *Trends in Ecology and Evolution* 9 (1994): 303–98; E. P. Mountford and G. Peterken, "Long-term change and implications for the management of wood-pastures: experience over 40 years from Denny Wood, New Forest," *Forestry* 76 (2003): 19–43.

24. Mountford and Peterken, "Long-term change and implications for the management of wood-pastures;" J. Bokdam, *"Nature conservation and grazing management. Free-ranging cattle as driving force for cyclic vegetation succession,"* PhD. Thesis, Wageningen: Wageningen University, 2003); E. P. Mountford, G. F. Peterken, P. J. Edwards, and J. G. Manners, "Long-term change in growth, mortality and regeneration of trees in Denny Wood, an old-growth wood-pasture in the New Forest (UK)," *Perspectives in Plant Ecology, Evolution and Systematics.* 2 (1999): 223–72.

25. Vera, *Grazing Ecology and Forest History*; F. W. M. Vera, E. S. Bakker, and H. Olff, "Large herbivores: missing partners of western European light-demanding tree and shrub species?," in K. Danell, P. Duncan, R. Bergström, and J. Pastor, eds., *Large Herbivore Ecology, Ecosystem Dynamics and Conservation* Conservation Biology 11 (2006) (Cambridge: Cambridge University Press): 203–31.

26. W. Szafer, "The Ure-ox, Extinct in Europe Since the Seventeenth Century: an Early Attempt at Conservation that Failed," *Biological Conservation* 1 (1968): 45–47; W. Pruski, "Ein Regenerationsversuch des Tarpans in Polen," *Zeitschrift für Tierzüchtung und Züchtungsbiologie* 79 (1963): 1–30.

27. Vera, *Grazing Ecology and Forest History*; H. Streitz, "*Bestockungswandel in Laubwaldgesellschaften des Rhein-Main-Tieflandes und der Hessischen Rheinebene,*" Dissertation Forstlichen Fakultät der Georg-August-Universität zu Göttingen (Hannover, Münden: 1967).

28. Cotta, *Anweisung zum Waldbau.*

29. Mantel, *Wald und Forst in der Geschichte*; W. Schubart, *Die Entwicklung des Laubwaldes als Wirtschaftswald zwischen Elbe, Saale und Weser,*" Aus dem Walde, Mitteilungen aus der Niedersächsischen Landesforstverwaltung 14 (1966).

30. Cotta, *Anweisung zum Waldbau*; Flower, "An Historical and Ecological Study of Inclosed and Uninclosed Woods in the New Forest, Hampshire;" Mayer, *Waldbau auf soziologisch-ökologischer Grundlage*; H. Mayer, *Waldbau auf soziologisch-ökologischer Grundlage*, 4., teilweise neu bearbeitete Auflage (Stuttgart: Gustav Fischer, 1992).

31. Mantel, *Wald und Forst in der Geschichte*; Schubart, *Die Entwicklung des Laubwaldes als Wirtschaftswald zwischen Elbe, Saale und Weser*; K. Vanselow, *Die Waldbautechniek im Spessart: Eine historisch-kritisch Untersuchung ihrer Epochen* (Berlin: Verlag von Julius Springer, 1926).

32. Bühler, *Der Waldbau nach wissenschaftlicher Forschung und praktischer Erfahrung*; Mantel, *Wald und Forst in der Geschichte*l; Vanselow, *Die Waldbautechniek im Spessart.*

33. Cotta, *Anweisung zum Waldbau*; Vanselow, *Die Waldbautechniek im Spessart*; A. Dengler, *Waldbau auf ökologischer Grundlage, Zweiter band. Baumartenwahl, Bestandesbegründung und Bestandespflege* (6th. ed., Hamburg and Berlin: Röhrig, E. and Gussone, H.A. Verlag Paul Parey, 1990).

34. Cotta, *Anweisung zum Waldbau*; Landolt, *Der Wald*; Tansley, *The British Islands and their Vegetation*; K. Vanselow, *Theorie und Praxis der natürlichen Verjüngung im Wirtschaftswald* (Berlin Neumann Verlag, 1949); K. Gayer Gayer, *Der gemischte Wald, seine Begründung und Pflege, insbesondere durch Horst- und Gruppenwirtschaft* (Berlin: Paul Parey, 1886).

35. Vera, *Grazing Ecology and Forest History*; R. Harmer, "Natural Regeneration of Broadleaved Trees in Britain: I. Historical Aspects," *Forestry* 67 (1994): 179–88.

36. Dengler, *Waldbau auf ökologischer Grundlage*; Harmer, "Natural Regeneration of Broadleaved Trees in Britain;" H. Grossmann, *Die Waldweide in der Schweiz*, Promotionsarbeit (Zürich: 1927).

37. Landolt, *Der Wald*; Vera, *Grazing Ecology and Forest History.*

38. J. H. von Hobe, *Freymüthige Gedanken über verschiedene Fehler bey dem Forsthaushalt, insbesondere über die Viehude in den Holzungen, deren Abstellung und Einschränkung* (Thal-Ehrenbreitstein, in der Gehraschen Hofbuchhandlund, 1805); Grossmann, *Die Waldweide in der Schweiz.*

39. M. G. Morris, "Oak as a habitat for insect life," in M. G. Morris and F. H. Perring, eds., *The British Oak, Its History and Natural History*, The Botanical Society of the British Isles (Berkshire: E.W. Classey, 1974): 274–97.

10 Regardening and the Rest

Chris Smout

REGARDENERS AND REWILDERS

According to Franciscus Vera's chapter, we might believe that nature exists independent of culture and was at its most perfect before humanity interfered by the intervention of farming. For him there was an "original-natural" state which forms a baseline that we need to understand if we are correctly to direct our attempts to restore natural processes and achieve a rich biodiversity. For him, nature conservation is about process. It is not about fiddling with little nature reserves to which one or two endangered species have retreated, or turning the clock back to one particular century. It is about understanding a time when human beings were without history, certainly without agriculture, in order to recover and release forgotten process. In the context of his own interests, it is about using large herbivores to create a landscape of groves, glades, and wood pasture which he believes was the true baseline for an "original-natural" Europe.

His chapter does not mention Oostvaardersplassen, the great Dutch reserve closely associated with his work and ideas. The Netherlands, the land of tulips and the home of gardening, does not have much wild land, however defined. In 1975, an area of 3600 hectares of land recently reclaimed from the sea was declared a nature reserve, the biggest in the country. An additional 2000 hectares was added in 1982. The land was not restored as tidal mudflat or seabed, but became the boldest experiment in terrestrial rewilding in Europe. In the words of the official book on the project:

> Here nature conservation got a new dimension. No longer conserving existing nature in mostly small reserves but allowing freedom to natural processes, and see what happens. From nature conservation to nature development. Since the seventies The Oostvaardersplassen is the flagship of a new direction. Many people from abroad look at it and are astonished about what appeared to be possible on only half an hour driving from the city of Amsterdam in such a densely populated country as the Netherlands.[1]

From reading this quotation, one might get the impression that the Dutch had enclosed the polder and gone away to leave it to nature without further intervention. Nothing could be further from the truth. Under the guidance of Vera, Oostvaardersplassen is now home to thousands of introduced Red Deer, Heck Cattle, and Konik Horses which roam the reserve untended, unculled, unfed, and unhindered. According to Vera's theory, which has proved immensely stimulating to conservation biology throughout northern Europe, this replicates a critical and forgotten natural process of the true original-natural past (i.e., before agriculture), that is to say, the benevolent impact on the land of numerous large herbivores.

So Oostvaardersplassen is at once regardened and rewilded. It was drained by man to create a terrestrial nature reserve on what was once a seabed and subjected by man to the impact of thousands of introduced wild and domestic animals. Then it was left utterly to itself and the forces of nature. Should it ever be shown that Vera's theory is wrong, and that such herds were not as numerous, large, or influential in the pre-agriculture natural state as he argues, the experiment of their introduction would be shown to have been (in its own terms) inauthentic, driven by a process not entirely "natural." But that would not for a moment diminish its objective value in creating and maintaining an astonishing biodiversity in this great reserve.

It is interesting to put all this in the context of Bill Adams's critique of modern British conservation policy. It was George Peterken who coined in 1981 the phrase "future nature," when he made a distinction between five "qualities of nature." He began with "original nature," the state which existed before significant human impact, Vera's "original-natural" baseline. He ended with "future nature," the state that would develop if human influence was completely removed now. It would not be the same as original-natural because soils and climate changed and are still changing, and because of the introduction of non-native species.[2]

Bill Adams has taken up the phrase in his book, *Future Nature: a vision for conservation*, where he criticizes British conservation policy as too often faint-hearted, fixated on small preserved sites, ignoring the wider countryside, and reluctant to "release the wild."[3] By this he means, like Vera, releasing natural processes to work without constraints, even if this means an element of unpredictability and the risk of losing some familiar forms of nature in the shape of present habitat and biodiversity. The gains would be greater, more natural, than the losses.

But this is vision. Let us take what actually happens in British conservation policy, and see how it uses, or fails to use, history, and then let us see how its confused attitude to history arises from a particular scientific view of the relationship between nature and humankind. Future nature, it will be proposed, has to arrive from acknowledgment of past culture.

HISTORY IN BRITISH NATURE CONSERVATION

In Britain, nature conservation has until now been overwhelmingly site based, despite some attempts by Scottish Natural Heritage in the early 1990s to break out of that straitjacket. There are signs that this might change with new agri-environment schemes to ameliorate the devastation of the wider countryside, but these are in their infancy. On protected sites, where most of what is valuable is concentrated, there are several current approaches that involve both restoration and an appeal to the past. Large-scale rewilding is still relatively unusual and mainly consists of projects for the future. In England, the most exciting involves the Wet Fens Partnership, a consortium of voluntary bodies, private landowners, and government organizations that plan to return 9000 hectares of farmland (not all contiguous) to fenland in East Anglia.[4] In Scotland, the most ambitious current project is to turn 9510 hectares of Glen Alladale in Sutherland into a fenced reserve, ultimately with reintroduced Elk (moose), Wild Boar, Lynx, Bear, and Wolves.[5] Even that would be dwarfed by Philip Ashmole's vision of returning montane birch and pine to over 6000 square kilometres of the Scottish mountains above the current timberline, but that is no more than a gleam in the eye.[6]

The term "rewilding" is undoubtedly in fashion in British conservation, however. The Wildland Network lists forty-seven current projects described as rewilding on its website, but in England and Wales out of thirty-two sites only four involve more than 1000 hectares, and several are under ten hectares. In Scotland, thirteen are above 1000 hectares, including several in the ownership of the John Muir Trust and other voluntary bodies.[7] But obviously there is less scope for rewilding in Britain than in countries like the United States, where, for example, a project is in train to reclaim for the "buffalo commons" 1.4 million hectares of the grasslands of the Midwest.

Most habitat restoration in Britain has perforce been of a "regardening" character, less ambitious, small scale management involving sites of a few dozen hectares or less. It is indeed sometimes derisively called "conservation gardening" by its detractors, who question whether much that is sustainable can be carried on at this scale. "Gardening" and not "rewilding," however, has been until now the staple approach of the wildlife trusts, of the Royal Society for the Protection of Birds, of most other conservation charities, and of most of the work on the ground by the three government conservation agencies, English Nature, Scottish Natural Heritage, and the Countryside Council for Wales. There is, however, a third approach gaining ground, sometimes called "landscape-scale change" that calls for a combination of the first and second. It consists of setting up networks of new or restored conservation habitat within a matrix of agricultural land, such as patches of woodland linked by hedgerows or of rough open grassland within insect-flight of each other. It is in some ways closer to rewilding than to gardening and some of the projects in the Wildlands Network database would fall more comfortably under this heading, such as the Border Mires project in the north of England, planned to associate 11,851 hectares of individual mires that vary in size from 2.5 to 400 hectares. The emphasis appears often to be on the abundance rather than the quality of what is created, eschewing micro-management in favor of releasing supposedly natural processes (for example in the work of the Woodland Trust and various agri-environment schemes), but at least for some insects there is a realization that exact quality of the patches is critical (exemplified in the various Marsh Fritillary recovery schemes in England and Wales). For the purposes of this chapter it is convenient to emphasize the contrasts between rewilding and gardening, without forgetting that landscape-scale change can combine the best of both, especially in a long-settled and long-farmed continent like Europe.

Historians and archaeologists are not routinely employed in nature conservation, yet all three approaches to habitat restoration imply a vision of the past, and appeals to the past are often explicit and elaborate. There could be no better example than the controversy around Vera's account of "original-natural" woodland and the experiment at Oostvaardersplassen, discussed earlier. This initiative and the theory behind it has had great influence and given rise to great disagreement. Palaeoecologists in Denmark and Ireland, for example, consider Vera's conclusions wrong, and prefer the older paradigm of the Mesolithic forest as predominantly continuous cover, broken by windthrow, small lawns, and bogs.[8] English Nature, led by its chief woodland ecologist, Keith Kirby, commissioned an impressive report from ecologists, palynologists, and specialists in macrofossil molluscs and invertebrates, and concluded cautiously with an English compromise:

> Parts of the Atlantic forest may have looked like a modern wood-pasture and there might have been some permanently open areas; but the majority seems likely to have been relatively closed high forest, but with a component of temporary and permanent glades.[9]

It is worth noticing a few things. First, the debate in this case has been conducted in rigorous historical terms, often by the most sophisticated means available. That it has been carried out in departments of science using the methods of laboratories does not make it less historical. It is simply that the Mesolithic did not generate documents, but it left a wealth of fossil pollen, snails and insects, and a few bones.

Second, there is often an unquestioned assumption that the Mesolithic past has the greatest relevance, since that was the last age that can be considered "original-natural." What we have around us today, the conservationists often seem to be telling us, is in degraded genealogical descent from a purer, noble, natural, wilder time. The English Nature report, however, is very conscious of the possibility that the wildwood might have been substantially changed by the hunter–gatherers of the time, raised by A. G. Smith as long ago as 1970 and reiterated recently by environmental historians such as Ian Simmons.[10] They had the capability, by ringbarking and fire, of modifying woodland to assist both hunting and gathering, by manipulating grassy spaces and increasing scrub and edge in a mosaic within the forest. If they were capable of doing it, they probably would do it, and if they did it on a large scale it would have implications for applying Vera's model today without additional human management.

Some rewilders, particularly in Scotland, are much less interested than English Nature or than Vera in environmental history, ancient or modern. Once they have convinced themselves of an original-natural state, typically based on a popular myth of the Caledonian Forest, they carry on with recreating what it represents, irrespective of 5000 years of climate change and human activity, or its mark on the landscape, and certainly irrespective of research. As suggested in this volume by Mairi Stewart in respect of Skye, and by Althea Davies in respect of West Affric, this can lead voluntary bodies and individuals to attempt inappropriate landscape change. What they will end up with will not be natural, but cultural landscapes that memorialize early third millennium conservation theories. These will be interesting landscapes in themselves, but probably not what their originators wanted.

Importantly, proponents of rewilding always (and rather paradoxically) deny any intention to recreate a particular point of time—in this they differ from those who seek to restore an historic building, who either want to recreate its appearance at a particular point (say, on construction) or to "conserve as found," a modern preference for retaining the patina and alterations up to the point of bringing the building into restoration management. Nature conservationists have two emphases in their response to the query about why they do not seek any point in time. Many say that they are trying not to recreate a period anyway, but to liberate a natural process: at Oostvaardersplassen, this process is naturalistic grazing, where ungulates impact on the landscape without human management or intervention. Others would say that you cannot exactly recreate a point in the past, partly because it is unknowable in sufficient detail but mainly because, with the best will in the world, many of the parameters are different now, such as the absence of wolves or the presence of non-native (alien) species, or simply a different climate where change may or may not have anthropogenic causes. Yet this does not prevent an obsession with the "original-natural," taken in Europe to be the Mesolithic, in America the pre-Columbian or at least the time before white colonization.

However, we can lift the stick from the other end of time. What is the point of harking back to an "original-natural" which in Europe is unknowable, irrecoverable, and 5000 years distant in time? It is simpler in America, where white colonization is sometimes barely 150 years old, as on the buffalo commons. The European landscape around us is not Mesolithic and original-natural but modern and cultural. It is impoverished and becoming more so, but the biodiversity we value, and the Red Data Book species we especially treasure, are largely the product of the cultural landscapes that immediately predated this one. Historians know, or could find out, a great deal about the cultural landscapes of the past 400 years or so, what they looked like and how they were maintained, and the file of contemporary information about its biodiversity thickens as we move from early days of the Royal Society in the seventeenth century, through the Linnean scholars of the eighteenth century, to the immense labors of the Victorian naturalists and their recent successors.[11]

It is the conservation "gardeners" who know this, although they seldom turn to professional historians for help. Because the "rewilders" work on a grand scale, they are often dismissive of managers looking after a reserve of limited size, considering them to be fiddling at the edges. In many cases they criticize them for actually working against the dynamics of nature by trying to hold back an ecological succession that would convert a dry heath or a raised bog into a birch wood, or a marsh into a rushy field. In the rewilders' ideal world, the area of ecological restoration would be large enough to accommodate any losses from ecological succession, as there would be space for nature to create new heaths and marshes from the free rein of natural processes.

To this the gardeners retort that we do not in Britain or Europe live in an ideal world, that there is little prospect of doing so, and that without their constant effort on small sites most of the highly valued assemblages of Red Data Book flora and fauna would become locally or nationally extinct. They could add, but seldom do, that the vast majority of British sites (and the existence of Red Data Book species in them) are only there because the sites are cultural, not natural, and have been maintained by past generations of farmers and others preventing the natural ecological changes that would have overwhelmed them. It is the character of ancient cultural landscapes to create niches, and the nature of biodiversity especially to appreciate them. Peter Marren described the newly formed UK Nature Conservancy in the 1950s as "fully aware that nature reserves would have to be managed if they were not to dry out, or become overgrown by scrub or rank grass, or harbour the 'wrong' sort of wildlife."[12] It is still true, and the management needs to be based on historical knowledge as well as on science.

One interesting corner of British conservation that almost by definition relates to conservation gardening rather than rewilding, is safeguarding the ancient weeds of arable cultivation. Many of these are not native plants but "archaeophytes," defined by botanists as alien species known to have been present since before 1500, and some of them believed to have been present since Neolithic times. As a group they have been severely threatened by changes in late twentieth-century agriculture, especially by herbicides, by intensive farming that obviates a need for a fallow break, and by effective seed-cleaning techniques.

There are 141 (by some accounts 167) species of archaeophytes, of which thirty-nine are among the 100 plants in the British flora to have shown the most marked relative decline since 1930. Eight out of the ten species to have shown most decline are archaeophyte weeds of arable fields.[13] The names of some of the most threatened reveal their long cultural association with people and farming: Lamb's Succory, Corn Cleavers, Corn Buttercup, Shepherd's Needle, Pheasant's Eye, Corn Chamomile, Stinking Chamomile, Corn Marigold, Black Bindweed, Corn Gromwell, Field Woundwort, Corn Salad, Good King Henry, Fat Hen, Venus's Looking Glass.

The rarest of them eke out their existence on the edges of fields in marginal places in the south of England, sometimes on the occasional farms that escaped, for one reason or another, the full blast of modern ideas. They are even threatened now in some places by ill-considered agri-environment schemes that favor planting field margins with hedges and trees, or allowing old fields to revert to scrub. And why not let them go, the purist may ask? What has their presence got to do with nature? They came with one farming regime, let them go with another.

That, however, and interestingly so, is not now the view of the UK Joint Nature Conservation Committee, whose *Red Data Book of Vascular Plants* (2005) lists a whole swathe of archaeophytes, though no neophytes (i.e., those plants believed to have arrived since 1500). It is easier than it was, but still much more difficult to get protection in Great Britain if you are an alien plant than an alien person, and heaven knows that is difficult enough.

Conscious of breaking new ground in protecting alien species at all on such a scale, the JNCC offered three justifications. First, archaeophytes are, as a group, stable or declining, often throughout the European range, and some of the arable weeds were under threat of extinction in the UK. Second, archaeophytes, unlike neophytes, tend to have a world distribution which is unknown or uncertain, and are often not regarded as native anywhere: to neglect them would be to risk their total loss from the planet. Third:

> Archaeophytes are of considerable historical and cultural interest. They have developed (and exploited) a close relationship with man which is, in effect, one of *commensalism*—many archaeophytes are, quite literally, 'followers of man.' The way in which humans now value these species is partly a consequence of having been so intimately associated with them over such a long time period.[14]

So culture as well as nature has here become explicitly a reason for nature conservation. It is to be hoped that a role will now be found for agricultural historians in learning how to tend and recover populations of these plants. And what is true of the plants of arable fields is no less true of the plants of traditional hay meadows and other grasslands, where the expertise of the documentary historian can again usefully be added to that of the ecologist.[15] Thanks to the work of Oliver Rackham, it is already true of the plants of traditional ancient woodland, where the informed maintenance or revival of coppicing can bring the woodland flora back to life, if levels of deer grazing permit.[16] These are all ways of gardening and they all have their critics on that account. But if the need for conservation gardening is accepted, so should be a role for the historian to recount the archaic practices that once maintained the rich cultural mosaic of ecological niches, just as the archaeologist has a role to pronounce on the history of the weeds themselves.

Peter Marren made a relevant point about this when he described the early days of Britain's National Nature Reserves in the 1950s:

> There was a great deal to be learned from traditional land husbandry and the craft industry, from coppicing and shelterwood systems in woodland to the commoning practices of rough grazing, turf digging and reed harvesting . . . [but] there is little evidence that the scientists talked to the woodmen, the commoners or the thatchers. Theirs was a hermetic world of seminar rooms and laboratories, and they thought in terms of the future rather than the past. As a result, they were in danger of having to reinvent the billhook.[17]

There is still a tendency to rely too much on science and too little on knowledge of past practice.

Many habitats in Great Britain have been entirely formed by cultural manipulation of nature, especially farming and hunting. Many more have come about through crude industrial activity, and the biggest threat to several species of butterflies and other rare insects in Britain is the building over of urban brownfield sites.[18] Among significant habitats, including small nature reserves, are a surprisingly large number involving disused historic mineral extraction sites, some of considerable interest to industrial archaeology. A quick trawl through the files of the journal *British Wildlife* in the last few years identified coal-mining relics in Yorkshire, and such reserves as the clay pits near Peterborough, with their nationally important population of Great Crested Newts, metal mines in Cornwall with their bryophytes, and peat diggings in Norfolk with their rare damsel flies.[19]

As Whitehouse and Davies emphasize in their contributions, there are two extensive habitats in upland Great Britain where the need for management and restoration is particularly clear, both of which call for deep understanding of their cultural past—heather moorland and raised bogs. Heather moorland was present on a small scale in the Mesolithic, although there is little evidence at present from palynology to suggest in the uplands any close equivalent to Vera's lowland savannahs. The great expanses of moor with which we are familiar today in Scotland and the north of England have been maintained as cultural landscapes free of the natural succession of tree cover only by, in the first instance, farmers burning and grazing the heather for millennia, and more recently by the managers of sporting estates burning and grazing in a subtly different way.[20] Grouse moors are critical habitat for a whole range of species from the Goldern Plover to the *jonellus* bee, and they are also the sites of great controversy between land-owners and conservationists concerning how far the maintenance of large numbers of predators such as Hen Harriers and Peregrines is compatible with large numbers of Red Grouse, on which the economic viability of the moors and therefore their continued management depends. If nature was to take its course, or the rewilders with their eye on 6000 square kilometres of potential montane woodland to succeed, both the game and the raptors would be lost, as well as all the incidental species. Even as it is, heather moorland is in steep decline throughout western Europe, and in northern Britain in particular there is a responsibility on us to maintain it both as biodiversity resource and cultural icon.

If it is to be treasured, it is an historical question as to how this habitat was managed in the past. It may be enough to say in this and similar situations, that we will maintain "traditional" management, which essentially means what the more old-fashioned managers are doing now. But it might be even better to try to find out (if it is possible) what had been the earlier management options, for example, pre-gun, pre-trap, pre-flame-thrower, with different types and sizes of animal, and see if you like the biodiversity of that time more than that of the present day. Both are a product of culture, of particular systems of management, and there is a question of choice (constrained of course by possibility, not least economic possibility) as to what biodiversity scene you may wish to have at the end of the day.

Raised bogs also show the paramount need for knowledge of past management. It is unclear how far peat bogs are entirely natural, but their formation accelerated with climatic change around the time of the onset of the Bronze Age, with more rain and wind, and human clearance of trees and cultivation may then and at other times have accelerated their growth. Many bogs overlie prehistoric field systems, but others are at least 9000 years old and predate agriculture. All were used for millennia for the provision of food, pasture, fuel, and building material, and it was this use that kept them in their distinctive form, especially the regular surface burning and grazing which prevented the resumption of tree growth and encouraged the continuous formation of the peat. When, at various times over the last half century, such bogs came under the care of conservation bodies, they were ignorant of all this. Misunderstanding what had kept them in being so long, they enclosed them against stock and banned the use of fire. The almost uniform result has been that bogs in the care of conservation bodies have become progressively more overgrown with wood, while those outside conservation care have fallen victim to commercial peat extraction (not, as in Ireland, to fuel power stations, but in Britain usually for garden centre material). Frequently in the twentieth century there was a judgment of Solomon, a division of the baby. Part of a great raised bog was used for commercial extraction and part was surrendered to conservation. Babies don't thrive from being cut in two. Extraction lowered the water table, to accelerate further the afforestation of the conservation sections. The reaction has been to try to stop or to limit commercial use, with lesser or greater success in the case of two extremely important sites, Thorne Moors in Yorkshire and Flanders Moss in Stirlingshire.

But what do you do about the trees once you have them? You might of course welcome them in the spirit of accepting a change that has probably happened before over the millennia, but most conservationists felt that tree cover would change the biodiversity they had inherited, and decided to extirpate them. Unable to believe that grazing and burning would provide a solution, the conservation managers often tried to destroy the wood by hand and spray, a battle against birch seed that surely will never succeed for long. Some managers, hearing that burning and grazing might be the answer, rushed in without asking what kind of burning and grazing had been practiced in the past: if they used a hot slow fire travelling against the wind towards unburned fuel, this could have a disastrous effect on the *sphagnum* moss of which the moss is composed, and if they used sheep that were unused to grazing on bogland, some of the poor beasts would drown. Had they learned from past management how to do it correctly, with fires set to travel rapidly downwind and beasts with more inherited local knowledge, they would have done better. A peat bog, like a grouse moor, is a cultural landscape that needs exact information about its past use.[21]

ON NATURE AND CULTURE

The terms "natural" and "cultural" have appeared often in this chapter because nature conservation scientists usually regard them as important distinctions. Generally speaking, the more something is "natural" or "wild" the more it is to be treasured, the more it is authentic.[22] If it is "cultural" or "anthropogenic," the less it is admired. Thus with alien species, the "natives" or non-aliens, have top conservation priority because their broad distribution can be taken to be "natural." (How this figures with swallows, collared doves, house sparrows, or other such commensal species is an open question: they are assumed to have arrived on their own wings, "naturally," but their habitat is totally determined by human culture).[23] Archaeophytes, or aliens, that arrived through man's cultural activity, and applied for a residency permit more than 500 years ago, can be accepted but are apologized for. Neophytes, aliens that have had an assisted passage since 1500, are beyond the pale, and never protected. Nevertheless, some of these may be both of cultural value and international conservation concern, like Lady Amherst's Pheasant in Bedfordshire, just as much as those threatened archaeophyte weeds.

But supposing an historian were to argue that maintaining the distinction between nature and culture is itself a cultural construct rather than a scientific fact? If man is part of nature then his culture must be part of nature too, and the distinction between "natural" and "wild" on the one hand and "cultural" and "anthropogenic" on the other, starts to change. And in what sense is man not part of nature? That it could be otherwise is part of a pre-Darwinian intellectual world, based on the Christian notions of the Divine separation of man from nature, of man's divinely sanctioned command over nature, of the corruption of man, and of the primal innocence of the rest of nature (this last a Romantic rather than a Christian belief). Then, all importantly, the fission between the human and the natural was confirmed independent of religion or sentiment (so untainted by either), by the rationalism of the Enlightenment. It was established as a category different from culture or nurture and in opposition to them, and secular modern science came also to be "built on a philosophical platform that makes nature separate from human society."[24]

Yet Darwinian science blows this apart. Darwin placed man as part of the tree of evolution, a twig no higher than any other, the outcome of a natural process that might have no creator at all. Huxley, no doubt to make evolution more palatable to his Victorian consumers, still intellectually fed by centuries of assuming the primacy and separateness of man, placed man as the top of the tree, the final triumphant bud at the end of the leading shoot to which all evolution pointed. This was a much less radical conclusion than Darwin's and much less disturbing to the hubris of contemporary Victorian and modern ways of thought.[25] The twentieth century, including conservation science, remained broadly content with Huxley's way of singling out man from the rest.

If we take a more thoroughly Darwinian view, man's actions appear scientifically as natural as a limpet's. But if our actions are entirely natural, are they then entirely excusable? Is to place a supermarket on a wildlife site as valid as saving the whale? How, then do we avoid ethical anarchy?

Here we should set aside science and consider the reason why culture has been separated from nature in common speech. We speak of nature because the otherness of the non-human is so impressive and so precious to us.[26] We may to a degree manipulate, but in the last resort we cannot control, the laws of the universe. Through taxonomy and biology we may study but never fully comprehend the lives of other species. We cannot place a thermostat on the sun or enter the mind of a swallow. Mystery is compounded by the compelling beauty of the other, forever real, forever in the eye of the beholder, seen at sunrise, in swallow-flight, and in every other contemplation of natural phenomena.

So if we do choose to refuse permission for the supermarket or fight to save the whale, it is not actually because of the scientific interest of butterflies and whales, though in our modern embarrassment about the unscientific we may pretend it is. It is actually because we value the otherness of other things.

The only way to avoid ethical anarchy is to return again to a spiritual view of stewardship through recognizing that, however it came about, man has exceptional powers both of command and of moral choice, and that this gives us a duty to protect the other—the rest of the natural world. Duty and self-interest turn out to be identical because culture needs nature: "I am part of nature and I damage myself when I damage it."[27]

In acting as stewards, however, we should think much more holistically and historically about what it is we should protect. Whether you are a rewilder or a regardener or something between, there is no logical reason whatever not to celebrate, preserve, and respect what came about, or was favored by, human cultural activity. I would go further, and argue that nature conservation will succeed much better in Britain when the assemblages of habitats and species are seen as part of our historical heritage, shaped by our culture, and valued as such. It will be better accepted when it becomes well integrated with historical interpretation of each site. Every ancient wood has its history and deserves a booklet for the local community and the visitor: the sheets of bluebells are an ancestral gift from generations of woodmen departed. The partnership dedicated to returning the fens to East Anglia will surely not wish to ignore the history of the hunter, the thatcher, the drainer, and the farmer, or omit to explain it alongside the history of the biodiversity, because the relationship between the two tells so much, and the explanation will gather public interest and support to its side.

"Rewilding," "conservation gardening," and "landscape-scale change" each exist in rich variety, as this book shows. They represent a spectrum of different approaches, all valid in context, if not all equally valid in every context. They are all cultural actions undertaken for our own satisfaction, and their best outcomes are great manifestations of the human spirit as much as manifestations of nature. They are what the natural world can no longer do unaided. We should not be apologetic about that, but we should celebrate and explain the intertwining of man with the rest of creation, or, if you prefer, with the rest of evolved nature.[28]

NOTES

1. Vincent Wigbels, *Oostvaardersplassen: New nature below sealevel* (Zwolle, Netherlands: Staatsbosbeheer, n.d., preface by Adrí de Gelder).
2. G. F. Peterken, *Natural Woodland: Ecology and Conservation in Northern Temperate Regions* (Cambridge: Cambridge University Press, 1966), 12–13.
3. W. M. Adams, *Future Nature: a vision for conservation* (London: Earthscan Publications, 2003, revised ed.).
4. See www.rspb.org.uk/Images/Wetfensleaflet_tcm5–91191.pdf.
5. See http//www.wildland-network.org.uk/glen_alladale.htm.
6. Philip Ashmole, "The lost mountain woodland of Scotland and its restoration," *Scottish Forestry* 60:1 (2006): 9–22.
7. See http//www.wildland-network.org.uk.
8. J.–C. Svenning, "A review of natural vegetation openness in north-western Europe," *Biological Conservation* 7 (2002): 290–96; F. J. G. Mitchell, "How

open were European primæval forests? Hypothesis testing using palaeoecological data," *Journal of Ecology* 93 (2005): 168–77.

9. K. H. Hodder, J. M. Bullock, P. C. Buckland, and K. J. Kirby, "Large herbivores in the wildwood and modern naturalistic grazing systems," *English Nature Research Reports* 648 (2005), 169.

10. A. G. Smith, "The influence of Mesolithic and Neolithic man on British vegetation: a discussion," in D. Walker and R. G. West, eds., *Studies in the Vegetational History of the British Isles* (Cambridge: Cambridge University Press, 1971): 81–96; I. G. Simmons, *The Environmental Impact of Later Mesolithic Cultures* (Edinburgh: Edinburgh University Press, 1996); I. G. Simmons, *An Environmental History of Great Britain from 10,000 Years Ago to the Present* (Edinburgh: Edinburgh University Press, 2001), 42–48; See also P. Mellars, "Fire ecology, animal populations and man: a study of some ecological relationships in prehistory," *Proceedings of the Prehistoric Society* 42 (1976): 15–45.

11. D. E. Allen, *The Naturalist in Britain: A Social History* (London: Allen Lane, 1976).

12. Peter Marren, *England's National Nature Reserves* (London: T. and A. D. Poyser, 1994), 23.

13. C. D. Preston, D. A. Pearman, and T. D. Dines, eds., *New Atlas of the British and Irish Flora* (Oxford: Oxford University Press, 2002), 37–38.

14. C. M. Cheffings and L. Farrell, eds., "The Vascular Plant Red Data List for Great Britain," *Species Status* 7 (Peterborough: Joint Nature Conservation Committee, 2005): 1–116.

15. See, for example, the work that John Rodwell is currently engaged on in the Dearn Valley, Yorkshire.

16. Oliver Rackham, *Ancient Woodland: its history, vegetation and uses in England* (Colvend: Castlepoint Press, 2003 ed.).

17. Marren, *Nature Reserves*, 24.

18. See, for example, Peter Harvey, "The East Thames Corridor: a nationally important invertebrate fauna under threat," *British Wildlife* 12:2 (2000): 91–98.

19. Peter Middleton, "The wildlife significance of a former colliery site in Yorkshire," *British Wildlife* 11:5 (2000): 333–49; Jeff Lunn, "Wildlife and mining in the Yorkshire coalfield," *ibid* 12:5 (2001): 318–26; Mark Crick, Pete Kirby, Tom Langton, "Knottholes: the wildlife of Peterborough's claypits," *ibid*, 16:6 (2006): 413–21; Adrian Spalding, "The nature-conservation value of abandoned metalliferous mine sites in Cornwall," *ibid*, 16:3 (2005): 175–83; Bob Gibbons, "Reserve Focus: Thompson Common Nature Reserve, Norfolk," *ibid*, 15:4 (2004): 240–43. See also Jonathan Briggs, "The saga of the Montgomery Canal," *ibid*, 17:6 (2006): 401–10.

20. C. M. Gimingham, "Heaths and moorland: an overview of ecological change," in D. B. A. Thompson, Alison J. Hester, and Michael B. Usher, *Heaths and Moorland: Cultural Landscapes* (Edinburgh: HMSO, 1995), 9–19; I. G. Simmons, *The Moorlands of England and Wales: an Environmental History 8000 BC–AD 2000* (Edinburgh: Edinburgh University Press, 2003).

21. R. Hingley and H. A. P. Ingram, "History as an aid to understanding peat bogs," in T. C. Smout, ed., *Understanding the Historical Landscape in its Environmental Setting* (Dalkeith: Scottish Cultural Press, 2002), 60–88. See also Parkyn, Stoneman, and Ingram, eds., *Peatlands*, and J. H. Tallis, R. Meade, and P. D. Hulme, *Blanket Mire Degradation: Causes, Consequences and Challenges* (Aberdeen: Macaulay Land Use Research Institute on behalf of the Mires Research Group, 1997).

22. But note the explicit acknowledgment of the cultural origins of heather moorland in Thompson, Hester, and Usher, eds., *Heaths and Moorland.*
23. See T. C. Smout, "The alien species in twentieth-century Britain: constructing a new vermin," Marcus Hall and Peter Coates, eds., *Landscape Research,* 28:1 (2003), 11–20.
24. Adams, *Future Nature,* 104.
25. Peter J. Bowler, "Darwinism and Victorian Values: Threat or Opportunity," in T. C. Smout, ed., *Victorian Values,* Proceedings of the British Academy 78 (1992): 129–48.
26. Adams, *Future Nature,* 105.
27. Mark Tulley, "Something Understood," broadcast on BBC Radio 4, 14 May 2006.
28. An aspect of British nature conservation not touched on here is the reintroduction of nationally extinct native species, which invariably involves archaeological and historical research on past status. See, as an outstanding example, Bryony Coles, *Beavers in Britain's Past* (Oxford: Oxbow Books, 2006).

11 Sidebar

Reforestation, Restoration, and the Birth of the Industrial Tree Farm

Emily K. Brock

Contemporary players in the corporate lumber industry present tree planting as a restorative response to logging damage that can renew trees while maintaining long-term forest health. The ultimate effects of these plantings are often viewed skeptically by mainstream restorationists, but the extent to which such plantings can represent good restoration should not be ignored—given the size, scope, and potential of this activity. It is worth reviewing the development of the industrial tree farm as an early form of restoration. These tree farms marked the arrival of a new rhetoric, a new scientific agenda, and a new economic approach to repairing degraded forests.

The first industrial tree farm was established on the Clemons tract of Weyerhaeuser Timber Company holdings in Grays Harbor County, Washington, USA. Intensive logging had been carried out there until the late 1930s when marketable timber was no longer available. Large fires then swept through logged-off areas of this parcel in 1938 and 1940, reducing the chances of natural reforestation on these acreages while requiring extra outlays of money for firefighting and recovery. In assessing the economics of fighting forest fires, the Weyerhaeuser Reforestation and Land Department determined that if the company had spent as much money on fire prevention as it had on fire recovery, most of the Clemons area could have been saved. Weyerhaeuser concluded that investing in fire prevention was the most economical way of preserving long-term forest production.

In order to make fire protection cost-effective, however, Weyerhaeuser accountants stipulated that protected areas needed to be fully stocked with Douglas fir. Moreover, the significant investments in fire suppression required significant payoff in lumber sales. Natural reforestation, although inexpensive, was too slow and too unpredictable to use in such a venture. The extensive degradation of the site also meant that this spontaneous reforestation was unlikely to occur over much of the area in a reasonable amount of time. To make fire protection cost-effective, intensive artificial reforestation was deemed necessary.

While tree plantations had existed in Europe for centuries, European antecedents differed greatly from their American counterparts in management philosophy, economic parameters, and ecological concerns.

Weyerhaeuser's newly conceived tree farm also differed from the Northeast and Midwest's scattered nineteenth-century tree plantations and 1930s Civilian Conservation Corps replanting projects. The Clemons was explicitly established for two, supposedly compatible long-term goals: production of industrial timber and refinement of artificial reforestation. Because the Clemons site had been completely logged out, and considered unproductive and worthless, company foresters were given free rein to experiment.

Although Weyerhaeuser never claimed to be re-creating an original forest at its Clemons site, it still saw immediate and long-term ecological benefits. The replanting of tree cover on clear-cut areas slowed soil erosion, protected streams from silt runoff, and rebuilt structure for wildlife habitat. In the eyes of its custodians, the landscape had not changed irrevocably in its passage from natural forest through clear-cut to tree farm, even if it had lost several key features. Creating a tree farm meant restoring some—but not all—ecosystem processes thereby reinstating elements valued by its land's users.

In both industrial reforestation and ecological restoration, certain species are favored, and the health of the rest of the system is judged through that prism. With ecological restoration the recovery of rare or emblematic species is often an indication of success, whereas on a tree farm the viability of a commercial lumber trees is generally the goal. One direct benefit for today's rewilders encountering Weyerhauser's reforested Douglas fir forests, for example, is that they are much less likely to find exotic tree species, a problem that restorationists often confront in other reforested regions. While most of today's industrial tree farms differ greatly from natural forests, it is still important to consider industrial restoration of natural resources within the full spectrum of restorative land use. Sometimes

Figure 11.1 Early publicity photograph of fire prevention measures at Weyerhaeuser Timber Company's Clemons Tree Farm in Grays Harbor County, Washington, c.1943. (Courtesy Weyerhaeuser Corporation)

voluntarily, sometimes mandated by law, the forest products industry has established tree farms on post-logging landscapes, and has developed techniques for growing trees quickly.[1]

NOTES

1. For further information on Weyerhaeuser's early tree farms, see Emily Brock, *Replanting the Douglas Fir Forest: Forest Science and Forest Practice in the Pacific Northwest, 1890–1945* (2004 dissertation, Princeton University), ch. 2, and Emily Brock, "The Challenge of Reforestation: Ecological Experiments in the Douglas Fir Forest, 1920–1940," *Environmental History* 9 (2004): 57–79.

Part III

Restore To What?

Selecting Target States

12 Informing Ecological Restoration in a Coastal Context

Anita Guerrini and Jenifer E. Dugan

In *The Mediterranean and the Mediterranean World in the Age of Philip II*, historian Fernand Braudel made the Mediterranean itself a central player in his narrative, which begins with a description of the geography of the region.[1] Another Braudelian classic, *Memory and the Mediterranean*, likewise relies on the *longue durée*, and stretches back to geologic time by beginning with the formation of the Mediterranean and the Paleozoic era.[2] This emphasis on the long view of history has been less influential on environmental historians—at least American environmental historians—than we might expect, and this neglect is also evident in the historical discourse of North American restoration.

The "problem of the baseline" is not unique to North American restoration. Timo Myllyntaus, Frans Vera, and Chris Smout note the problem of defining a European baseline. What historical moment do we choose when we attempt to turn the clock back? For North American ecologists, this question has generally not been problematic: The baseline has been Columbus. But this notion is increasingly being challenged. In the past few decades, American environmental historians have begun to look beyond the great Columbian divide to the *longue durée* of North American history. In works such as William Cronon's *Changes in the Land*, they have found that North America was far from untouched by humans before Columbus.[3] This emphasis on pre-Columbian human impacts fits well into Braudel's framework. He viewed the landscape as acted upon by humans, but paid less attention to the dynamic qualities of landscape itself. By looking at the modern Mediterranean, he claimed, we can see "a landscape and sky like those of long ago."[4] But landscapes in general, and a coastal landscape in particular, would have experienced many changes between ancient times and now, not only man-made but also natural.

Change and the loss of historical and ecological landscapes can be particularly evident in the coastal zone, where a large proportion of the world's population dwells. The disappearance of wetlands, native vegetation and wildlife, and the alteration of natural processes have greatly affected the ecology of remaining coastal zone habitats, while the historical introduction of numerous species of plants and animals has transformed much of the open landscape. Cultural and historic structures and land uses have vanished through modern development, change, and decay. Ecological restoration and historic preservation efforts are often separate processes lacking a synthetic context and are constrained to fragments or remnants of coastal landscapes.

Coastal environments are inherently dynamic. This dynamism leads to questions about restoration, in particular the establishment of a baseline, but also the whole idea of rewilding, of restoring to a particular state without human presence. As the goal of historic preservation has generally been to fix a historical moment rather than to acknowledge change, so restoration ecology has sought to restore a landscape to a particular single state or reference condition. The idea of a dynamic landscape at the interface with the sea complicates this goal considerably. Coastal human populations also are dynamic, engaging in a variety of interactions with the environment. The human attraction to the ocean is of long standing, but what it now means is an increasingly dense population in coastal zones.

We are assembling the ecological and cultural history of a small fragment of the southern California coast with the goal of developing a model approach for informing restoration efforts. Located on the shores of the Santa Barbara Channel, northwest of the city of Santa Barbara, this area is now owned by the University of California at Santa Barbara. Known as the West Campus, this area is largely occupied by a natural reserve, part of the University of California Reserve System, with the unlovely but descriptive name of Coal Oil Point. Along with the adjacent nearshore reefs, kelp forests, and ocean channel, it forms part of the Santa Barbara Coastal Long Term Ecological Research site. It also includes a school for the disabled, a preschool, horse stables, art studios, and university faculty housing. The coastline offers a wide variety of habitats including sandy beaches, dunes, rocky tidepools, cliffs, a saltwater slough, coastal mesas, and freshwater marshes and pools. The setting is on the edge of a highly urbanized and quickly developing area, a liminal space that is a borderland both literally and symbolically.

The history of the area includes early human occupation, ranching by Spanish and American settlers, and a large estate. This area is part of a major land-use agreement that is being drawn up between Santa Barbara County, the University, the young city of Goleta, and private developers. Known as the Ellwood–Devereux plan, this agreement would leave much of the coastal area free of new development in exchange for development rights farther inland. Ecological restoration is part of this plan and is also a major objective of the Coal Oil Point natural reserve.

Toward the goal of providing cultural and ecological perspectives that can inform the restoration process, we are using a number of different kinds of evidence to examine the human history of the West Campus and Coal Oil Point, which dates back at least 8000 years, and the history of this dynamic coastal landscape itself, which has undergone many changes

Figure 12.1   The view in 2008 of the south-facing coastline of the study area (UCSB campus shoreline) showing the nearshore reef, kelp forest, sandy beach and coastal strand, and bluff habitats typically present now. The cypress and eucalyptus trees were planted by the Campbells in the 1920s and 1930s. (Photograph by Jenifer Dugan)

over time. We wish to increase the range of evidence considered in restoration while also contributing to a new and more inclusive definition. Like Lynne Westphal in an upcoming chapter, we believe that restoration cannot proceed without recognition of human impacts and cultural significance, and as Eric Higgs notes, historical fidelity is as important as ecological integrity.

The coastal zone of the Santa Barbara Channel is particularly dynamic because it lies at the sharp oceanographic and biogeographic transition zone between the cold waters of the California current and warmer southern California waters.[5] Point Conception, at the western end of the channel, represents a major biogeographic boundary where the southernmost or northernmost limit for numerous species intermix. This leads to very high diversity in many taxa (birds, marine mammals, fish, invertebrates, and algae) as well as to the potential for major shifts in species composition in response to climate and ocean currents.[6] For example, a gradual increase in sea surface temperature in this region over the past century has been accompanied by lower production at many levels of the food chain. Ecosystem changes associated with these shifts can be detected in fish, birds, and other taxa.[7]

The Santa Barbara Channel is also a site of dramatic physical and biological responses to El Niño Southern Oscillation (ENSO) events. These events occur on a semi-decadal scale with seventy-five events recorded in the last 500 years and intense events in 1833, 1877, 1878, 1889, 1973, 1983, and 1998. [8] ENSO events are accompanied by abnormally high rainfall, warmer water, sea level rise, intense storms, and strong waves, all of which contribute to coastal erosion and change and result in strong biological responses. El Niño events have played a major role in shaping the beaches and shoreline as well as the human history of the study area. [9] For example, a seventy-year record of the study shoreline indicates the 1982 to 1983 El Niño reduced beach widths and habitats by more than 50 percent and most of the beaches had not recovered to pre-1982 to 1983 widths by 2003. [10]

The disappearance of keystone predators such as sea otters has also affected the ecological dynamics of this coastal region. Sea otters once lived around the Pacific from northern Japan to central Baja California, but were extirpated throughout most of their natural range (including southern California) through hunting by the beginning of the twentieth century. After decades of protection, sea otters have begun slowly to reappear in the region, but face an ecosystem altered by climatic and human influences. Marine protected areas are now being implemented in California waters as an approach to restore these nearshore marine ecosystems. [11]

To begin with the oldest evidence, we consider long-term data on sea temperatures, based on analyses of records from sediment cores in offshore basins from our colleague, geologist Jim Kennett. The composition of organisms in these sediments can be used to indicate sea temperature changes in the ocean. This evidence shows shifts in water temperatures from cool to warm in approximately 10,000 year intervals. [12]

The rock types and fossils we see today in our study area also reflect the long term nature of coastal landscape dynamics. The marine terraces of the West Campus area are composed of the Sisquoc Formation, a gray, diatomaceous shale or claystone formed in offshore marine conditions during the Pliocene that underlies much of the coastal plain. It contains fossilized whale bones and is estimated to be 12 to 6 million years old (myo). At a higher sea level stand, this formation was cut by waves into a Pleistocene marine terrace, which was then overlaid by a layer of clay and later uplifted above wave action. The Sisquoc Formation lies conformably (without an erosional surface) on the Monterey Formation, which is a light buff to pale brown, silaceous shale composed largely of diatoms deposited in deep ocean conditions, interbedded with softer shale and limestone that is Miocene in age, c. 20 to 15 myo. These marine terrace deposits and the underlying shales form the dominant soil types. [13]

The marine terrace deposits exposed on the sea bluffs above the Sisquoc formation contain 40,000-year-old marine invertebrate fossils, including many species of clams and snails. The species of shellfish found here suggest a landscape of a protected or shallow muddy bay with fringing marshes. This landscape would be in marked contrast to the exposed rocky shore, sandy beach, and non-tidal lagoon of the present time. During the Pleistocene, the sea level was at times much higher than today, and it is estimated that the present shoreline was one to two miles inland.[14] Evidence of lagoon and dune sand deposits underlie what is now shrubby grassland with vernal pools at least one kilometer from the present shoreline. The present shoreline position and height therefore reflects major sea level change, erosion, and uplift.

Violent uplift associated with earthquakes may be characteristic of the region, meaning that this uplift can be punctuated rather than gradual. The present-day coastal landscape of bluff-backed beaches, coastal mesas, dunes, and structural basin estuaries formed over the last 40,000 years with several uplift events (~10) raising the mesas by perhaps a meter at a time to their current heights.[15] The earthquake of 1812 was one of the largest recorded in the state, reported as a 10 on the Rossi/Forel scale. Aftershocks continued for months. The last major earthquake in the region was the 1925 Santa Barbara earthquake, measured at a 6.3, although there have been several earthquakes of lesser magnitude since then, including a 5.1 in 1978 that caused significant damage. Tsunamis also affect this coast; records from the 1812 earthquake, which had an offshore epicenter, noted that a ship was pushed up a coastal canyon and that the tsunami wave reached a mile inland at Gaviota.[16]

The fossil records also show that about 10,000 years ago, around the time of the first human habitation in the region, conifers such as Monterey pine and Douglas fir grew near the coast; they do not now.[17] Estimates made using archaeological records of a beach-dwelling clam suggest that wide, fine sand beaches were present ~7000 years ago in the study region during a period when sea level was not changing rapidly; these beaches then declined 5000 to 4000 years ago.[18] What these diverse data demonstrate is the inherent instability and highly dynamic nature of a coastal environment. It is always in process, never static. Within living memory, the ocean outlet of Devereux Slough has shifted some one hundred and fifty meters to the west.[19]

Our research shows that human activities and impacts on the study site have been substantial over the last 8000 to 9000 years. Archaeologists Michael Glassow and Dustin McKenzie analyzed artifacts and faunal remains collected from the site's midden deposits, the latter including numerous shells of open-coast and estuarine species. They have determined by carbon dating that humans were present at least 8500 years ago. Arrowheads and grinding tools ("manos") indicate an active economy and a diverse diet. Shifts in diet and tool use are apparent over the archeological record for the site and may be tied to environmental change. For example, certain types of shellfish disappear from middens for a time or are replaced by other species, results which could indicate a crash of targeted shellfish populations from overfishing and/or environmental changes associated with ENSO or climate. The Chumash settlements along the Santa Barbara Channel were among the most populous native

American settlements in North America and had an extensive trading network.[20] Obviously, these pre-Columbian inhabitants had an impact on the landscape.

Although Spanish explorers traveled up and down the California coast in the sixteenth and early seventeenth centuries, Europeans only settled in California in the eighteenth century. Accounts from early explorers provide some clues about the landscapes they observed, as well as confirmation of substantial native settlements. Restoration ecologists working on the site have asserted that when the Spanish first arrived, they saw a coastal plain without trees. Yet eighteenth-century lore stated that it was possible to walk from Refugio—ten miles north of the study site—to Santa Barbara—fifteen miles south—in the shade of oaks. Which story is correct? While the current flora of eucalyptus and olive trees has only existed for the last century or so, it is not at all clear what preceded them.

The Spanish left several written accounts. How can we best use these? The ecological philosopher Paul Shepard warned about the limitations of human accounts:

> we must stand apart from the conventions of history, even while using the record of the past, for the idea of history is itself a Western invention whose central theme is the rejection of habitat. It formulates experience outside of nature and tends to reduce place to location. . . . History conceives the past mainly in terms of biography and nations. It seeks causality in the conscious, spiritual, ambitious character of men and memorializes them in writing.[21]

The first European explorer of the California coast, Juan de Cabrillo in 1542, described a rich and densely settled land:

> All these pueblos are between the first pueblo of Las Canoas, which is called Xucu, and this point. They are in a very good country, with fine plains and many groves and savannahs. The Indians go dressed in skins. They said that in the interior there were many pueblos, and much maize three days' journey from there. They call maize "Oep". They also said that there were many cows; these they call "Cae". . . . All this coast which they have passed is very thickly settled. . . . The country appears to be very fine.[22]

Sixty years later, Sebastian Vizcaíno confirmed Cabrillo's account of densely populated fertile land:

> In this place there are great numbers of Indians, and the mainland has signs of being thickly populated. It is fertile, for it has pine groves and oaks, and a fine climate, for although it gets cold it is not so cold as to cause discomfort.[23]

These accounts seem to confirm archaeological evidence indicating a densely populated area, and point to the methodological issue of needing to use sources other than written ones. But these accounts give little specific information about land use or appearance. The major written sources for early California history come from over 150 years after Vizcaíno, in the various accounts of the Portolà expedition of 1769. Everyone on that trip seems to have kept a diary, and the engineer Miguel Costanso and the priest Juan Crespi described the landscape they encountered. Costanso noted "verdure and trees," a land "exceedingly pleasing, with an abundance of pasture, and covered with live-oaks, willows, and other trees." Crespi described a coastal plain of "dark, friable soil and very grown over with very tall broad grasses. It had been burned off in some spots and not so in others; the unburnt [grasses] were so tall that they topped us on horseback by about a yard." The presence of burning indicates some sort of active landscape management by the Chumash peoples. Crespi also described seeing many large live oaks on the tablelands and expanses of wild roses on the plains.[24]

Both accounts describe a land of great fertility and considerable population. But Costanso's description of "verdure" and Crespi's account of tall grass must give us pause: this is, after all, southern California in August, when the landscape is not green at all. If we believe these sources, could the landscape have changed so much between then and now? Perhaps 1769 was an El Niño year, with so much rain that the land was still green in August. Work on historical El Niños indicates that there may have been one in the late 1760s.[25] It could also be that the native perennial grasses were better adapted to seasonal rainfall and were therefore still green. Perhaps the overall climate was wetter; the Little Ice Age was still in play in the eighteenth century. But if Costanso and Crespi were accurate in their descriptions, where does it leave the question of restoration? Further substantiating the existence of a number of trees is the 1793 account of Archibald Menzies of the Vancouver expedition, who stated that at Santa Barbara, wood "was easily procured at no great distance from the beach as there were some large trees of a kind of evergreen oak."[26] Similarly, forests have a dynamic character in Finland, as Timo Myllyntaus notes in his contribution to this volume.

Following the "European incursion" of the Spanish missions, the area probably became much less populated than it had been, as the Chumash people began to die out and their way of life to disappear. The area was part of a Spanish land grant given to an Irishman named Nicolas Den in 1842. The "diseño" of the land from around 1850 gives some interesting details but does not tell us much about things like vegetation or climate. The survey made at this time is equally vague. But the 1861 "official" survey is little better, since it entirely omits the major landscape feature of the Devereux Slough.[27] These maps also tell us little about land use. We know from some other written sources that Den and his children farmed the land, but we aren't sure what they grew. The Coast Survey topographical map from 1871 gives the best indication of land use at that time.

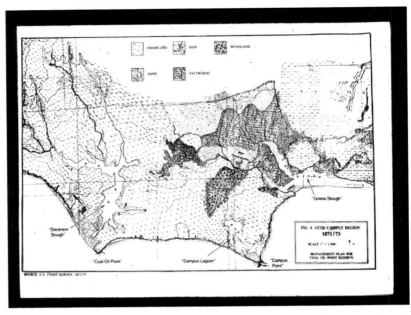

Figure 12.2   The Topographical Map of the study area (UCSB campus), taken from the U.S. Coast and Geodetic Survey's 1870 map survey, the earliest georeferenced map available for use in developing this site's restoration goals. The map indicates the diversity of landscapes and major vegetation communities present in the area as well as the historical extent of the wetlands and watershed. More than fifteen miles of irrigation canals had been dug in the vicinity of the Goleta Slough prior to 1850. By the time of this survey, the area had supported extensive livestock grazing for many years, and was largely covered by grassland. The legend and identifying labels are recent additions. (Courtesy National Oceanic and Atmospheric Administration)

We have more information for the period from 1919–1940, when Nancy and Colin Campbell owned the land. The Campbells planted grain and raised chickens as well as building a mansion, dredging the Devereux slough for a yacht harbor, and developing coastal nearshore oil claims. In addition, Mrs. Campbell gave the ecologist Frederic Clements permission to establish test plots on the coastal plain, which he used from 1934 until his death in 1946. These were apparently quite extensive. In three test gardens on the site, Clements introduced as many as 200 species of plants for his experiments on adaptation. Some of these plants were very common to the region, while others were obviously exotic. Donald Burnette's work on Clements and on these plots in particular shows that at least some species thought to be unique to the area were in fact introduced by Clements.[28]

Such a long human presence on this site, coupled with the inherent dynamics of the coastal zone, leads to a questioning of the very definition of restoration. In this coastal context, it can only mean restoring the ecological processes, not a particular point in time. This approach may become increasingly relevant in the face of anticipated impacts of global climate change on coastal ecosystems.[29] Larger answers to the challenge of developing restoration goals for the narrow, heavily impacted but highly productive, and ecologically and culturally diverse coasts of the world will require a synthesis of physical and ecological dynamics and processes, anthropology, history, sea level change, natural and cultural resources, and human population growth and needs. As Eric Higgs notes, this is a self-consciously social process, based on human concepts of design, ecological integrity, focal practices, and historical fidelity. The impact of climate change adds a further dimension of dynamism and uncertainty to this environment. Settling on restoration goals represents a formidable task for historians and ecologists alike and illustrates the need for an interdisciplinary approach that considers all available sources, both historical and scientific.

ACKNOWLEDGMENTS

We especially thank Marcus Hall for motivating and encouraging this effort and the participants of the Restoring or Renaturing workshop for stimulating discussion. We gratefully acknowledge Michael Glassow, David Hubbard, Dustin McKenzie, Jill Jensen, Donald Burnette, Kathy Kwong, Karinna Hurley, Peter Neushul, Deborah Bahn, James Kennett, David Revell, Beverly Schwartzberg, and Michael A. Osborne for their input and intellectual contributions to the many aspects of this research. Our research was supported by a Collaborative Programs Grant to A. Guerrini and J. Dugan from the National Endowment of the Humanities, as well as grants from the Pearl Chase fund and the Research across Disciplines program of the University of California at Santa Barbara. J. Dugan was also supported in part by the Santa Barbara Coastal LTER (National Science Foundation OCE #99–82105).

NOTES

1. Fernand Braudel, *The Mediterranean and the Mediterranean World in the Age of Philip II*, trans. Siân Reynolds, 2 vol. (New York: Harper and Row, 1972), vol. 1 Part 1.
2. Fernand Braudel, *Memory and the Mediterranean*, trans. Siân Reynolds (New York: Knopf, 2001), Chapter 1.
3. William Cronon, *Changes in the Land* (New York: Hill and Wang, 1983). There is a growing literature on this topic: see, for example, Paul H. Gobster and R. Bruce Hull, eds., *Restoring Nature: Perspectives from the Social Sciences and Humanities* (Washington, DC: Island Press, 2000); J. Baird Callicott, "Choosing appropriate temporal and spatial scales for ecological restoration," *Journal of Bioscience* 27: (2002), 409–20; Eric Higgs, *Nature*

by design: people, natural process, and ecological restoration (Cambridge, Mass.: MIT Press, 2003).

4. Braudel, *Memory and the Mediterranean*, 3.
5. M. P. Otero and D. A. Siegel, "Spatial and temporal characteristics of sediment plumes and phytoplankton blooms in the Santa Barbara Channel," *Deep-Sea Research* II, 51 (2004):149.
6. J. W. Valentine, "Numerical analysis of marine molluscan ranges on the extratropical northeastern Pacific shelf," *Limnology and Oceanography* 11 (1966): 198–211; J. C. Briggs, *Marine zoogeography* (New York: McGraw–Hill, 1974).
7. J. A. McGowan, D. R. Cayan, and L. M. Dorman, "Climate-ocean variability and ecosystem response in the Northeast Pacific," *Science* 281 (1998): 210–217; S. J. Holbrook, R. J. Schmitt, and J. S. Stephens, Jr., "Changes in an assemblage of temperate reef fishes associated with a climate shift," *Ecological Applications* 7 (1997):1299–1310; J. R. Smith, P. Fong, and R. F. Ambrose, "Dramatic declines in mussel bed community diversity: response to climate change," *Ecology* 87 (5) (2006): 1153–1161.
8. S. Brönniman et al., "ENSO influence on Europe during the last centuries," *Climate Dynamics* 28 (2007): 181–97.
9. R. E. Flick, "Comparison of California tides, storm surges, and mean sea level during the El Niño winters of 1982–83 and 1997–98," *Shore and Beach.* 66 (3) (1998); J. C. Allan and P. D. Komar, "Climate controls on US west coast erosion processes," *Journal of Coastal Research* 22, 3 (2006): 511–29; D. L. Revell and G. B. Griggs, "Beach width and climate oscillations along Isla Vista, Santa Barbara, California," *Shore and Beach* 74 (3) (2006): 8–16; P. K. Dayton and M. J. Tegner, "Catastrophic storms, El Niño, and patch stability in a southern California kelp forest," *Science* 224 (1984): 283–85; D. M. Hubbard and J. E. Dugan, "Shorebird use of an exposed sandy beach in southern California," *Estuarine, Coastal, and Shelf Science* 58S (2002): 41–54; Otero and Siegel, "Spatial and temporal characteristics"; M. S. Edwards and J. A. Estes, "Catastrophe, recovery, and range limitation in NE Pacific kelp forests: a large-scale perspective," *Marine Ecology Progress Series* 320 (2006): 79–87; D. J. Kennett and J. P. Kennett, "Competitive and cooperative responses to climatic instability in coastal southern California," *American Antiquity* 65:2 (2000): 379–95; F. P. Chavez et al., "From anchovies to sardines and back: Multidecadal change in the Pacific Ocean," *Science* 299:5604 (2003): 217–21; A. L. Weinheimer, J. P. Kennett, and D. R. Cayan, "Recent increase in surface-water stability during warming off California as recorded in marine sediments," *Geology* 27:11 (1999): 1019–1022; P. M. Masters, "Holocene sand beaches of southern California: ENSO forcing and coastal processes on millennial scales," *Palaeogeography, Palaeoclimatology, Palaeoecology* 232 (2006): 73–95.
10. Revell and Griggs, "Beach width."
11. G. R. Van Blaricom and J. A. Estes, eds., *The Community Ecology of Sea Otters.* Ecological Studies Vol. 65 (New York: Springer Verlag, 1988); K. W. Kenyon, "The sea otter in the eastern Pacific Ocean," *North American Fauna* 68 (1969): 1–352; J. E. Dugan and G. E. Davis, "Applications of fishery refugia to coastal fishery management," *Canadian Journal of Fisheries and Aquatic Sciences* 50: (1993), 2029–2042; C. Roberts et al., "Ecological criteria for evaluating candidate sites for marine reserves," *Ecological Applications* 13(1 Suppl S) (2003): S199–S214.
12. T. R. Baumgartner et al., "The recording of interannual climatic change by high-resolution natural systems: Tree-rings, coral bands, glacial ice layers

and marine varves," *AGU Geophysical Monographs* 55 (1989): 1–14; R. J. Behl and J. P. Kennett, "Brief interstadial events in the Santa Barbara basin, NE Pacific, during the past 60 kyr," *Nature* 379: 6562 (1996): 243–46.

13. R. M. Norris and R. W. Webb, *Geology of California* (New York: J. Wiley, 1976); L. Gurrola and E. A. Keller, eds., *Friends of the Pleistocene Pacific Cell Field Trip Guidebook, Santa Barbara Fold Belt and Beyond* (privately printed, 2004).

14. E. A. Keller and L. D. Gurrola, *Earthquake Hazard of the Santa Barbara Fold Belt, California.* USGS Final Report (Washington, DC: USGS, 2000).

15. Edward Keller, personal communication.

16. F. J. Weber, Jr. and E. W. Kiessling, "Historic earthquakes: effects in Ventura County," *California Geology*, 31:5 (May 1978); H. O. Wood and N. H. Heck, *Earthquake History of the United States—Part II, Stronger earthquakes of California and western Nevada* (Washington, DC: U.S. Dept. of Commerce, 1966); Keller and Gurrola, *Earthquake Hazard;* Southern California Earthquake Data Center website, (for the 1925 earthquake) http://www.data.scec.org/chrono_index/santabar.html (for the 1978 earthquake) http://www.data.scec.org/chrono_index/santab78.html.

17. D. I. Axelrod, "Outline history of California vegetation," in M. Barbour and J. Major, eds., *Terrestrial Vegetation of California* (New York: J. Wiley, 1977), 139–220; D. I. Axelrod, "History of the maritime closed-cone pines, Alta and Baja California," *University of California Publications in Geological Sciences* v. 120 (1980); L. E. Heusser, "Rapid oscillation in western North American vegetation and climate during oxygen isotope stage 5 inferred from pollen data in the Santa Barbara Basin (Hole 893A)," *Palaeogeography, Palaeoclimatology, Palaeoecology* 161 (3–4) (2000): 407–21.

18. David Hubbard, personal observation; Gurrola and Keller, "Friends of the Pleistocene"; Masters, "Holocene sand beaches."

19. David Revell, "Evaluation of Long Term and Storm Event Changes to the Beaches of the Santa Barbara Sandshed," (PhD dissertation, University of California, Santa Cruz, 2007).

20. Philip L. Walker and John R. Johnson, "For everything there is a season: Chumash Indian births, marriages, and deaths at the Alta California missions," in A. C. Swedlund and D. A. Herring, eds., *Human Biologists in the Archives* (Cambridge: Cambridge University Press, 2003), 53–77; D. McKenzie, "West Campus Archaeology," in A. Guerrini and J. Dugan, eds., *Historicizing Ecological Restoration* (forthcoming); Kennett and Kennett, "Competitive and cooperative responses"; Michael Barbour et al., *California's Changing Landscapes: Diversity and Conservation of California Vegetation,* (Sacramento: California Native Plant Society, 1993).

21. Paul Shepard, *Nature and Madness* (San Francisco: Sierra Club Books, 1982), 95.

22. Herbert E. Bolton, ed., *Spanish Exploration in the Southwest, 1542–1706* (New York: Charles Scribner's Sons, 1916), 487.

23. Ibid., 90.

24. F. J. Teggart and Manuel Carpio, eds. *The Portola Expedition of 1769–1770: Diary of Miguel Costanso* (Berkeley: University of California Press, 1911), 39–41; Juan Crespi, *A Description of Distant Roads: Original Journals of the First Expedition into California, 1769–1770* ed. and trans. Alan K. Brown (San Diego: San Diego State University Press, 2001), 329; on burning see Barbour et al., *California's Changing Landscapes.*

25. Brönneman et al., "ENSO influence."

26. Archibald Menzies, "Archibald Menzies' Journal of the Vancouver Expedition," *California Historical Society Quarterly*, II, (1924); George Vancouver,

A Voyage of Discovery to the North Pacific and Round the World (London: John Stockdale, 1801).

27. *Diseño del Rancho Los Dos Pueblos, California* (San Francisco: US District Court, 185?); J. E. Terrell, *Plat of the Rancho Los Dos Pueblos, finally confirmed to Nicolas A. Den: Surveyed under instructions from the U.S. Surveyor General, November 1860* (San Francisco: US District Court, 1860); Jasper O'Farrell, "Survey of Los Dos Pueblos Rancho, Santa Barbara County," ca 1845, O'Farrell collection, MS OF 10, L12 B9, Huntington Library, San Marino, California.

28. On the Campbells, see Anita Guerrini, "The Campbell Era, 1919–1940," in Guerrini and Dugan, eds., *Historicizing Ecological Restoration*; on Clements's work at Coal Oil Point, see Donald Burnette's forthcoming PhD dissertation on Clements (History Department, UCSB).

29. C. D. G. Harley et al., "The impacts of climate change in coastal marine systems," *Ecology Letters*, 9 (2006): 228–41; *California Coastal Erosion Response to Sea Level Rise—Analysis and Mapping*, Report for the Pacific Institute (San Francisco: Philip Williams and Associates, March 2009).

13 South Yorkshire Fens
Past, Present, and Future

Ian Rotherham and Keith Harrison

The demise of the great fenlands of eastern and northern England has been described in detail by Ian Rotherham. The land-use such as peat cutting which drove this has also been discussed, and the social, cultural, and economic tensions which resulted have been closely observed by Rotherham and McCallam.[1]

Oliver Rackham believes that "about a quarter of the British Isles is, or has been, some kind of wetland." Chris Smout adds that "There are many thousands of hectares of what is now prime arable land, especially in northern England, that were in the seventeenth century, fen and mire" and "it is surprising how . . . Yorkshire. . . . fenlands have evaporated from general memory." This chapter explores the implications of these statements and offers a reconstruction of historic wetlands, before generating new observations and conclusions about the wetlands and the scale of anthropogenic change.[2]

By the early 1900s one of England's biggest fenlands, over 100,000 hectares in South Yorkshire of central England and adjacent counties, was virtually gone. It suffered long-term impacts of drainage by Dutch engineers followed by intensive farming. The incentive for these massive reclamation schemes was increased revenue, especially for the Crown, along with the suppression of political unrest. In 1600, Parliament passed "An Act for the recovery and inning of drowned and surrounded grounds and the draining of watery marshes, fens, bogs, moors and other grounds of like nature." This drainage idea was developed with James I and implemented by Charles I. Before drainage, 36,420 hectares of the Humberhead Levels was "A continual lake and a rondezvous of ye waters of ye rivers."[3]

Our research involved the mapping and 'virtual' re-construction of the former wetlands across the study region. Fieldwork and archival research generated GIS computer maps of the wetlands from pre-Roman to the present day. A range of sources was accessed to suggest the variety and abundance of likely fauna of these wetlands. Area-based case studies identified the key political and economic drivers for change, and help expose the consequences and impacts. Mapping began by referencing key maps of earlier researchers.

In 1933, Wilcox produced two maps showing the prehistoric marsh, moss, and fen: one is based on geology, topography, and climate; the other on early literature. Surveying the whole of England, they considered only major lowland floodplains to be historic wetland, an approach we also adopted.[4]

Our own project builds on a study undertaken by Darby and Maxwell for the Domesday Geography series. These researchers mapped the region's wetlands and related features such as peat, alluvium, fisheries, and mills to begin creating an accurate image of this extensive former wetland.[5]

This research is aimed at helping reconstruct the wetlands in our era of post-industrial and post-agricultural landscapes. Farmers are diversifying their products to include wetland crops; tourists and recreational visitors are taking advantage of re-wetted areas. The rationale for wetland restoration includes water management for alleviating floods and droughts, compliance with the EU Water Framework Directive, and halting regional declines in biodiversity. Carbon sequestration through wetland creation offers additional justification for re-wetting. Adding urgency to the project is the UK government's policy on climate change and expanding regional developments such as the new international airport at Finningley.

The South Yorkshire Fens have been surveyed in incredible detail. Van de Noort and Davies, and Van de Noort and Ellis, for example, offer major reviews of archaeological studies across one subsection called the Humberhead Levels. Dinnin has described the drainage of the Humberhead Levels in great detail. Buckland as well as Whitehouse, et al., have undertaken major and ground-breaking studies of the palaeoecology of the region. There is also the meticulous entomological research sponsored by the Doncaster Museum, together with the valuable *Thorne and Hatfield Moors Papers* as summarized and reviewed by Peter Skidmore. Lastly, a five-year program of the Countryside Agency called "Value in Wetness" promoted and co-funded detailed studies of land use, economics, ground water potential, hydrology, and potential carbon sequestration through re-wetting. The compilation of all this information offers an unusual depth of understanding about the state of this remarkable landscape, now and in the past.[6]

Yet there has been little attempt to join up this intensive scrutiny of the lowlands with the wider region, or to re-construct the landscape and ecological history of the area. In particular, there has been little attention to the relationships between the lowland fens, bogs, and heaths with the mid-ground of the Coal Measures rivers and wooded valleys, or with the extensive upper catchments of the south Pennines moors. There has, moreover, been no attempt at synthesizing this information to interpret the landscape's historic ecology. Such omissions make efforts of rewilding difficult. Barriers to change, challenges of short-term priorities, and lack of corporate vision are discussed against this dearth of synthetic information.

METHODOLOGIES

We began our research by consulting early maps and accounts. A recon-struction of the wetlands across the Humber region would utilize topo-graphic, geological, and pedological information. We combined Wilcox as well as Darby and Maxell's methods with contemporary tools such as air photographs, remote sensing, and *MapInfo Geographical Information System* to create layers of landscape history. Such layers were designed to reveal key changes in wetlands and demonstrate reconstruction potential.

Rivers were indicated by *Ordnance Survey Geological Maps* of Drift (Ordnance Survey 1949–1969), and maps of the sixteenth to nineteenth centuries. Alluvium was taken to imply original wetland over 100,000 hectares (1,000 sq. km). Preliminary results were discussed with interested researchers modified when appropriate.[7]

Obvious anthropogenic change stemmed from construction of water-mills, drainage for agriculture, urban expansion, industrial development, dumping of mine spoil, and the building of reservoirs and transport sys-tems. In specific areas, detailed case studies were chosen to collect informa-tion on a range of topographies and drivers of change (identified from the historic review). Computer-generated reconstruction of former wetlands was then attempted to produce time-sliced maps. The final overview map joined the lowland and upland wet landscapes, uniting the Humber fens with the Coal Measures valleys and the south Pennine upland mires. This holistic regional view offers the best picture for any genuine rewilding proj-ect. Combining this graphic with early descriptions, travel itineraries, and other sources such as feast menus and game books provided special insight into the area's ecological history, and a glimpse at the historic wilderness.[8]

Additional research supported by the Countryside Agency assessed the potential for wetland-based or heritage-based tourism and recreation. This was to guide and promote transition to a wetter landscape through the engagement of farmers and landowners in policy-driven, economically facilitated diversification.

RESULTS

The study generated computer early maps of River Don and Went water-courses before the Roman Turnbridge Dyke, the first major anthropogenic change to the rivers. Capturing the Went it diverted part of the Don into the River Aire. The locations of Thorne Mere and two nearby meres were mapped; other smaller, now forgotten lakes, meres, and carrs were ignored. In the wider landscape, extensive wetlands were indicated by maps of allu-vial deposits and peat and imposed onto maps of watercourses. Rivers and wetlands were revealed in 1600, before the first major drainage by Ver-muyden and colleagues. A study of the uplands reveals extensive wetlands

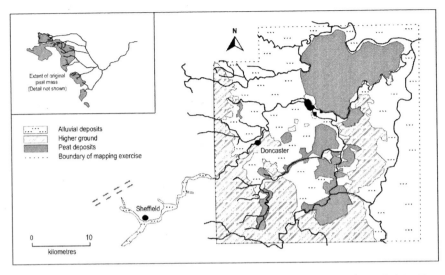

Figure 13.1 A composite map of research findings show the South Yorkshire Fens and surrounding area before the era of draining and canalling.

and subsequent medieval and later drainage. The composite map of this information is presented in Figure 13.1, which shows the probable linkage between lowland peatlands and upland mires.[9]

Through the initial drainage schemes, Thorne Mere was lost and the earlier course of the River Don changed, diverting a number of major watercourses across the region. The main driver for this was agricultural improvement and subsequent profits from land sales. There may have been other political motives oriented toward subduing unrest and non-conformism. The extensive fens were refuges for those seeking independence and distance from the law.[10]

By 1800, the Turnbridge Dyke north of the Dutch River had disappeared, along with Potteric Carr, south of Doncaster. Nineteenth-century drainage removed almost all the wetlands apart from remnants of peat on Thorne and Hatfield Moors. In the wider landscape intensive, mostly arable farming further desiccated the already dry lands. Agri-industrial farming relied on groundwater extraction for irrigation in a climatically dry region along with extensive drainage to remove surplus water from this flatland. The result was extensive desiccation of land.[11]

A Former Wetland Lost From Memory

Anecdotal evidence from a range of sources suggests former abundant wildlife that included hundreds of thousands of birds and mammals. At the same time, human predation was massive, peaking in the eighteenth and nineteenth centuries. Early hunting techniques and technologies limited the taking of game, especially in open, wet landscapes. The advent of Dutch decoys and effective fowling guns changed that. The nineteenth century saw fewer hunters, but almost total loss of wild stock and widespread habitat destruction. Imported food became more available as wild harvests became less economical, and subsistence use diminished. However, despite a decline in commercial and sport hunting, the dramatically changed landscape precluded recovery. Some 98 percent of the historic wetland was gone; the remaining 2 percent altered beyond recognition.

The southwest Doncaster carrs were described as outliers of the great fen originally extending north to the River Humber, east to the Trent lowlands, and south to Nottinghamshire, including the Isle of Axholme, Thorne Waste, Marshland, and Hatfield Chase Fen. The 1600 hectares of Pottrick Carr included vast numbers of duck, bittern, ruff and reeve, black-tailed godwit, marsh harrier, great crested grebe, and water rail until Smeaton's drainage in the late 1700s. Well-documented and anecdotal evidence suggests that the pre-drained landscape rivalled today's Coto Doñana in its rich birdlife.[12] As Chris Firth declares, "*the destruction of the wetland habitats would, by today's standards, be regarded as an ecological disaster of enormous proportions . . . equal in proportion to the present day destruction of rainforests.*" In just the southern part of the Ouse–Trent confluence, 28,000 hectares of Hatfield Chase were inundated annually before Vermuyden and his associates began to drain them in 1626.[13]

Records and memories of the region's wildlife give some impression of the vanished landscape. Bitterns were sufficiently common to find a place in local folk-rhymes: "When on Potteric Carr the Butter Bumps cry, the women of Bulby say summer is nigh." In the early 1900s, elderly people around Beverley recalled hearing bitterns. But two hundred years of draining meant that this wet wilderness was converted and "improved" into a rich agricultural land, as rich as any in all England outside the Cambridgeshire Fens. The improvers increased land value at Hatfield Chase from 6d per acre to 10s. However, this value to farmer and landowner doesn't account for the former advantages and benefits of living and working around a richly productive and biodiverse wetland.[14]

A Productive Landscape

For nearby communities, this bountiful wetland provided fish, reed and rushes for thatching, flooring and candles as well as peat fuel, brushwood (from the carrs) for fuel and light construction work, and pasture for cattle. Indeed, wetlands across northern Britain provide economic benefits. Contrary to opinions of many landowners and politicians in the days of drainage, these wetlands were rich and productive landscapes, offering a variety of land uses and sources of productivity. Such areas were also important hunting grounds in early times, as evidenced by such features as Conis-brough Castle Park and nearby Hatfield Chase. Shortly after the Norman Conquest, 70,000 low-lying hectares of Hatfield Chase were the private forest of the de Warennes of Conisbrough, reverting to the Crown in 1347. According to one report, a thousand deer (Red and Fallow) roamed these areas in 1607, and were as common "as sheep upon the hills," being "so unruly that they almost ruined the country." At the last major hunt in 1609, the royal party used a hundred boats to pursue 500 deer across Thorne Mere. Famous for fisheries and swans, Hatfield Chase was deforested and then drained in the 1600s.[15]

John Leland (Henry VIII's antiquary) described the feast in 1466 for the enthronement of George Neville, Archbishop of York. Although the following list of delectable offerings may be an exaggeration, since much was taken from the Derwent Washlands of south of York and across the South Yorkshire Fen, it gives insight into the diversity of wildlife, domesticated, and semi-domesticated stock found in the region:

"Oxen 104; Wild Bull 6; Muttons 1,000; Veales 304; Porkes 304; Piggs 3,000; Kidds 204; Conyes 4,000; Staggs, Bucks and Roes 504; Pasties of venison cold 103; Pasties of venison hot 1,500; Swans 400; Geese 5,000; Capons 7,000; Mallard and Teal 4,000; Plovers 400; Quails 100 dozen; Fowles called Rayes 200 dozen; Peacocks 400; Cranes 204; Bytternes 200; Chickens 3,000; Pigeons 4,000; Hernshawes (young herons) 400; Ruff 200; Woodcock 400; Curlews 100; Pheasants 200; Partridges 500; and Egritts 1,000."[16]

As described by Ian Rotherham, other household accounts confirm cranes, herons, snipe, bittern, quail, larks, dotterel, and bustards for the table (1526), peacocks, cranes, and bitterns (1530), and twelve spoonbills at 1s each, and ten bitterns at 13s and 4d (1528). Many are wetland birds and mammals from forest or chase, demonstrating the wide and productive extent of these landscapes. Little bittern, night heron, and purple heron probably survived in English wetlands until the 1600s, with breeding cranes and spoonbills extant in England for around 300 years. Until the 1820s, ruff bred at Hatfield Chase, and were taken in nets, fattened in captivity and sold on the table for 2s each.[17]

Before modern guns, new techniques in wildfowling were crucial for people living near these wet landscapes. The Dutch duck decoy mentioned previously debuted in the 1600s, presumably with the Dutch engineers, ushering in commercial exploitation. Thousands of wildfowl were captured annually from the South Yorkshire fens, with the Doncaster Corporation investing in decoys for upkeep of the poor. Construction of a special three-quarter-mile access embankment, Decoy Bank, cost £160. A circular 6½ acre (2½ hectare) decoy pond utilized six 'pipes' to collect the ducks; leased in 1662 for twenty-one years, its annual rental of £15 fell in 1707 to only

£3, possibly reflecting the impacts of drainage and wetland contraction. In 1707 the lessee specialized in pochards, good ducks for the table. Settling on the water, pochards were captured by nets raised with pulleys on poles, and the duck pipes were still there in 1778, with the last decoy man dying in 1794. By the late 1800s, as described by Rotherham, the *Great Northern Railway* ran straight through the site.[18]

VALUE IN WETNESS

Today, this drained region is managed under the *Integrated Land Management Project*, one of twelve across England. Research outcomes from targeted studies are to inform proposals for reconstructing wetlands from post-industrial and post-agricultural lands. The goal is to diversify the farming landscape and economy by favoring wetter conditions and sustainability. Water-related tourism and recreation are potential economic drivers that would benefit local employment and business; local people are therefore encouraged to support re-wetting.

The coming changes are influenced by a matrix of drivers and factors, and an historic perspective allows longer-term appraisal of likely outcomes. Current goals include managing water for preventing floods and droughts, complying with the EU Water Framework Directive, promoting rural economic diversification, and reversing regional biodiversity declines. Carbon sequestration is an additional benefit of creating wetland and wetter landscapes. UK climate change policy and construction of a nearby international airport have made the potential for a carbon sink more urgent.

A variety of meticulous research supports re-wetting, the recent five-year "Value in Wetness" program promoting detailed studies of land-use, economics, soil water potential, hydrology, and potential carbon sequestration. This work synthesized earlier studies and provided an integrated vision for land-use change. Surprisingly the agency provisions for major wetland conservation areas, such as Thorne and Hatfield Moors, offer little acknowledgment of tourism potential, the need to provide visitor facilities, or requirements for tangible an economically and environmentally sustainable future. There is a serious lack of communication, common language, and purpose between conservation managers and regional economic planners. Similar problems occur around the major coastal areas of the Humber Estuary, which includes the 400-hectare Alkeborough. Opportunities to understand and integrate re-wilding are being lost at several levels in the local community and sustainable, long-range thinking gives way to intensive, profit-driven farming.

Despite rising appreciation for wetland habitat, it remains surprising that there is no bigger restoration vision that joins focused lowland studies with regional landscape and ecological histories. There is now localized habitat creation but no genuine, large-scale rewilding vision which, informed by the past might help re-construct the lost landscapes.

CONCLUSIONS

With integrated approaches and emerging awareness of historic drivers, we can formulate effective ways to respond to two thousand years of human disruption. Historical ecological research demonstrates the former richness of the region. However, there are many challenges arising from short-term policy approaches and lack of corporate vision. Rotherham has demonstrated the importance of landscape-change in current flood-risk scenarios and the need to work at a regional catchment level to remediate damage. Drawing on the work of others, this study confirms the scale of regional wetland loss and presents evidence for former wetlands and wet landscapes across the area. It demonstrates wetland continuity throughout the catchment basin, with temporal and spatial links between lowland river valleys and upland blanket mires.

Some of the wetland loss has been partially redressed in the middle and upper catchment, as through remnant millponds, later reservoirs, and, recently, wetland nature reserves. There are now subsidence flashes, flood management washlands, and purpose-built sites. Large wetlands are being developed across parts of the region with the Royal Society for the Protection of Birds (RSPB) Dearne Valley and Yorkshire Wildlife Trust (YWT) Potteric Carr Nature Reserve. But these areas represent tiny fractions of the once-extensive and varied wetland. There is no evidence of a wider approach or any vision of the scale of historic loss, and the enormity of the task ahead.

History suggests that concerns about floods, droughts, and water-shortages, are politically short-lived. Crises punctuate socio-economic and political realms, shaping environmental resources and resulting in flurries of funded, directed research and policies. These gather significant information, promote modest achievement on the ground, and often wither when key officers move on to new posts and government agencies re-shuffle. There is little sign of large-scale change necessary to avert further, serious, long-term declines in the resource. Furthermore, there is no indication from politicians or leading bodies such as Development Agencies that they understand or wish to address the challenges. It is generally "business as usual," short-term economic focus, and jobs. There is passing reference to *sustainable development* and *quality of life*, the latter acceptable if it doesn't affect the former!

This does not give cause for optimism combating global climate change by rejuvenating wetland sites. Hogan and Maltby considered the potential for carbon sequestration in the Humberhead wetlands, noting carbon sequestration is an effective measure for decreasing atmospheric carbon dioxide. They suggested strategic wetland conservation and management as effective approaches to delivering social, economic, and environmental benefits. For strategic planning, they advocated that freshwater restoration across key areas would make the biggest contribution to carbon sequestration.[19]

Clearly much contemporary, intensive farming in the region's lowland areas is environmentally unsustainable. To resolve deep-seated issues resulting from historical drainage and "improvement" requires farmers to embrace new approaches and without their co-operation, progress will be slow. A holistic approach combining farming with recreational visitors should be informed by knowledge of historic landscape-scale change.[20]

Finally, there remain critical questions about the scale and potential of remediation. How much land can we expect to rewild, and how might resulting landscapes alter the balance of nature and culture through time? These seem academic questions, removed from the practical achievement of creating sustainable landscapes, but this study suggests otherwise. Decisions need to be made in the full light of ecological and cultural histories of this region. Since economic drivers once encouraged water removal and wetland destruction, to remediate the losses, the drivers must change.

NOTES

1. I. D. Rotherham, *The Lost Fens of Cambridgeshire, Lincolnshire and Yorkshire: The untold story of England's greatest ecological disaster* (Ashbourne, Derbyshire: Landmark Publishing Limited, 2009); I. D. Rotherham, *Peat and Peat Cutting* (Oxford: Shire Publications, 2009); I. D. Rotherham and D. McCallam, *Peat Bogs, Marshes and Fen as disputed Landscapes in Late eighteenth-Century France and England*; L. Lyle and D. McCallam, eds., *Histoires de la Terre: Earth Sciences and French Culture 1740–1940* (Amsterdam: Rodopi, 2008), 75–90.
2. O. Rackham, *The History of the Countryside* (London: J M Dent & Sons, 1986); C. Smout, *Nature Contested: Environmental history in Scotland and Northern England since 1600* (Edinburgh: Edinburgh University Press, 2000).
3. R. Van de Noort, *The Humber Wetlands: The Archaeology of a Dynamic Landscape* (Macclesfield: Windgather Press, 2004); A. De La Pryme, "Letters" [1639] as quoted in M. Dinnin, "The drainage history of the Humberhead Levels," in R. Van de Noort, and S. Ellis, *Wetland Heritage of the Humberhead Levels: An Archaeological Survey* (Humber Wetlands project, Hull: University of Hull, 1997).
4. H. A. Wilcox, *The Woodlands and Marshlands of England* (London: University Press of Liverpool, Hodder & Stoughton, 1933).
5. H. C. Darby, and I. S. Maxwell, eds., *The Domesday Geography of Northern England* (Cambridge: Cambridge University Press, 1962).
6. Van de Noort and Davies, *Wetland Heritage*; Dinnin, "The drainage history of the Humberhead Levels;" Noort and Ellis, *Wetland Heritage of the Humberhead Levels*; N. J. Whitehouse, M. H. Dinnin, and R. A. Lindsay, "Conflicts between palaeoecology, archaeology and nature conservation: the Humberhead Peatlands SSSI," in I. D. Rotherham and M. Jones, eds., *Landscapes ... ... Perception, Recognition and Management: reconciling the impossible? Proceedings of the conference held in Sheffield, UK, 2–4 April, 1996. Landscape Archaeology and Ecology*, 3 (1998) 70–78; P. C. Buckland, *Thorne Moors: a palaeoecological study of a Bronze Age site; a contribution to the history of the British insect fauna* (Department of Geography

Occasional Publication No. 8, Birmingham: University of Birmingham, 1979); P. Skidmore, M. Limbert, and B. C. Eversham, "The Insects of Thorne Moors," *Sorby Record* 23, Supplement (1985).

7. C. Saxton, *Map of the County of Yorkshire* (1577); C. Saxton and W. Goodman, *Map Of Pottrick Carr near Doncaster* (Local Archives, Doncaster MBC Libraries, 1616); C. Vermuyden, *Map of Hatfield Chace Before The Drainage* (1626); Burdett, *Map of the County of Derbyshire* (Derbyshire County Libraries: Matlock, 1767); Jeffrys, *Map of the County of Yorkshire* [1772] (Harry Margary: Lympne, Kent, 1973); J. Colbeck, *Plan of the rivers cuts drains and watercourses subject to the direction of the Trustees and which drain and preserve certain Lands within the parishes Townships and Hamlets of Doncaster Balby Carhouse High* (1782); Director General of the Ordnance Survey (1949–1969) *Geological Survey of Gt Britain (England and Wales)*: Ordnance Survey, *Sheets 77, 78, 79, 86, 87, 88, 99, 100, 101*; P. C. Buckland, "South Yorkshire Wetlands: The palaeo-ecologist's view," One-day workshop on wetland biodiversity and management, Sheffield Hallam University, November 2002, *South Yorkshire Wildlife Review* (in press); J. Leland, *The Itinerary of John Leland the Antiquary* (Third Edition, Published from the original manuscript in the Bodleian Library by Thomas Hearn, 1769).

8. This is described more fully in Rotherham, *The Lost Fens.*

9. I. D. Rotherham, P.A. Ardron, and O.L. Gilbert, *Factors determining contemporary upland landscapes . . . A re-evaluation of the importance of peat-cutting and associated drainage, and the implications for mire restoration and remediation*, in *Blanket Mire Degradation. Causes, Consequences and Challenges*. Proceedings of the British Ecological Society Conference in Manchester, 1997 (British Ecological Society and the Macaulay Land Use Research Institute: Aberdeen, 1997), 38–41; I. D. Rotherham, P.A. Ardron, and O. L. Gilbert, "Peat-cutting and upland landscapes: case-studies from the South Pennines," in *Landscapes . . . Perception, Recognition and Management: Reconciling the impossible?* Proceedings of the conference held in Sheffield, UK, 2–4 April, 1996, *Landscape Archaeology and Ecology*, 3 (1998): 65–69.

10. M. Taylor, *Thorne Mere and the Old River Don* (Ebor Press: 1997).

11. Burdett, Map of the County of Derbyshire (1767); Jeffrys, Map of the County of Yorkshire [1772] (1973).

12. Cornish, C. J., *Wild England Today* (Seeley and Co.: London,1895); Jeffrys, Map of the County of Yorkshire [1772] (1973); E. Clarke, "Potterick Carr," *The Field* (November 26th, 1887).

13. C. Firth, *900 Years of the Don Fishery: Domesday to the Dawn of the New Millennium* (Leeds: Environment Agency, 1997).

14. W. Cobbett, *Rural Rides* (Expanded edition of the 1830 issue with appendices, London: 1853); Smout, *Nature Contested.*

15. Smout, *Nature Contested*; M. Jones, "Deer in South Yorkshire: An Historical Perspective," in M. Jones, I. D. Rotherham, and A. J. McCarthy, eds., "Deer or the New Woodlands?," *The Journal of Practical Ecology and Conservation, Special Publication*, 1 (November 1996).

16. J. Leland, *The Itinerary of John Leland the Antiquary* (Third Edition, Published from the original manuscript in the Bodleian Library by Thomas Hearn, 1769).

17. Rotherham, *The Lost Fens*; Thomas Pennant, *A Tour in Scotland in 1769* (John Monk: Chester, 1771).

18. I. D. Rotherham, "Landscape, Water and History," *Practical Ecology and Conservation* 7 (2008): 138–52.

19. D. V. Hogan and E. Maltby, *The potential for carbon sequestration in wetlands of the Humberhead Levels* (Royal Holloway Institute for Environmental Research, Royal Holloway London: University of London, 2005); Anon., *Reprint of the First Edition of the One Inch Ordnance Survey of England & Wales*, (Doncaster: David & Charles Publishers Ltd, 1990) Sheet 22.
20. I. D. Rotherham, "South Yorkshire's Wetlands: Their obscure past and uncertain future. South Yorkshire Wetlands," One-day workshop on wetland biodiversity and management, Sheffield Hallam University, November 2002, *South Yorkshire Wildlife Review* (in press).

14 Uneasy Relationships Between Ecology, History, and Restoration

Jan E. Dizard

For decades, the dominant trope in environmental writing has been elegy. Optimism, to the extent that it was in evidence, was largely defensive: save the redwoods, declare an unspoiled area "wilderness," list a species as endangered. But then in the 1980s, a shift of perspective began to take hold among environmentalists and, more particularly, among a growing group of biologists and ecologists who carved out a new sub-field they called "conservation biology." Acknowledging loss was still at the center of concern (indeed, conservation biologists promoted the idea of a biodiversity crisis) but the point was less to save the few remaining remnants of unspoiled nature than it was to enlist fellow biologists and the public in an effort to recover what we have lost, to undo where possible the damage we have done. Conservation biology quickly gave birth to another sub-field, restoration ecology, and a formal society, the Society for Ecological Restoration.[1]

Of course the idea of restoration was not new. Private groups, often in cooperation with federal and state agencies, have been practicing restoration for decades. Ducks Unlimited has been restoring breeding habitat for waterfowl in the United States and Canada since the 1950s. Trout Unlimited has been restoring habitat for the various trout species, both native and introduced, for nearly as long. And there have been many small projects aimed at returning a place to biotic vitality, if not to its once pristine condition. But these efforts, even when successful, were either small in scale or focused on a single species. Conservation biologists and ecological restorationists began arguing for much more ambitious goals—nothing less that the "rewilding" of very large swaths of land.

The case for large-scale rewilding is, at first blush, compelling. Whatever the virtues of small-scale restorations, they simply cannot make amends for the sweeping effects of the landscape conversions that have been wrought by agriculture, industrialism, and urban sprawl. Only restorations at the landscape scale can hope to serve as something of a counterweight to the relentless forces of development. As Donald Waller has argued:

> Because habitat loss, fragmentation, and other forms of degradation threaten such a large fraction of our biota, we must pursue strategies

that can succeed not only in conserving populations now but also in perpetuating the ecological conditions that will sustain their evolutionary future. . . . Only large areas support larger, more viable, and interconnected populations of rare and threatened species and perpetuate the ecological processes that sustain other elements of biodiversity.[2]

In this same spirit, Michael Soule and Reed Noss have argued for large rewilded reserves because only large areas can support the large carnivores they claim are key to maintaining biodiversity. They write:

> By the early 1980s biologists recognized that large carnivores . . . require extensive, connected, relatively unaltered, heterogeneous habitat to maintain population viability. . . . These became the animals used to justify large nature reserves. . . . The assumption in this approach is the large, wide-ranging carnivores offer a wide umbrella of land protection under which many species that are more abundant but smaller and less charismatic find safety and resources.[3]

Extending the logic of Soule and Noss to its furthest reach, some conservation biologists have argued that rewilding should take as one of its goals repopulating parts of North America with surrogates of the megafauna that went extinct in the late Pleistocene.[4] Needless to say, such an undertaking would require an enormous expanse of land. Whether the goal is driven by the desire to have tracts large enough to contain a wide range of floral and faunal communities or by the desire to make room for megafauna, the result is the same: Very large areas will need to be set aside. Projects on this scale pose obvious and daunting challenges. But practical challenges can be surmounted if the goal is worthy. We need to ask ourselves if the goal of rewilding has merit.

THE PAST PROBLEM: BASELINES

Restoration in general and rewilding in particular derive much if not all their appeal from one major premise: At some point in the past, nature functioned in the way it was meant to function. The condition, for the sake of argument let's call it the "original condition," was marked by oscillations of varying lengths as species adjusted to one another as well as to perturbations of drought, storms, pathogens, and fire, to name only the most common factors that kept systems from achieving complete stability. Niles Eldredge captures this view nicely in describing the unique Okavango ecosystem in Botswana. The Okavango River had its outlet to the sea blocked by an earthquake some 5000 years ago, thus creating a unique landlocked delta-like system. Eldredge describes in detail the intricately linked relationships that weave myriad organisms into a system that is "astonishingly intact, incredibly complex, yet at the same time evincing a kind of simplicity that only systems as old, as well broken in, can possibly show. The Okavango is the real thing."[5]

What makes something the "real thing"? Eldredge provides us with a point of departure. Though he was arguing in behalf of preserving the Okavango from mounting pressure from cattle herders and others with designs on the region, he could as easily have been describing the goals of large-scale restorations: reproducing an intact, complex system in which myriad species are intricately articulated to ebbs and flows and, taken as a whole, are "well broken in." Until another earthquake or the press of humanity causes the system to change. Diamonds, discovered near the Okavango Delta, may be forever, but any specific state of nature is temporary.

Restoration requires choosing a baseline, a time when the "real thing" is presumed to have existed. But baselines are arbitrary, even misleading, as F. Vera has made clear.[6] The Okavango, as wondrous as it is, is both accidental and transient: It is, in effect, awaiting another event that will rearrange things in ways we cannot predict very well. Change may be imperceptible on the scale of a human's lifetime, but it is occurring nonetheless. The attempt to return a landscape to some past condition is necessarily an attempt to hold at bay the forces of nature that inevitably militate against stability.

The authors in this volume all acknowledge this, in one way or another, though few are as candid about the implications as Anita Guerrini and Jenifer Dugan.[7] Perhaps this is because they and their colleagues are working with a coastline—coastlines are about as volatile as nature gets, even at the scale of a human's lifetime. If nature is in flux, what criteria should inform the choice of one state or condition over another? Appeals to "original condition" are complicated for at least two reasons. First, as Vera argues, our understanding of the past in inextricably bound up with cultural assumptions that often bear little relationship to what the past actually looked like.[8] The second reason to worry about baselines is that the closer we look, the more evidence we find of significant early human impacts on the environment. Beginning with Martin's claim that Paleo-Indians were responsible for the extinction of North America's mega fauna,[9] the evidence of indigenous people having had a profound impact on both the flora and fauna of the Americas has steadily risen.[10] If the landscape has been shaped by humans for thousands of years, seeking the "original condition" becomes akin to searching for the Holy Grail.

At the very least, we need to ask how important it is to erase the footprints of our ancestors, especially the ancient footprints bound up in extinctions, fire regimes, and early agriculture. We need to be wary of dubbing the present profane and the distant past sacred. The attempt to reconstruct ecological history must give way to an appreciation for how intimately intertwined nature and culture have been for a very long time. Even places now revered as "unspoiled" are revealing to a new generation of archeologists evidence of extensive human impacts dating back several thousand years or more. Guerrini and Dugan make clear that the Santa Barbara coastline with which they work has been modified by humans for at least the past 8500 years.[11] The abundance of fruit- and nut-bearing trees in the Amazon, long thought to be "natural," now seems more likely the result of active cultivation—orchards, if you will.[12] And the Okavango is yielding evidence of horticulture and domesticated flocks dating back at least a 1000 years.[13] The role of humans in shaping the landscape can be lamented, but Kidner's skepticism about the degree to which humans altered what for us have become iconic symbols of untouched nature is fast being overwhelmed by the evidence of sweeping human impacts on North America long before Europeans even dreamed of sailing west.[14]

Myllyntaus also acknowledges the difficulties associated with establishing baselines. For him, as for Vera, baselines are entangled in versions of the past that are more products of cultural beliefs or scientific theories that have become threadbare, if not completely out of date. This said, Myllyntaus as well as Rotherham and Harrison are confident that we can know what a predisturbed forest or fens was like. But neither is sanguine about the prospects of recovering (rewilding) the forests of Finland or the fens of Britain on a scale that would make restoration more than a symbolic gesture, a demonstration project, if you will. Baselines are, as Myllyntaus puts it, "moving targets." Picking one moment as the way nature should be distorts nature and obscures the fact that humans have been interacting with and transforming nature for a very long time.[15]

FROM BASELINES TO PROCESS

Donlan and Greene attempt to sidestep the problem of human involvement by taking their reference point for North America the time before the first humans set foot on the continent, roughly 13,000 years ago.[16] That surely is a moment when things were indeed the result of purely natural forces. They also resist the temptation to argue for a Pleistocene version of Jurassic Park. Instead, their proposal is to restore the "evolutionary potential" that was lost when the megafauna were extirpated. Not only have we lost the evolutionary potential of the creatures that went extinct, Donlan, et al., argue that "our activities have curtailed the evolutionary potential of most remaining large vertebrates [in North America]."[17] Donlan and Greene press the point, saying that " . . . from a continental perspective, the major missing component of ecosystems today, compared to the Pleistocene . . . is megafauna. Today, nature is an anomaly because it misses what we can infer are critical cogs in the wheels—megafauna."[18] The implication is that we have thrown evolution off course and rewilding will, hopefully, return cogs to the wheel and return us more nearly to the trajectory they were on before we messed things up. This echoes Waller's point, quoted earlier, that only large wild areas will "sustain [species'] evolutionary future."

There can be no gainsaying that North America, or any other place for that matter, would look very different had *Homo sapiens* not entered the scene. Our appearance irrevocably changed things, just as the arrival of another novel species would have. But it is impossible to know what evolution would have produced absent humans. Invoking an interrupted evolutionary potential comes perilously close to advocating a teleological view of evolution. The oft-heard phrase, "the way nature intended," attributes intentionality when none exists. Moreover, no organism's evolutionary potential can be known since what it becomes is the result of an extraordinarily complex interaction between the organism, the other organisms with whom it co-exists, and the natural forces with which it contends. Evolutionary potential is unknowable precisely because evolution is so heavily contingent that no outcome can be said to be more consistent with evolution than any other outcome, including extinction. By the same token, there is no way, at least given present knowledge, to determine whether or not a species is fulfilling its evolutionary potential. It is impossible to say how contemporary flora and fauna have been altered by virtue of the Pleistocene extinctions. All we can be confident of is that North America would be different were wooly mammoths and saber-toothed tigers still with us.

North America was not meant to be what it was in the Pleistocene any more than it was meant to be what it is today. It was the way it was 13,000 years ago because of physical and organic properties which we can now pretty accurately describe, at least in general—mean temperatures, rainfall, plants, mammals, etc. We are the way we are for the same reasons. Neither condition can be privileged or called "the real thing." Each are both real—incredibly complex systems brimming with energy and ripe for change. Thinking of nature as a particular ensemble of "cogs" not only risks the temptations of teleology, it also comes uncomfortably close to a secular version of Louis Agassiz's theory of special creation, which he invoked to dispute Darwin (in the same spirit as contemporary advocates of "intelligent design" inveigh against evolution). If "cogs" are missing from places where they belong, we should put them back where they were meant to be. Rather than claiming that systems were designed by a supreme being, we instead speak of "intact ecosystems," and the "integrity" and "health" of ecosystems. These metaphors are predicated on an untenable assumption: that everything has its place and if it goes missing, the system of which it was a part is no longer whole.

These are tempting metaphors, judging by how frequently they are used. They are also deeply misleading. As Keulartz helpfully reminds us, metaphors must be carefully weighed.[19] Almost all of the common metaphors used in restoration are rooted in the ecology of Clements and Odum, both of whom posited an orderly process through which ecosystems pass after a disturbance. In the end, the process concludes when the pre-disturbance or "climax" condition is attained. But the Clements–Odum paradigm is in tatters and the metaphors arising from this tradition are misleading.[20]

To be sure, the once dominant theory—and the metaphors derived from it—still has its defenders. This is so largely because no widely agreed upon theory has taken its place. In a sense, restoration ecology is in a state that Kuhn described as a crisis: The king (Clements–Odum) is dead and there is no clear successor to the throne.[21]

Needless to say, this lack of a widely agreed upon theory of how biotic communities develop and change compounds the difficulties of agreeing upon a condition to which an area should be returned. In this sense, Donlan and his fellow Pleistocene rewilders are appropriately circumspect, regarding the rewilding as an experiment rather than a full-scale undertaking. But even with a careful matching of the proxies with the original species, the result would at best be a very partial approximation of the Pleistocene menagerie. This makes the larger claims—replacing the "cogs" and thus restoring evolutionary potentials—highly tenuous. The effort will also require heavy and on-going management and monitoring, precisely, I dare say, because it is virtually impossible to predict what the impact of the imported proxies will be on native flora and fauna, organisms which have adapted quite well to the absence of huge carnivores and other very large mammals. Whatever might come of the experiment, it won't be "the real thing."

Rewilding to restore ecological services, as opposed to evolutionary trajectories, is another way of avoiding or at least reducing the problem of establishing a baseline. Rotherham and Harrison see restoration as a way to make nature work for us in the way it once did. Restoring the fens would foster carbon capture, provide flood control, and produce habitat for at least some of the wildlife that now cling to small bits of their former range. Guerrini and Dugan are reluctant to use the term "restoration" for the coastal settings they are interested in precisely because they are, as I noted earlier, so intrinsically volatile. But they do want to see the coast function in a way that sustains both marine and terrestrial biodiversity. Their historical analysis of coastal ecology of the Santa Barbara region highlights the need to focus on ecological processes rather than baselines.

Shifting the focus to ecological processes and services interestingly brings us back to where we began—the need for very large-scale undertakings. If carbon capture is one goal, it makes no sense to think small. Similarly, if we want to restore estuaries' biotic productivity and deltas their protection against storms, we are talking about very large undertakings. In the world we inhabit, the larger the scale, the more toes get stepped on. This means that restoration projects, if they are to provide more than token services or go beyond demonstration projects, are inherently political.

WHO DECIDES WHAT TO RESTORE?

Environmental philosopher Eric Katz has sharply criticized ecological restorations for being nothing more than human artifacts, much as parks, subdivisions, and shopping malls are human artifacts.[22] For reasons that should by now be apparent, I tend to agree with him. But unlike Katz, I do not think this makes restorations any less worthy of our regard than those vanishingly rare "untouched" spots of nature that Katz honors. That said, we need to ask just what sorts of artifacts restoration projects are. More pointedly, we need to ask *whose* artifacts they are. Are they artifacts of environmental activists' ambitions? Are they artifacts of professional biologists and ecologists who claim to know what nature should be. Are they artifacts of the convictions or hobbies of the very rich or of well-endowed organizations like The Nature Conservancy or the Turner Foundation? A number of studies have shown that the public is of many minds about what sort of nature they prefer.[23] Everyone can agree that clean water and air are good things. Saving endangered species also is broadly appealing. But when we move beyond what may be called "core values," the picture becomes much less clear. Some like it wild, some like it tame; some think tall grass prairie is wonderful, others find it monotonous. As Helford makes clear, people resent being told by experts what sort of nature they should appreciate.[24]

The problem of legitimacy and expertise gets more intense and complex as the scale of restoration increases. This is certainly a key reason so many large restoration/rewilding proposals have depended heavily on private funding and private landholdings. The idea for Pleistocene rewilding was hatched at a retreat hosted by the Turner Foundation on one of Ted Turner's large ranches in New Mexico (a ranch that is now home to the first Bolson tortoise to live north of the Mexican border in thousands of years). The Turner Endangered Species Fund has also been working closely with the Peregrine Fund on a project to reintroduce aplomado falcons on Turner's Armendaris Ranch in New Mexico.[25]

It ought to give us pause when restoration projects proceed largely or exclusively under private auspices. Whether it's wealthy individuals or conservation organizations like The Nature Conservancy or the World Wildlife Fund, an agenda is being set that is essentially removed from public debate and consent. To be sure, it is their land (with apologies to Woody Guthrie) and they are more or less free to use it as they wish. If all goes well, these private efforts may generate broad public support that may lead to public support for large-scale restoration, but it is easy to imagine the opposite outcome: public resentment of an environmental policy over which they have little or no say. There are some cautionary tales in this regard.

Roxanne Quimby made a small fortune with a cosmetics line, Burt's Bees, and has devoted much of her wealth to promoting a proposal to set aside much of the Northern Forest in Maine, New Hampshire, and Vermont. She has encountered bitter opposition from locals who resent her designs, which include ending access to her properties for hunting, fishing, and snowmobiling.[26] Even more disturbing is the growing opposition in the Third World to American conservation organizations. Mark Dowie writes: "Today, the list of culture-wrecking institutions put forth by tribal leaders on almost every continent includes not only Shell, Texaco, Freeport, and Bechtel, but also more surprising names like Conservation International, The Nature Conservancy, the World Wildlife Fund, and the Wildlife Conservation Society."[27]

As Kidner points out, rewilding requires a significant change in the way humans view themselves and their relationships with nature.[28] I agree, but so far the evidence is that rewilding ourselves is a much harder and much more complicated proposition than moving wolves from Canada to Yellowstone or, perhaps, moving elephants and lions from Africa to the Great Plains. In any case, large reserves will bring large problems. The reserves will function in much the same way Martin and Szuter argue the "no-man's zones," the areas between hostile Indian nations, functioned. Game were heavily hunted outside these de facto refugia. Game was replenished in these no-man zones. The species encouraged to flourish within the rewilding sites, no matter their expanse, will not stay put. To the extent that rewilding is successful on its own terms, species will fan out and those who leave the reserves will either have to be trapped or shot or otherwise kept from migrating to places where they will be neither safe nor welcome. We should learn from the headaches that park rangers at Yellowstone have every year with Bison wandering across Park boundaries. Managing Bison is easier than managing the public. Imagine the public outcry should we be faced with culling elephants to keep them from damaging crops or competing with elk.

CONCLUSION

Before humans appeared on earth, nature could be characterized as self-regulating and self-replicating because there was no one around who cared about any particular ebb or flow. It is only when we gained a secure toehold that the otherwise erratic shifts in nature began to have meaning. Thus began our long career of attempting to modify nature to suit ourselves. We may now lament this history, though I think this represents an unappealing tilt toward self-loathing: "The world would be better off without people." But it is too late to turn back. Large-scale rewilding will require constant management and this will, in turn, require shifting resources and energy away from other and arguably more pressing environmental challenges.

For better or worse, the closer we look, the more difficult it becomes to reconcile ecological and human history in a way that gives us a clear line of sight to whatever target we might prefer. The conceptual ambiguities as well as the political difficulties rewilding faces have led some restorationists to abandon the search for an ideal to which we should return and, instead, to simply admit that restoration is about creating, not recreating, landscapes that meet our needs—in the broadest and most inclusive sense of the word. Allison concludes his reflections on ecological restoration by suggesting that it may be more accurate to label what restorationists do "ecological gardening."[29] Perhaps the point is not to replicate the past but to understand the past so that we can be smarter than we have thus far been about what we want for the future.

NOTES

1. For accounts of these developments, see David Takacs, *The Idea of Biodiversity: Philosophies of Paradise* (Baltimore: The Johns Hopkins Press, 1996); Timothy J. Farnham, *Saving Nature's Legacy: Origins of the Idea of Biological Diversity* (New Haven: Yale University Press, 2007); and Eric Higgs, *Nature by Design: People, Natural Process, and Ecological Restoration* (Cambridge, MA: MIT Press, 2003).
2. Donald M. Waller, "Getting Back to the Right Nature: A Reply to Cronon's 'The Trouble With Wilderness'," in J. Baird Callicott and Michael P. Nelson, eds., *The Great New Wilderness Debate* (Athens, GA: The University of Georgia Press, 1998), 554.
3. Michael Soule and Reed Noss, "Rewilding and Biodiversity: Complementary Goals for Continental Conservation," *Wild Earth* (Fall, 1998), 21.
4. See C. Josh Donlan et al., "Re-wilding North America," *Nature* 436 (18 Aug 2005): 913–14; Paul S. Martin, *Twilight of the Mammoths: Ice Age Extinctions and the Rewilding of America* (Berkeley, CA: The University of California Press, 2005); also Donlan, Chapter 26, this volume.
5. Niles Eldredge, *Life in the Balance: Humanity and the Biodiversity Crisis* (Princeton, NJ: Princeton University Press, 1998), 18.
6. F. W. M. Vera, *Grazing Ecology and Forest History* (New York: CABI Publishing, 2000); and Vera, Chapter 9, this volume.
7. This volume.
8. This volume.
9. Paul S. Martin, "Prehistoric Overkill," in P. S. Martin and H. E. Wright, Jr., eds., *Pleistocene Extinctions: The Search for a Cause* (New Haven, CT: Yale University Press, 1967): 75–120.
10. William Cronon, *Changes in the Land* (New York: Hill and Wang, 1983); Shepard Krech III, *The Ecological Indian: Myth and History* (New York: W. W. Norton, 1999); Charles C. Mann, *1491: New Revelations of the Americas Before Columbus* (New York: Knopf, 2005).
11. This volume.
12. Mann, *1491*.
13. Edwin N. Wilmsen, *Land Filled With Flies* (Chicago: The University of Chicago Press, 1989).
14. David Kidner, this volume.
15. Timo Myllyntaus, this volume; Ian Rotherham and Keith Harrison, this volume.
16. C. Josh Donlan and Harry Greene, this volume.
17. Donlan et al., "Re-wilding North America," 913.
18. Donlan and Greene, this volume.
19. Josef Keulartz, this volume.
20. For discussions of this, see Daniel Botkin, *Discordant Harmonies* (New York: Oxford University Press, 1990); Donald Worster, "The Ecology of Order and Chaos," in Donald Worster *The Wealth of Nature: Environmental History and the Ecological Imagination* (New York: Oxford University Press, 1993); Dana Phillips, *The Truth of Ecology: Nature, Culture, and Literature in*

America (New York: Oxford University Press, 2003); and R. Bruce Hull, *Infinite Nature* (Chicago: The University of Chicago Press, 2006).

21. Thomas Kuhn, *The Structure of Scientific Revolutions* (Chicago: The University of Chicago Press, 1962).

22. Eric Katz, "Another Look at Restoration: Technology and Artificial Nature," in Paul H. Gobster and R. Bruce Hull, eds., *Restoring Nature: Perspectives from the Social Sciences and Humanities* (Washington, D.C.: Island Press, 2000): 37–48.

23. Willett Kempton et al., *Environmental Values in American Culture* (Cambridge, MA: MIT Press, 1995); Reid M. Helford, "Constructing Nature as Constructing Science: Expertise, Activist Science, and Public Conflict in the Chicago Wilderness," in Gobster and Hull, eds., *Restoring Nature*, 119–42; Jan E. Dizard, *Going Wild: Hunting, Animal Rights, and the Contested Meaning of Nature* (Amherst, MA: The University of Massachusetts Press, 1999); Deborah Lynn Guber, *The Grassroots of a Green Revolution: Polling America on the Environment* (Cambridge, MA: MIT Press, 2003); Ann Herda-Rapp and Theresa L. Goedeke, eds., *Mad About Wildlife: Looking at Social Conflict Over Wildlife* (Leiden, The Netherlands: Brill, 2005).

24. Helford, op. cit.

25. Jack Hitt, "One Nation Under Ted," *Outside* (December, 2001), 74–76 ff; Stephanie Paige Ogburn, "Bred for Success," *High Country News* 38:21 (13 November 2006), 8–13, 19; Donovan Webster, "Welcome to Turner Country," *Audubon* (January–February, 1999): 48–56.

26. See Jonathan Finer, "Park Debate is a Battle Over the Future of Maine," *The Washington Post*, November 28, 2004, A3.

27. Mark Dowie, "Conservation Refugees: When Protecting Nature Means Kicking People Out," *Orion* 24:6 (November–December 2005): 16–27.

28. David Kidner, Chapter 23, this volume.

29. Stuart K. Allison, "What Do We Mean When We Talk About Ecological Restoration?" *Ecological Restoration* 22:4 (December, 2004): 281–86.

15 Sidebar

Designing a Restoration Mega-Project for New York

Mark B. Bain

Large-scale ecosystem restoration projects have emerged in the last decade around the world: the Fouta Djallon Highlands of Africa, the grasslands of Central Asia, the Danube Delta of Europe, and Patagonian rangelands of South America. Well-known cases from North America include the Everglades, Sacramento–San Joaquin Delta, and Chesapeake Bay. These restoration efforts can be termed mega-projects because they encompass one or more ecosystems, expansive landscapes, multiple levels of government, and varied considerations spanning environment, society, and economy. Restoration mega-projects are appealing because they address complex problems; act as symbols of decisive action; bring together many governments and private institutions; employ a systems approach that connects diverse issues; and utilize many sources of financing for a common purpose. However, their record of accomplishment remains to be established. Moreover, there is a pressing need for effective planning of restoration mega-projects. Restoration planning at large scales needs to identify fundamental approaches, methods for selecting goals and objectives, and ways to report progress.

The case under question here is the Hudson–Raritan Estuary (HRE) restoration program authorized by the Congress to enhance the environment of the waterways and harbor surrounding New York City. The Hudson–Raritan Estuary, also called the New York–New Jersey Harbor Estuary, has been extensively modified for ship navigation, includes extensive port facilities, and was designated an "Estuary of National Significance" by the Environmental Protection Agency. The waterways of the estuary cover 500 km^2 with a geomorphology stemming from Pleistocene glaciers and subsequent flooding by rising seas. The resulting islands, rivers, channels, and bays form a complex waterway with strong tidal currents of fresh and marine waters. Water quality has improved from generally poor conditions in the 1970s to consistently good quality today. Now the harbor almost always meets quality standards for oxygen, but sediment contamination remains high and bacterial levels impede safe seafood production. Habitats have been dramatically altered during the development of the urban cores: 80 percent of coastal wetlands have been lost, Manhattan Island has been enlarged by 25 percent, Newark Bay has been reduced by 25 percent with average depth increasing by 50 percent. Yet public interest in the waterways has risen and many new developments are oriented toward waterfronts. By necessity or desire, this estuary has become increasingly vital to metropolitan life, continuing its role in making New York City into a global economic and cultural center.

The HRE restoration plan was designed by an interdisciplinary team of ten members supplemented by periodic consultation with implementing agency representatives and outreach liaison. Team deliberations defined the fundamental properties of the restoration problem, its approach, goal, and measurable objectives. Additional experts were used to justify and document specific plan elements, as well as generate initial restoration ideas. The approach and program goal were broadly defined so that key plan elements and measurable objectives could follow.

New York City's metropolitan region constrains potential future states of the ecosystem. A consideration of these constraints yielded four assumptions that defined the context and scope of feasible HRE restoration objectives: the ecosystem is human dominated; it has been irreversibly changed; it is dynamic and will change further; and environmental enhancements can be made through science and technology. From these assumptions, a goal was developed to guide HRE restoration for its human and non-human inhabitants: *A mosaic of habitats will be created that provide society with new and increased benefits from the estuary environment.* Enhancing or creating a mosaic of habitats would support greater biodiversity, diversify ecosystem functions, promote resilience and persistence of flora and fauna, reverse habitat loss and degradation, and increase public enjoyment of the aquatic environment.

Measurable objectives of restoration programs have been called endpoints, characteristics, indicators, and targets. The term *Target Ecosystem Characteristic* (TEC) was adopted as a specific, measurable ecosystem property or feature providing environmental benefits valued by society. TECs were developed with a process that sought broad input of many disciplines and perspectives, while allowing review by independent scientists and management specialists. The final list of TECs was limited to the number judged to be manageable but sufficiently diverse for guiding the restoration program. Once documented and justified, the targets were reviewed by independent scientists and agency managers, and final adjustments made using review comments. The final list of eleven ecosystem targets includes: Shorelines and Shallows, Islands for Waterbirds, Enclosed Waters, Oyster Reefs, Fish and Lobster Habitat, Eelgrass Beds, Tributary Connections, Maritime Forest, Public Access, Sediment Quality, and Coastal Wetlands. Each of these targets was described and given qualitative criteria. For example, the TEC "Shoreline and Shallows" is shown in Figure 15.1.

Making a restoration plan for the Hudson–Raritan Estuary was difficult and required the expert team to adopt one vision and one method for proceeding. How the major challenges were handled provide some lessons

Figure 15.1 This existing Shoreline and Shallows site (Caven Point Beach, Jersey City) is in the intensively developed shoreline of the Upper Harbor. This site contains a vegetated riparian zone with channels and marsh, a vast mud flat extending several hundred meters out from the beach at low tide, and expansive shallow waters. (Courtesy of Julian Olivas / Air-to-Ground)

for planning other restoration mega-projects. Initial planning discussions clearly revealed that expert team members joined the effort with very different perspectives on restoration. Public benefits were determined to be critical for enhancing and improving this world cultural center. The ecosystem is far from its original condition and returning it to that past condition is not realistic or desirable. Moreover, envisioning restoration as a return to a fixed state was also seen as impractical and undesirable. The adopted goal directs HRE restoration efforts at producing ecosystem elements or habitats that promote both human and natural benefits. Perhaps the most difficult step in the planning work was accepting this dual approach across the interdisciplinary expert team.

The practice of providing restoration objectives in the form of precisely stated and quantified targets was the central method for designing this restoration mega-project. Use of a consistent language and precise terminology helped move the planning work along and communicate to outside groups and agencies. We found the planning work proceeded quickly once the team adopted the TEC concept and viewed the definition of ecosystem targets as the primary work to be accomplished. This target setting method succeeded in making a clear HRE restoration agenda that has gained broad support from government agencies and independent conservation organizations. Working toward the goal of promoting a mosaic of new and enhanced habitats will diversify the physical complexity of the ecosystem, increase biodiversity, and create positive ecosystem values for the people of New York. We realize that public interests will also change, as will the ecosystem. Based on public and scientific insight, specific restorative activities should now proceed and we are confident that achieving any of the targets will improve the Hudson–Raritan Estuary.

Part IV

What To Restore?

Selecting Initial States

16 Reflooding the Japanese Rice Paddy

David Sprague and Nobusuke Iwasaki

The South Yorkshire Fens, once England's third largest fenland, and an abundant hunting ground for waterfowl, was ultimately drained and transformed into farmland. More recently, the South Yorkshire Biodiversity Research Program has embarked on projects to map the extent of the original fens, and restore the fens (Rotherham and Harrison, Chapter 13, this volume). This train of events is astonishing to an Asia-based observer, first, at the initial self-denial of wetland resources by the past residents of Yorkshire. Astonishment continues at the rediscovery of the fen by their descendants, and their growing efforts today at restoring this thoroughly dry land to a wetland for future English peoples.

However, most compelling to the Japan-based authors of this chapter is our self-realization that if this fen had been located in Japan, farmers would probably have turned it into rice paddies. This realization triggers a train of inquiry for us about how to compare Europe and Japan, as well as how we are to join an already complex transatlantic dialogue about restoration ecology.[1] We will not plunge here into a detailed comparison of the South Yorkshire Fen, a Japanese rice paddy, and perhaps a North American wetland. Rather, our intent in this chapter is to identify key value scales for understanding decisions about land use based on the valuation of land, the products and services harvested from them, and the rediscovery of forgotten land-values, which are then judged to require preservation or restoration.

We concentrate on the value of rural landscapes as habitat for wildlife, and explore how the Japanese rice paddy can be placed on a scale developed primarily in Europe, designed to identify rural landscapes with high-nature value. If the Japanese rice paddy can be placed along this scale, then a transcontinental comparison would allow researchers to help define and identify ecological restoration and preservation. The rice paddy has a two thousand year history in Japan, and its modern version occupies a large proportion of Japan's farmland. How they are managed will have profound effects on the distribution and content of biodiversity in Japan.

This value scale assigns rural areas their own unique ecological status and defines the agro-environment (or more often "agri-environment" in

Europe) as a distinct area of research and policy. The agro-environmental perspective appreciates the myriad forms of farming, pastoralism, agroforestry, and their intricate interactions with land and biodiversity. The semi-natural biota that survived under agriculture is considered to be a unique form of biodiversity worthy of preservation or restoration in its own right. This perspective assumes that the "natural" baseline has little practical value, since it serves only to show that the majority of the original biodiversity has disappeared. Historical analysis chooses baselines defined in time and land use regime.

AGRO-ENVIRONMENTS AT THE CENTER

The agriculture-centered view of nature conservation has both a practical and idealistic side. On the practical side, agriculture is, simultaneously, one of the most important drivers of environmental destruction, preservation, and restoration. On the idealistic side, the "traditional" rural landscapes are thought to express the Arcadian values embodying a harmonious coexistence of humans within nature.[2] More importantly, for the purpose of this discussion, we feel that the agriculture-centered view forces us to recognize that decisions about environmental preservation and restoration require us to face the trade-offs between various use and non-use values of land.

Humans must make a living off of the land, and the valuation of land often involves some sort of usage value. Since food has been one of the most important products obtained from land, the values of agriculture and pastoralism have been imposed upon much of the landscape in regions with long histories of human occupation. Agriculture continues to be important today. After all, even when agriculture accounts for ever-smaller percentages of our economies, it still provides all of our food, and occupies a large proportion of our land.

The fate of biodiversity is intimately linked to the use of land for agricultural production.[3] Agricultural intensification is cited as a prime cause for the loss of biodiversity in European nations, including the collapse of the farmland bird populations.[4] At the same time, abandonment is blamed for the loss of extensive, semi-natural grasslands maintained by traditional farming systems, where abandoned grasslands are overgrown with shrubs or afforested.[5]

Agricultural issues have emerged onto policy arenas in international environmental organizations. The Convention on Biological Diversity established a thematic program on agricultural biodiversity. The Food and Agriculture Organization is particularly concerned with the genetic diversity of crops and livestock. Agriculture takes center stage at the World Trade Organization, as disagreements over agricultural policies paralyze talks on international trade (as of 2006), while scholars and policy makers debate the effects on rural environments of liberalizing international trade in agricultural commodities.[6] Agro-environmental schemes, designed to encourage environmentally beneficial farming practices, are spending larger proportions of government budgets.[7] The Common Agricultural Policy of the European Union spent approximately 5 percent of its budget between 2000 to 2002 on various types of agro-environmental payments, while the United States spent about 8 percent of total budgetary agricultural spending on agro-environmental payments.[8]

Agriculture occupies the crucial middle ground in many scales of environmental value. The agro-environment lies conceptually between the totally artificial and the wilderness. Geographically, the rural lies between the cities and the deep mountains, or the coast and the outback. On a historical scale, the Arcadian value system invokes the near-past, a time period preceding the advent of modern intensive agriculture but well past the clearing of wilderness by the earliest human settlers.[9] The twin pressures of land use intensification, on one side, and abandonment followed by unmanaged vegetation succession, on the other, threaten high-nature value rural landscapes.[10] In government policy, farm areas are bounded on one side by policies encouraging more intensive land uses, and on the other side by nature protection policies prohibiting human activity, sometimes leading to boundaries drawn on maps or visible on the ground. On a scale of nature destruction, agriculture can be the destroyer of more tropical rain forests, or the buffer holding back urbanization.

THE JAPANESE RICE PADDY IN TRI-CONTINENTAL CONTEXT

We anticipated many bewildering twists and turns in comparing Japanese rice paddies with wetlands of Europe or North America. For example, the rice paddy would share with the former English fens the present state of prime arable land, but differ in that rice paddy depends on maintaining a wetland, and thus retaining to some extent the wetland functions of allowing water flow and supporting wetland biodiversity. Both fen and paddy can be subjects of environmental restoration, but differ in that the ultimate goal of South Yorkshire fen restoration may be "rewilding" (Rotherham and Harrison, Chapter 13, this volume), while rice paddy restoration might involve "renaturing," the restoration of non-crop biodiversity into the tightly managed modern farm setting. In some cases, this may be achieved by returning to a "traditional" rice paddy,[11] embodying the Japanese version of the Arcadian values extolling a harmonious coexistence of humans within nature. Jumping to North America, some ornithologists tout California rice fields as aids to avian conservation.[12] Presumably, however, Californians could not propose a return to a "traditional" rice paddy, although their advocates in southern Europe may argue that "traditional" rice fields exist in Spain or Italy, or native Americans may propose a return to a rural environment based on their "traditional" subsistence practices (Tomblin, Chapter 17, this volume; Casagrande and Vasquez, Chapter 18, this volume). Already, the juxtapositions reach an unmanageable complexity.

AGRO-ENVIRONMENTS ENCAPSULATED IN INDICATORS

The currents of agriculture, biodiversity, landscape, transcontinental valuation, and government policy, all converge in the ambitious undertaking by the Organization of Economic Cooperation and Development (OECD) to formulate a comprehensive set of indicators to measure the effects of agricultural policies on the environment.[13] A crucial aspect of this effort is that the proposals encapsulate a theory of environmental valuation with a value scale for rural wildlife habitats that may allow us to locate the rice paddy in a transcontinental framework.

The OECD habitat indicators use a three-class scale of the agro-environmental landscape: (1) intensively farmed agricultural habitats, (2) semi-natural agricultural habitats, and (3) uncultivated natural habitats.[14] Intensively farmed areas are defined as areas used to produce arable crops and improved grasslands, commonly treated with fertilizers and pesticides, and subject to plowing, sowing, weeding, and harvesting. Semi-natural agricultural habitats are areas of farmland relatively undisturbed by farming practices, where farm chemical-use is reduced or absent. A list of semi-natural habitats includes extensive grassland and pasture, fallow land, extensive margins in cropped land, "low intensity" fruit orchards and olive groves, grazing marshes or water meadows, pastoral woodland, and alpine pastures. Uncultivated natural habitats refer to habitats that are on, crossing, or bordering agricultural areas and include natural woodlands and forests, small ponds, lakes, rivers, and unexploited wetlands and bogs. Some countries put man-made features among semi-natural habitat or landscape features, although these, such as hedges and stonewalls, shelterbelts, ditches and woodland planted on farms, can be used as landscape indicators.[15]

The proposed indicators apply this three-class scale of land use. The centrally located indicator is based on semi-natural habitat: the share of the agricultural area covered by semi-natural agricultural habitats.[16] This indicator has a potentially powerful impact on the evaluation of rural environments, since the loss of semi-natural areas is considered by many researchers to be a prime cause for biodiversity-loss in Europe, and the focus of many restoration efforts.[17] The same indicator is listed among biodiversity related indicators for agriculture by the European Environment Agency.[18] However, the habitat classification, as seen in the examples of semi-natural habitats, seems to be based on traditional European, extensive pastoral land uses. Many European representatives refer to the semi-natural habitats in their papers for the OECD expert meeting on agro-biodiversity.[19] The three-class spectrum is very similar to one proposed by Swart and colleagues.[20] European NGOs and advocacy groups emphasize the importance of traditional pastoral landscapes (e.g., EFNCP).[21]

The Euro-centrism of the proposed indicators presents a problem for ecologists working in non-European nations. A paper on Australia presented to the OECD expert meeting on agro-biodiversity did not use the word semi-natural at all.[22] A paper on the U.S. attempts to find a match for the semi-natural habitat within the U.S. definition of rangeland.[23] According to this paper, most U.S. rangelands would fall under the OECD definition of semi-natural agricultural habitat since rangeland is land used by grazing animals where the management consists of manipulating the vegetation primarily by adjusting grazing extent, or by prescribed fire, and other methods generally without cultivating the soil. However, pasture may be classified as intensively farmed agricultural habitats if intensively managed by planting desired vegetation and applying soil amendments to increase productivity.

A similar problem arises for placing the Japanese rural landscape under this classification. Japan, like Europe, has a long history of human occupation, and it is possible to go in search of the traditional Japanese rural landscape in the contemporary landscape as well as in history. However, Japan simply did not experience the same degree of pastoralism. How to fit the Japanese rural landscape to the OECD classification requires detailed examination of the land uses carried out by Japanese farmers.

EVALUATING THE TRADITIONAL JAPANESE RURAL LANDSCAPE

A school of Japanese ecology emphasizes the high-nature value of the traditional Japanese rural landscape that combines aquatic, wooded, and grassland elements, under the term *satoyama*, a word that can be translated as "the rural foothills."[24] The rice paddy, and its associated ponds and waterways, plays a central role in the *satoyama* by providing the aquatic elements. Rural woodlands of pine and coppice broadleaf trees provide the wooded elements. Grasslands existed as areas where farm communities collected various grass resources, and is closer to meadow in English, since the cut grass was carried back for use in the village and farmers did not graze livestock in large numbers.[25] Each of these elements can provide habitat for a variety of wildlife species adapted to moderate disturbance regimes. Rice paddy, in particular, can support organisms that spend part of their life cycle in aquatic environments, such as dragonflies, frogs, and fishes.

The temporal baseline for measuring change in the *satoyama* can be set at 1950, as suggested by the OECD, since agricultural modernization speeded up in the post-World War II period as it did in many other parts of the world. We have suggested that the baseline should be taken back to the 1880s.[26] This is reasonable, first, because this period is just after the change in government in 1868 that ended feudalism and set Japan on a course for modernization in agriculture as well as all other aspects of technology and economy. Another practical reason is that the first series of Japanese topographic maps provide detailed land use information, at least for the Kanto Plain surrounding Tokyo. The maps show that the full-set of rice paddies, woodlands (mostly pine or coppice), and grasslands had existed in the Kanto Plain, as shown in Figure 16.1 for southern Ibaraki Prefecture. Measured by a naturalistic baseline, this landscape is utterly unnatural. The potential natural vegetation in this region is mostly broad-leaf forest.[27] Rural grasslands were maintained by the repeated cutting and firing that kept vegetation from succeeding to woodland.[28] Thus, the landscape seen in Figure 16.1 was largely the product of human effort. Nevertheless, this landscape supplied the natural resources supporting agricultural production for the farm communities in this region.

Given this background, locating the Japanese rural landscape on the three-class scale appeared straightforward. The *satoyama* school of ecology is accustomed to thinking of rural Japan as secondary nature providing semi-natural habitat for wildlife. The farm practices creating the *satoyama* are traditional ones of the near-past prior to the modernization of agriculture. The rural woodlands, grasslands, and rice paddies can be included within the semi-natural category; dry field would not. The uncultivated natural habitats exist in the form of forests in deeper mountains, or wetlands that had not been converted to rice paddy. Modern intensive agriculture could be identified with the use of farm chemicals, or the re-engineering of paddy fields for strict water control that allows paddies to be thoroughly dried when desired.[29]

However, classifying the traditional Japanese rice paddy as a semi-natural habitat conflicts with the European definition of semi-natural. First, the rice paddy is arable field dominated by a single crop species. Second, the paddy field is subject to regular disturbances of the soil. Third, the paddy field is, in itself, an artificial construction, although this point may be less critical if hedgerows or stonewalls can be counted as semi-natural habitat in some European nations.

Sprague proposed, nevertheless, that the traditional Japanese rice paddy can be counted as semi-natural habitat, based on three premises.[30] First, rice paddies and their associated waterways, can, in fact, provide habitats for a wide variety of organisms. Second, rice agriculture mimics part of the disturbance regimes of natural rivers and wetlands caused by the seasonal flood, flow, and ebb of water through the paddies, ponds, and waterways of the rice paddy system. Third, the loss of traditional Japanese rice paddies would lead to a radical restructuring of wildlife habitats in a Japanese rural region.

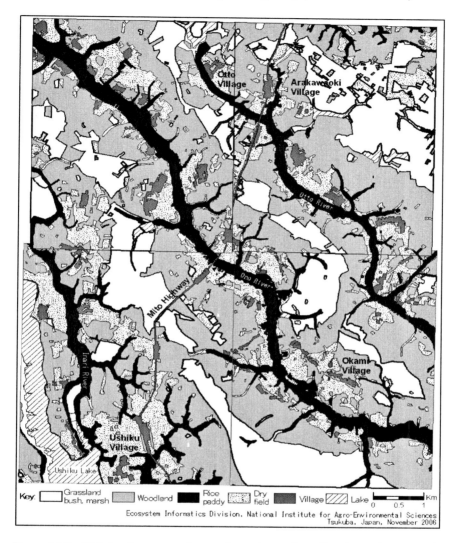

Figure 16.1 This land use map of a rural area in southern Ibaraki Prefecture in the Kanto Plain, Japan, is based on four Rapid Survey Maps surveyed in 1881 (lower half) and 1883 (upper half).

CONCLUSION: DECLARING VALUES

In proposing the semi-natural status of the traditional Japanese rice paddy, we are not asking European ecologists to alter their habitat definitions as applied in their own rural landscapes. We are, rather, declaring our value scale, and identifying our place on that scale in our valuation of the traditional Japanese rice paddy. We feel that our scale has much in common with the European value scale. The Japanese *satoyama* and European extensive grassland concepts share a common historical vision of the predicament faced by rural landscapes today, that the rural landscape is threatened on two sides by land use intensification and abandonment. T. Takeuchi, a prominent proponent of the *satoyama* concept, stated:

> Today, the word *satoyama* . . . generally indicates a natural environment that is being managed and, therefore, its basic element can be represented as secondary nature. Secondary nature is easily lost in the process of large-scale urban development but, on the other hand, if it remains untouched it will be thoroughly transformed by natural vegetation succession. In order to conserve *satoyama* landscapes, adequate management is essential, as has been observed in traditional agricultural activities.[31]

In its details, the Japanese and European scales necessarily differ, since the histories of agriculture and biogeography differ between Japan and Europe. The problem faced by Japan-based agro-ecologists is that a strict application of a European pastoral-based definition of semi-natural might leave Japan with little semi-natural habitat. Japan has very little extensive grassland left today. If we are to ascribe semi-natural status to both extensive grassland and traditional rice paddy, we are faced with the task of seeking a definition that is not dependent on any particular form of agriculture, or even any particular form of biodiversity. This task cannot be avoided, and ecologists in each region of the world need to provide the ecological basis to habitat classifications and content.

The importance of a habitat scale is in declaring a value system to be applied on the landscape. The use of a scale recognizes that values are not unitary or static. The scale recognizes a variety of values and allows one to identify one's own value point along it in relation to other values. The scale allows for movement along the scale, and thus can be used to define the meaning of environmental preservation or restoration along the scale.

Even the simple, three-class habitat definition used in this chapter allows ecologists to interpret many forms of ecological management. The semi-natural advocate may declare that restoration is the movement of a habitat into the semi-natural category from the modern intensive or uncultivated natural categories. A more naturalistic ecologist may define restoration as the movement of habitat into the uncultivated natural categories from the semi-natural category or straight from modern intensive category. Some ecologists may feel that categorizations are too arbitrary. By strictly biological measures, there may be no need to decide on a habitat category. An ecologist can count the number of fishes in a rice paddy regardless of whether it is classified as a semi-natural or modern intensive habitat. If so, restoration can be defined as quantitative movement in a desirable direction along the scale.

Various forms of river and wetland restoration occur in Japan because much of both have been rebuilt for flood control or agriculture.[32] Japanese agriculture, recognizing the effect of modern farm practices on biodiversity, census rice paddy biodiversity, and implement projects to restore some of the biodiversity lost from rice paddies. The two Ministries of Agriculture and Environment jointly organize an annual rice paddy biodiversity census. The 2005 census found ninety-four species of fish and seventeen species of frogs in the rice paddies and associated waterways.[33] The Ministry of Agriculture funds projects to rebuild waterways to allow for the survival of aquatic organisms, such as by removing large steps preventing the movement of aquatic organisms through the waterways, or removing one, two, or even all three concrete sides of the concrete canals often built into modern Japanese rice paddies.

One of the simplest ecological restoration techniques is to keep a rice paddy flooded over the winter. Water control is so effective in modern rice fields that a paddy can be left completely dry for long periods, a difficult condition to achieve in older rice paddies originally built in wetlands. Winter reflooding can be considered a form of renaturing for paddy fields that otherwise may be devoid of wetland biodiversity over the winter. While many farmers use winter reflooding mainly as part of low-till or organic farm management, reflooding projects involving local government or NGOs more often cite biodiversity benefits among the reasons for winter reflooding.[34] One of the most famous reflooding projects is being carried out in the Kabukuri-numa wetland region of Miyagi Prefecture of northern Japan, a RAMSAR site, in association with a project to enhance the quality of a winter habitat for the greater white-fronted goose (*Anser albifrons*).[35] To reduce population pressure on a natural wetland, neighboring rice paddies are flooded over the winter to allow the geese to feed in the rice paddies. Winter reflooding alone may not lead to a rice paddy that can be classified as traditional or semi-natural, but it does move these rice paddies in the direction towards an enhanced habitat for various wetland organisms.

Will North American or Australian ecologists wish to declare a scale that includes the semi-natural concept? North American ecologists may prefer to apply the pre-European settlement state as the baseline.[36] The Conservation Reserve Program implemented by the U.S. Department of Agriculture has the objective of returning farmland permanently to natural wetland.[37] Similar values may apply in Australia. Abensperg-Traun and colleagues compared the agro-environmental policies of Austria and Australia to conclude that ecological restoration in Western Australia needs to address sustainable agriculture as well as the restoration of ecosystems as they were prior to European colonization, while biodiversity conservation in Austria needs to emphasize the maintenance or restoration of traditional landscape patterns and associated land use practices to favor a return to higher biodiversity patterns.[38]

In conclusion, after observing the transcontinental debate on rural habitat classifications, we feel that values matter. We invite readers to declare their own value scale and place themselves on it. With these value scales, we will have a basis for discussion, negotiation, and planning. A great variety of preservation and restoration projects are carried out around the world. They range in size from individual gardeners in their garden, to the EU agro-environmental schemes. Each is based on a value system that should be declared. After values become clear, we will then have the basis for negotiations about exactly what sort of biodiversity to give priority to, or the restoration programs that should be encouraged.

NOTES

1. M. Hall, *Earth Repair: A Transatlantic History of Environmental Restoration* (Charlottesville: University of Virginia Press, 2005).
2. J. A. A. Swart et al., "Valuation of nature in conservation and restoration," *Restoration Ecology* 9 (2001): 230–38.
3. F. Berendse et al., "Declining biodiversity in agricultural landscapes and the effectiveness of agri-environmental schemes," *Ambio* 33 (2004):499–502; L. G. Firbank, "Striking a new balance between agricultural production and biodiversity," *Annals of Applied Biology* 146 (2005): 163–75; E. H. A. Mattison and K. Norris, "Bridging the gaps between agricultural policy, land-use, and biodiversity," *Trends in Ecology and Evolution* 20 (2005): 610–16.
4. T. G. Benton et al., "Linking agricultural practice to insect and bird populations: a historical study over three decades," *Journal of Applied Ecology* 39 (2002): 673–87; P. F. Donald et al., "Further evidence of continent-wide impacts of agricultural intensification on European farmland birds, 1990–2000," *Agriculture, Ecosystems and Environment* 116 (2006): 189–96.
5. S. Petit et al., "MIRABEL: models for integrated review and assessment of biodiversity in European landscapes," *Ambio* 30 (2001): 81–88.
6. OECD, 2005. Agriculture, Trade and the Environment: The Arable Crop Sector (Paris: OECD).
7. Berendse et al., 2004; Firbank, 2005; Mattison and Norris, 2005.
8. OECD, 2003a. Agricultural Policies in OECD Countries (Paris: OECD).
9. J. Pfadenhauer, "Some remarks on the socio-cultural background of restoration ecology," *Restoration Ecology* 9 (2003): 220–29; Swart et. al., 2001.
10. Firbank, 2005; Petit et al., 2001.
11. K. Takeuchi et al., eds., *Satoyama: The Traditional Rural Landscape of Japan* (Tokyo: Springer-Verlag, 2003).
12. C. Elphick, "Functional equivalency between rice fields and seminatural wetland habitats," *Conservation Biology* 14 (2000): 181–91.
13. OECD, 2001. *Environmental Indicators for Agriculture*, vol. 3: Methods and Results (Paris: OECD); OECD, 2003b. *Agriculture and Biodiversity: Developing Indicators for Policy Analysis* (Paris: OECD).
14. OECD, 2001, 333–347.
15. OECD, 2001, 365.
16. OECD, 2001, 342.
17. F. Klotzli and A. P. Grootjans, "Restoration of natural and semi-natural wetland systems in central Europe: progress and predictability of developments," *Restoration Ecology* 9 (2001): 209–19.

18. EEA, 2003. *An Inventory of Biodiversity Indicators in Europe, 2002* (Copenhagen: European Environment Agency).
19. OECD, 2003b.
20. Swart et al., 2001.
21. EFNCP, 2001. *Recognizing European Pastoral Farming Systems and Understanding Their Ecology: A Necessity for Appropriate Conservation and Rural Development Policies*, Proceedings of the 7th European Forum on Nature Conservation and Pastoralism. European Forum on Nature Conservation and Pastoralism, Argyll, U.K.
22. J. Walcott et al., "Indicators of agri-biodiversity-Australia's experience," in OECD, 2003b: 222–40.
23. S. J. Brady and C. H. Flather, "Estimating wildlife habitat trends on agricultural ecosystems in the United States," in OECD, 2003b: 156–67.
24. H. Moriyama, *What Does Rice Paddy Protection Mean?* (Tokyo: Nosangyoson Bunka Kyokai, 1997) [in Japanese]; Takeuchi et al., 2003.
25. S. Sato, "Perspectives on the grazing system in Japan," *Grassland Science* 51 (2005): 27–31.
26. D. S. Sprague and N. Iwasaki, "The Rapid Survey Maps: Japan's first modern topographic maps and the GIS analysis of historical land use in the Kanto Plain," in T. Uno, ed., *Reading Historical Spatial Information from Around the World*, Proceedings of the 24th International Symposium, International Research Center for Japanese Studies (Kyoto: 2006) 165–76; D. S. Sprague et al., "Measuring rice paddy persistence over a century with Japan's oldest topographic maps: georeferencing the Rapid Survey Maps for GIS analysis," *International Journal of Geographical Information Science* 21 (2007): 83–95.
27. T. Miyawaki, *The Vegetation of Japan: Kanto* (Tokyo: Shibundo, 1986) [in Japanese].
28. K. Ohkuba and K. Tsuchida, "Conservation of semi-natural grassland," in M. Numata, ed., *Handbook of Nature Conservation* (Tokyo: Asakura-shoten, 1998) 432–476 [in Japanese].
29. S. J. Lane and M. Fujioka, "The impact of changes in irrigation practices on the distribution of foraging egrets and herons (Ardeidae) in the rice fields of central Japan," *Biological Conservation* 83 (1998)· 221–30; D. S. Sprague, "Monitoring habitat change in Japanese agricultural systems," in OECD, 2003b: 168–79.
30. Sprague, 2003.
31. K. Takeuchi, "*Satoyama* landscapes as managed nature," in Takeuchi et al., 2003, 9–16 (10).
32. K. Nakamura et al., "River and wetland restoration: lessons from Japan," BioScience 56 (2006): 419–29.
33. MAFF, 2005. *Joint Ministry of Agriculture, Forestry, and Fisheries, and Ministry of Environment Rice Paddy Biodiversity Census 2005*, press release March 24, 2006 [in Japanese].
34. H. Kurita et al., "Environmental potentials of winter-flooding rice fields for wetlands restoration," *Journal of the Japanese Society of Irrigation, Drainage, and Reclamation Engineering* 74 (2006): 713–17 [in Japanese].
35. T. Itoh, "Attempts to incorporate biodiversity into rice agriculture," *Kagaku* 76 (2006): 309–313 [in Japanese].
36. Hall, 2005.
37. Brady and Flather, 2003.
38. M. Abensper-Traun et al., "Ecological restoration in the slipstream of agricultural policy in the old and new world," *Agriculture, Ecosystems and Environment* 103 (2004): 601–11.

17 American Indian Restoration

David Tomblin

Indigenous people's cultural traditions challenge Western notions of ecological restoration. When Europeans arrived on the North American continent, they saw "wilderness"—a landscape free of systematic human interventions. At the same time, however, indigenous North Americans perceived themselves as living in a managed, cultural landscape. Collectively, Native Americans had created a variety of "semi-natural" agro-ecosystems: what Sprague and Iwasaki describe in their chapter on traditional Japanese rice paddies as being situated "conceptually between the totally artificial and the wilderness." Although the position along the artificial/natural continuum may be different for Native Americans, conflicting perspectives on the extent to which pre-European landscape was wilderness, nonetheless complicate Euro–American restorationist aspirations for "rewilding;" that is, restoring wild conditions. Indeed, many indigenous restoration efforts take on the characteristics of regardening projects commonly found in Europe and Japan. This chapter focuses on American Indian visions of ecological restoration while emphasizing how their practice contests, complicates, and enriches Western definitions of restoration. Furthermore, the interaction between indigenous restorationists and Western science embodies a continuation of the Native American historical struggle against the expansion of Western civilization. A primary result of this struggle is the creation of eco-culturally hybrid landscapes that integrate features from both the past and present, and Western and native traditions.[1]

RESISTANCE AND EXCHANGE

Cultural resistance to and exchange with Western traditions has long been part of American Indian history dating back to the first contact with Euro–American colonists and explorers. As Europeans moved inland, spread disease, introduced new technologies and Christianity, waged wars, and disrupted traditional American Indian political and social systems, American Indians had to adapt. As a matter of survival, these societies adopted cultural, political, and technological elements from both the traditional remnants of numerous fragmented tribes and European colonizers. This story of cultural survival and hybridization continues in the twentieth century with the evolution of natural resource management on American Indian reservations. During the twentieth century, cultural resistance responded to federal policies of cultural assimilation and land dispossession. Since the 1940s, however, many tribes have gradually obtained greater political autonomy and reclaimed control over natural resources on their lands. Associated with this political resurgence were a growing number of ecological restoration projects on reservations. These projects helped buttress the foundation of this political movement. And as American Indians began to restore their cultures and natural resources, they recognized the necessity of negotiating both "time's arrows" and "time's cycles" that David Lowenthal mentions in his chapter. As a consequence, ecological restoration for American Indians is not solely about returning to the past; it is about dealing with the present and creating a sustainable future.[2]

Fidelity to the past is an important part of American Indian restoration projects, but within certain physical and historical constraints. A survey of current American Indian natural resource management strategies reveals no plans to take their societies back to 1492. Nonetheless, it is not uncommon to find tribes participating in rewilding projects with federal agencies and environmental NGOs, but for different cultural reasons than those held by Western restorationists. For instance, large vertebrates are important characters in many tribal legends, serve spiritual and ceremonial roles, or were important sources of food. Most American Indian restoration projects, however, focus on restoring disrupted ecosystem processes, remediating hazardous waste contamination, and reestablishing plant and animal species important to the production of cultural artifacts (e.g., medicinal plants, plant fibers for basket making, timber, etc). Through a variety of restorative approaches and even preservation, American Indians on a regional and continental scale are collectively embarking upon a landscape-scale eco-cultural restoration project. This growing movement is adapting ecological restoration for the local needs of each tribe. Furthermore, the cultural appropriation of restoration is one of the many ways that American Indians are coping with rapid social change while attempting to restore a semblance of cultural and ecological stability on their lands.[3]

The rest of this chapter surveys the diversity of goals and values that drive restoration projects implemented by the White Mountain Apache Tribe on the Fort Apache Indian Reservation. The Apache's restoration work exemplifies how American Indian tribes in general are blurring the boundaries of the modern and the premodern, indigenous and Western traditions, and visions of past landscapes with new landscapes.

WHITE MOUNTAIN APACHE RESTORATION PROJECTS

The White Mountain Apache Tribe is one of several bands (San Carlos, Cibecue, Northern Tonto, and Southern Tonto) that belong to the Western

Apache group. Situated in a ponderosa pine ecosystem in east-central Arizona, the Tribe is dependent upon timber, grazing, hunting, fishing, ecotourism, a ski resort, and a casino to support their economy. Ecological restoration has become an important tool for sustaining the Tribe's eco-cultural landscape. Restoring the Apache natural resource base is also essential to the survival of the Tribe. As Ronnie Lupe (tribal chairman 1966–1970, 1974–1986, 1990–1998, reelected in 2006) asserted in a lecture at Harvard University in 1992: "The White Mountain Apache Tribe is fortunate to have a land base that is rich in natural beauty and bounty. Our resources provide us with the potential and opportunity for sustainable development . . . In today's world, you must continue to grow or face stagnation and eventual failure." Implicit to this message is that the goal of any restoration project on the reservation has to integrate the past with present needs. Apaches won't survive if they don't move forward. They must renature the landscape—but not exactly to the same configuration as it was prior to significant European disruption. As the Apache regained control over their natural resources in the 1980s and 1990s, they began creating a hybridized eco-cultural landscape by integrating the past with the present, and Apache traditions with Western traditions.[4]

The Fort Apache Indian Reservation has a rich history of both American Indian and Euro–American restoration projects. This stems from a long record of poor federal management of reservation lands. Fire suppression policies left forests vulnerable to destructive wildfires that recently culminated in the Rodeo–Chediski fire of 2002; the U.S. Forest Service and Bureau of Indian Affairs (BIA) promoted policies that encouraged the overexploitation of timber and overgrazing. And in an attempt to slow soil erosion caused by overgrazing and to provide more water to Euro–Americans in the Salt River Valley, the BIA implemented a watershed project that exacerbated erosion problems and negatively impacted plant communities along targeted waterways. Coincident with federal agencies assuming management responsibilities of tribal lands in the late nineteenth century were assimilation policies that disrupted Apache culture and their land management traditions. Because of the tight link between ecological degradation and cultural disruption, ecological restoration became important to the Apache community for both cultural and ecological reasons.[5]

The BIA directed most of the early restoration projects on the reservation. In the late 1940s, the BIA's forestry division began experimental prescribed burns in ponderosa pine forests. Harold Weaver, a vociferous critic of federal fire suppression policies, led these experimental forays into forest restoration. Within his papers, Weaver acknowledged the significance of American Indian burning to the maintenance of a healthy ponderosa pine ecosystem. At the same time, members of the White Mountain Apache Tribe began the restoration of depleted Apache trout populations. The Tribe closed streams to fishing and with the help of the federal government built fish hatcheries, restored stream habitat, and erected barriers to limit the migration of non-native trout into Apache trout habitat. Currently, the Apache trout is well on its way to recovery, providing a source of revenue for the Tribe through fishing licenses. Even the BIA's ill-fated watershed project in the 1960s involved restoration. Although the project extensively altered the vegetation structure of the Cibecue watershed, one of the main goals of the project targeted the restoration of overgrazed land. However, the competing goal of providing water to downstream users worked against restoring an ecologically sustainable watershed.[6]

By the 1990s, the White Mountain Apache had reclaimed political autonomy over natural resource management and cultural affairs on the reservation. By the year 2000, the Fort Apache Indian Reservation swarmed with a wide variety of restoration projects ranging from those that focused mostly on cultural concerns to those focused solely on ecological concerns. Most projects, however, incorporated a combination of ecological and cultural values. The subsequent survey reveals that the Apache considerably varied the degree of cultural integration and the fidelity to the past in their restoration work.

Watershed Restoration—The Tribal Watershed Program originated from funds awarded to the White Mountain Apache Tribe in a 1995 court settlement with the federal government. The Tribe filed the lawsuit in 1950 to sue the federal government for mismanagement of their natural resources prior to 1946. The Apache received $22 million, reserving 20 percent of this settlement for a permanent land restoration fund. A Tribal Council resolution mandated that the fund support the restoration of "tribal ecosystems to a condition that better reflects their condition prior to suffering damage from the mismanagement [by federal agencies] and to fund the education of Tribal members in the disciplines related to natural resource management." The Council gave funding priority to projects that developed comprehensive watershed planning activities, incorporated community-based efforts, and included "activities that promote traditional cultural practices, the Apache language, and the education of tribal members." From the beginning, unlike BIA projects, the Council explicitly articulated eco-cultural concerns for the tribal community.[7]

Watershed restoration is culturally significant to Apaches, with water embodying both ecological and spiritual qualities. Ronnie Lupe explained, "The Apache way is to bless ourselves with water . . . Water is *tuu*. It's a language that my brother and I have. It's the Apache word for water, the life blood of my people. It has sustained us from time immemorial. It nourishes and sustains the plants, the animals, and the ecosystem—to use a modern word—upon which we live and survive." Thus, uncontaminated, free flowing water remains imperative to Apache cultural and spiritual survival.[8]

For instance, the Cibecue Bridge Project combined elements from the past and present to restore riparian plant communities and water flows disrupted by the accumulation of sediments in the streambed. This project took into consideration the local needs for passable roads, livestock grazing, traditional plant use (e.g., replanting cattails—important for curing ceremonies and girl's puberty ceremonies), and hydrologically sound stream flow. The project required heavy earthmoving equipment to clear channels and reduce sedimentation. Restorationists erected solar-powered electric fences to protect wetlands from cattle and elk grazing. Local Apache students cleared overgrowth and planted native plants along the streams near the bridge. This newly constructed landscape, therefore, took aspects of the past (functioning stream flow, traditional plants) and fused them with modern features of the landscape (cattle grazing, bridges, electric fences). With other projects, blending cultural elements with natural processes was less obvious. The restoration of Soldier Spring, a sacred place important for ceremonial performances, illustrates this point. Restorationists employed more subtle interventions to prevent erosion and restore stream flow. In order to accommodate Apache customs, instead of using metal, a material often used to construct erosion abatement structures in stream restorations, tribal members installed check dams and riffle structures constructed out of local rock and plant material. The tribal community appreciated this innovation as metal compromises the spiritual integrity of ceremonially significant streams.[9]

On the surface, the Watershed Program seemingly constitutes a landscape-scale restoration effort to erase past BIA-spawned ecological degradation. However, upon closer inspection, Apache restorationists incorporated a diversity of goals, adapting each project to address local concerns. Soldier Spring and Cibecue Bridge took on characteristics of what Chris Smout (Chapter 10, this volume) calls "conservation gardening," small-scale management projects that restore culturally managed species assemblages. How one characterizes a restoration project, therefore, depends very much on the scale at which a project is observed.

Forest Restoration—Like the Watershed Program, the Apache's restoration efforts after the Rodeo–Chediski fire also resembled a landscape-scale restoration project. However, in this case, because of the significant damage to the Apache's economic base, flood hazards to several communities, the size of the job, and limited funds, the level of detail given to local concerns was less than with the Watershed Program. The task involved the restoration of 276,000 acres of forested reservation lands devastated by the largest conflagration in Arizona's history (476,000 acres in total). To facilitate this effort, the Tribal Council requested and received federal assistance through the Burned Area Emergency Response program. The program initially entailed a cooperative action between the White Mountain Apache, BIA fire management office, the Forest Service, and the Ecological Restoration Institute of Northern Arizona University. In 2002 they began restoring watersheds made vulnerable to flooding and burned timber lands. Although restoration efforts primarily focused on economic resources and community safety, project managers nonetheless consulted tribal members about damaged sacred sites, burial grounds, and areas of archeological importance. Because of the culturally sensitive nature of the work, the Tribe employed numerous Apache to protect and restore the integrity of these sites.[10]

Wildlife Restoration—The White Mountain Apache also participate in a number of rewilding projects. Most visible to the non-Indian public is the Mexican Wolf Recovery Program led by the National Wildlife Federation and the U.S. Fish and Wildlife Service (FWS). In 1998, the Tribal Council voted to allow two Mexican wolves that wandered onto the reservation to stay. Since then, under the condition that they have full control of the reservation wolf population, the Tribal Council approved the release of captive-bred wolves onto reservation lands. As of 2006, the FWS estimated that over 20 wolves either reside on or move through the reservation. While this is largely a rewilding project, this program also has relevance to Apache cultural restoration efforts. Prior to cultural disruption and the Mexican wolf's extirpation from the United States, the Apache held a spiritual connection to this species. According to Krista Beazley, a White Mountain Apache wildlife biologist who monitors the wolves on the reservation, Apache warriors in the past sometimes asked for "wolf power" prior to hunts and battles. They would perform a ceremony that involved imitating wolf-stalking behavior while singing a ritual song to wolves. After completing the ceremony, Apache warriors could borrow a wolf's hunting ability.[11]

Cultural Restoration—The Western Apache Placenames Project further illustrates the difficulty of separating the Apache's natural and cultural worlds. Apache place-names and their stories hold vital information about how the Apachean landscape has changed over time, making this information useful in restoration projects. These stories also contain moral lessons that instruct the Apache people about the right way to treat each other and the land itself. Reviving these stories for future generations has implications for reestablishing long fragmented relationships with the land and cultural traditions.[12]

Passive Restoration—Even the goal of *preserving* wilderness is evident in White Mountain Apache natural resource management issues. As with Western "wilderness" ideals, permanent disturbances of nature are prohibited in sacred places. The White Mountain Apache are among several Western Apache tribes involved in protests against University of Arizona's plans to construct telescopes on Mount Graham in southeastern Arizona. Apaches have long held this mountain as a sacred place, so that any technological intervention here is considered sacrilegious, thus inhibiting their religious freedom. To the Apache, the desecration of Mount Graham is equivalent to tearing down a Christian church to build a shopping mall. The Apache, therefore, justify the preservation of "wilderness" on cultural grounds, which differs from Euro–Americans who typically justify preservation on ecological grounds. Furthermore, Apache eco-cultural restoration efforts along with their desire to prevent construction of telescopes on

Mount Graham demonstrate cultural flexibility toward acceptable technological intervention in natural processes. As in Western society, the agree-
able level of intervention ranges from low to heavy.[13]

THE NECESSITY OF SEMI-PERMEABLE MEMBRANES

Although it is evident that the White Mountain Apache combine many
Euro–American and Apache traditions when conducting restoration projects, they also erect cautious boundaries between such traditions. The
Apache are usually quite critical of Western capitalism, Christianity, and
technological determinism that they see as undermining their traditional
way of life, which is more in tune with natural processes. As Apache restorationist Benrita Burnette asserts, "Apaches see the cause of unhealthy
ecosystems not as the result of traditional philosophies, but as the disruption of those philosophies by external influences."[14]

Apaches are suspicious of restoration goals that overlook economic,
political, and cultural issues. In the early 1990s, for example, the FWS proposed a critical habitat designation under the Endangered Species Act for
the loach minnow, a species of fish that inhabits the streams of the reservation. This action delayed the construction of a project designed to deliver
water to drier parts of the reservation. Apaches found irony in this situation.
Legislation conceived to save endangered species from extinction was possibly hastening the extinction of their people. Thus, what sets Apache restoration projects apart from their non-Indian counterparts are that Western
environmental perspectives often don't account for local community needs.
Such philosophical differences meant that numerous partnerships emerging
between the White Mountain Apache and outside institutions (BIA, Forest Service, FWS, National Wildlife Federation, etc.) remained on delicate
ground. As a result, a dynamic tension between cultural resistance to and
exchange with outside cultures remains a force that shapes the Apache culture and the lands they manage.[15]

CHALLENGES TO EURO–AMERICAN CONCEPTIONS
OF ECOLOGICAL RESTORATION

Beyond their presence as active managers in the pre-European settlement
landscape, current American Indian restoration efforts complicate, challenge, and potentially enrich Western notions of restoration in at least
five ways. First, and perhaps most obvious, is the integration of cultural
and ecological concerns in almost all American Indian restoration projects. Second, the eco-cultural emphasis on restoration has led to a flexible,
context-specific planning approach. Moreover, American Indian restoration efforts demonstrate the necessity of locally designed restoration goals.

For instance, the White Mountain Apache transcend battles over which historical landscape to restore, consequently combining a variety of goals and values to produce eco-culturally hybrid landscapes. The diversity of projects occurring on the Fort Apache Indian Reservation, all approved by the Tribal Council, illustrates the possibility of developing comprehensive, landscape-scale restoration projects that incorporate local, regional, and national needs along with considerations for nonhuman nature. On a local scale, goals and values associated with rewilding, conservation gardening, and wilderness protection are all evident among White Mountain Apache restoration projects. On a scale encompassing the entire reservation—some 2,627 square miles (slightly larger than Delaware)—the patchwork of local restoration projects, regardless of the goals, collectively act to erase past federal mismanagement of their homeland, incorporating eco-cultural elements from both the past and present.

Third, the incorporation of different cultural perspectives is important to the success of restoration projects on reservation lands. In all of the projects mentioned in this chapter, Western science and technology are fused with indigenous knowledge and technology. This has been a potentially fulfilling endeavor within a reservation setting. However, beyond the reservation, skepticism about the efficacy of indigenous knowledge is very common. This skepticism exists despite the methodical efforts of such scholars as Kat Anderson and Nancy Turner who advocate incorporating indigenous knowledge into Western-based land management strategies. Encouragingly, the National Park Service, the Forest Service, the FWS, conservation biologists, and restoration ecologists have begun to acknowledge the importance of indigenous knowledge. But progress towards incorporating such knowledge into management plans is slow and is often met with political resistance. This is apparent in the recent unprecedented move by the FWS in December 2004 to turn half the management duties of the National Bison Range over to the Confederated Salish and Kootenai Tribes (CKST), a group with a respectable record of natural resource management on the Flathead Indian Reservation. Resistance to this action, which began well before the transfer of responsibilities, continues today from within the ranks of the FWS, environmental groups, science watchdog groups, and conservatives seeking to undermine tribal sovereignty. Many of the formal complaints question the scientific competency of the CKST, but these criticisms only mask the territorial nature of the dispute. Equally evident are concerns about job loss for federal employees, loss of federal control over natural resources, and the special status of tribes as sovereign nations. Still, the successful integration of indigenous and Western knowledge on some reservations suggests that resistance to importing indigenous knowledge into the mainstream of Western science is less about epistemological incompatibility than power struggles and economics.[16]

Fourth, American Indian restoration efforts are explicitly political. Historical and contemporary "external threats" to sovereignty and cultural survival often motivate restoration projects on tribal lands, as Casagrande and Vazquez make evident in the next chapter. In mixing science with politics, American Indians realize that the process of building sustainable societies cannot rely too completely on restoring the past. They must take into consideration what has transpired to create their current condition and the constraints it has placed on any plans for cultural or ecological restoration. This mindset towards restoration follows what Casagrande and Vazquez suggest: "Restoration can be thought of less as the quest for a future steady state, and more as a process of continuous communal re-interpretation of conditions from the immediate and long-term past." This process not only involves blending past and present physical features of the landscape but different cultural traditions as well. In creating these landscapes, American Indians employ Western and indigenous techniques and knowledge to shape eco-cultural landscapes in accordance with their history and contemporary physical and political constraints. In doing so, they cautiously attempt not only to protect their cultural identity, but to transcend rigid historical barriers (e.g., the wilderness ideal) and cultural barriers (e.g., Western versus non-Western knowledge systems). As a consequence, American Indian resistance to assimilation ultimately entails cultural exchange with Western society, which leads to the creation of eco-culturally hybrid landscapes on reservation lands.

Fifth, the political nature of American Indian restoration projects also points to the importance of restoring or developing a sense of place. For American Indians, restoring a sense of place is essential to recovering eco-cultural sustainability and cultural survival. This is a truly innovative aspect of indigenous restoration projects—they draw from the past to reconstruct eco-cultural ties to the land. This component of indigenous restoration projects often remains absent from Euro–American projects, which tend to focus solely on ecological or economic issues. Perhaps if Western restoration efforts centered on "becoming indigenous" and promoting eco-cultural survival, these projects might have a lasting public appeal. This is, in fact, beginning to happen in many places, especially in the western United States in areas ravaged by timber over-exploitation.[17] Like American Indian restoration projects, these community-based efforts are ultimately tied to cultural survival. More restoration projects must follow this pattern if they are going to have a lasting effect on solving environmental problems. Restorationist Robin Wall Kimmerer, of American Indian descent, nicely expresses this sentiment of blending indigenous and Western land management goals:

> Traditional ecological knowledge is not unique to Native American culture. It is born of long intimacy and attentiveness to a homeland and can arise wherever people are materially and spiritually integrated with their landscape. The writings of such luminaries as Aldo Leopold in "The Land Ethic" and others in the Western tradition express

this imperative most powerfully. The goal should not be to appropriate the values of indigenous peoples. As an immigrant culture, non-Native Americans must start to engage in their own process of becoming indigenous to this place.[18]

NOTES

1. M. Kat Anderson, *Tending the Wild* (Berkeley: University of California Press, 2005); Marcus Hall, *Earth Repair* (Charlottesville: University of Virginia Press, 2005).
2. Richard White, *Roots of Dependency* (Lincoln: University of Nebraska Press); Charles Wilkinson, *Blood Struggle* (New York: W&W Norton, 2005).
3. Anderson, *"Tending the Wild;"* Ann Garibaldi and Nancy Turner, "Cultural Keystone Species: Implications for Ecological Conservation and Restoration," *Ecology and Society* 9(3) (2003): 1. [online] URL: http://www.ecologyandsociety.org/vol9/iss3/art1 accessed 27 March 2009.
4. Keith Basso, "Western Apache," in *Handbook of North American Indians* (vol. 10), Alfonso Ortiz, ed. (Wash., D.C.: Smithsonian, 1983), 462–88. Quote from Ronnie Lupe, "The Challenges of Leadership and Self-Government: A Perspective from the White Mountain Apaches (Speech)," (John F. Kennedy School of Government, Harvard University: Harvard Project on American Indian Economic Development, October 1992), 2.
5. Arthur R. Gomez, "Industry and Indian Self-Determination: Northern Arizona's Apache Lumbering Empire, 1870–1970," in *Forests under Fire*, C. J. Huggard and A. R. Gomez, eds. (Tucson: University of Arizona Press, 2001), 3–40; Jonathan W. Long, "Restoring the Land and Mind," *Journal of Land, Resources, and Environmental Law* 18 (1998): 51–61; Harold Weaver, "Fire as an enemy, friend, and tool in forest management," *Journal of Forestry* 53 (1955): 499–504.
6. Weaver "Fire as an enemy;" Laura Tangley, "Restoring a Lost Heritage," *National Wildlife Magazine* 41(1) (December/January 2003); Jonathan W. Long, "Cibecue Watershed Projects: Then, Now, and in the Future," in *Land Stewardship in the 21ˢᵗ Century: The Contributions of Watershed Management, Proceedings RMRS-P-13*, P. Ffolliot et al., technical coordinators (Fort Collins, CO: U.S. Department of Agriculture, Forest Service, Rocky Mountain Research Station, 2000), 227–33.
7. Russell A. Calleros and Anna Ling, *Maximizing the Impact of the Restoration Fund: Policies the White Mountain Apache Tribe Can Use to Manage the 22-H Fund, PRS 97-4* (John F. Kennedy School of Government: Harvard Project on American Indian Economic Development, 1997); White Mountain Apache Tribal Council, *White Mountain Apache Land Restoration Code*, available at www.wmat.us/Legal/LandRestorationCode.html.
8. Ronnie Lupe, "Comments of the Chairman of the White Mountain Apache Tribal Council," in Ted Olinger, ed., *Indian Water—1997 Trends and Directions in Federal Water Policy* (Boulder, CO: Report to the Western Water Policy Review Advisory Commission, 1997): 38–44.
9. Long, "Cibecue Watershed Projects."
10. Gail Pechuli, "White Mountain Apache tribe uses BAER to restore charred land," *Indian Country Today* (3 February 2003).

11. Tangley, "Restoring Lost Heritage."
12. John R. Welch and Ramon Riley, "Reclaiming Land and Spirit in the Western Apache Homeland," *American Indian Quarterly* 25 (Winter 2001): 5–12; Keith Basso, *Wisdom Sits in Places* (Albuquerque, NM: University of New Mexico Press, 1996).
13. John Dougherty, "Making a Mountain into a Starbase: The Long, Bitter Battle over Mount Graham," *High Country News* (July 24, 1995).
14. Jonathan W. Long, Aregai Tecle, and Benrita Burnette, "Cultural foundations for ecological restoration on the White Mountain Apache Reservation," *Conservation Ecology* 8(1) (2003): 4. [online] URL: http://www.consecol.org/vol8/iss1/art4 accessed 26 March 2009.
15. Lupe, "Comments of the Chairman."
16. J. Ford and D. Martinez, "Traditional Ecological Knowledge, Ecosystem Science, and Environmental Management," *Ecological Applications* 10(5) (2000): 1249–1250; Dave Egan and M. Kat Anderson, eds., "Theme Issue: Native American Land Management Practices in National Parks," *Ecological Restoration* 21(4) (2003): 245–310; Perry Backus, "Report Critical of Bison Range Management," *Missoulian* (July 7, 2006).
17. Freeman House, *Totem Salmon: Life Lessons from another Species* (Boston: Beacon Press, 1999).
18. Robin Kimmerer, "Native Knowledge for Native Ecosystems," *Journal of Forestry* 98 (August 2000): 4–9, 9.

18 Restoring for Cultural–Ecological Sustainability in Arizona and Connecticut

David G. Casagrande and Miguel Vasquez

The Hopi villages in northeastern Arizona are far removed from the West River neighborhood in New Haven, Connecticut. It is not at first obvious how much these two communities have in common or how important their commonality is. One is a remote Native American tribe in the American Southwest, and the other a mostly African American, inner city community on the East Coast. Both however are economically marginalized communities that share a vision that may be crucial for global sustainability. Members of each community see ecological restoration as a tool for improving social conditions by restoring the natural links between ecological and human cultural processes. We define 'renaturing' as an intentional and reflective attempt to restore human relationships with natural processes of ecosystems in addition to the more common focus on restoring the biophysical health of ecosystems. This chapter compares case studies from these two communities to illustrate why renaturing is a form of ecological restoration that can contribute to sustainability.

Ecological restoration is an intentional attempt to restore damaged ecosystems, and most practitioners look to the past to evaluate the present. Usually they consider humans as inherently destructive agents. Humans however are part of the ecosystem, and restoration provides opportunities to create mutually beneficial relationships between ourselves and the non-human landscape.[1] We advocate a re-definition of 'nature' in which humans and non-human entities are equal partners in a process of continuous co-evolution. Throughout history, social systems have collapsed when cultural reproduction and bio-ecological regeneration have taken separate trajectories.[2] By integrating ecological regeneration and cultural reproduction, 'renaturing' could become a form of ecological restoration that looks to the past and present to evaluate potentially sustainable futures.

In the evaluation of restoration projects we should be asking, "How does this project contribute to sustainability from a long-term cultural perspective?" Many restoration projects that are successful from a biological standpoint or have strong community support may not be sustainable on a long-term basis if they reify the human–nature dichotomy. As environmental anthropologists, we conceptualize culture as the social relationships, moral guidelines, and historical trajectories of identity and practice that provide a template for continuous experimentation in how to live correctly. Cultural reproduction refers to how social relationships are reproduced as people conduct these experiments. Examples of processes of cultural reproduction include cultural transmission of ecological knowledge through moral narratives, intergenerational bonding through ritual interaction with landscape elements, patterns of subsistence, and shared norms of reciprocity and redistribution of resources. Human populations experience social dysfunction, crisis, or extinction when these processes break down.

The Hopi have a long historical continuity with their landscape, although they have experienced a gradual degradation of their sustainable agricultural lifestyle since World War II. Recent efforts to restore the practice of terraced gardening provide an excellent opportunity for exploring how processes of cultural reproduction and bio-ecological regeneration might be integrated in a sustainable manner. We compare the Hopi case to a post-industrial, urban, salt marsh restoration project in New Haven, Connecticut, which included efforts to engage the economically marginalized people who live next to the restoration site. Our comparison suggests that important components of sustainable renaturing include common property, local leadership, the use of local history to respond to external threats to cultural identity, active public engagement with the landscape, and the role of external facilitators.

In both cases, important landscape features no longer exactly fulfilled their historical roles, but nevertheless provided templates for possible future trajectories. People were motivated by tying ideas to a commonly shared landscape feature that was important to cultural identity. Active public engagement with the landscapes helped close the gap between cultural and ecological regeneration. In the case of the Hopi, the gap is not so large. The project in New Haven presents more of a challenge.

HOPI TERRACE GARDENING

The Hopi present a useful and interesting case because of their cultural association with a particular bio-ecological region over millennia. They developed sustainable desert farming by adapting what one author has characterized as a *theology of place*—"the complex of relationship, symbolism, attitude, and a way of interacting with the land" that has derived from their long-term engagement with the surrounding landscape.[3] As one Hopi elder explained, "This is the place that made us." Hopi traditional knowledge demonstrates a profound understanding of the agro-ecological niches of this seemingly inhospitable environment. Over centuries, the Hopi have developed a culture of the commons, based on the cultivation of corn, beans, and other crops, both traditional and modern.

In recent years however, agricultural production has declined precipitously and given way to a relatively impoverished mixed economy of wage work, artisanry, and subsistence ranching, with only limited farming or gardening in most communities. This disconnect from local ecological processes and the subsistence base that is the core of Hopi cosmology, economy, and social organization has resulted in social and economic dysfunction, a decline in human health, and degradation of the surrounding biogeophysical environment.

Restoration of centuries-old terrace gardens by the Hopi provides an excellent example of integrating cultural and ecological restoration, and the revitalization of a once-vibrant cultural commons. In re-engaging youthful Hopi participants with the landscape and elders willing to work with them, ties between local ecological and cultural processes have been strengthened, and intergenerational cultural reproduction of the commons has been enhanced. This has enabled Hopi and their collaborators from the Northern Arizona University Department of Anthropology (NAU) to demonstrate a culturally appropriate paradigm for healthy, mutually beneficial relationships between themselves and the natural landscape.

Decline in Traditional Agriculture and Hopi Cultural Identity

With the pressures of modern lifestyles most modern Hopi have given up challenging the natural limits of the area, dry farming the dunes and washes of surrounding desert and steppe lands, and most who do are over sixty years of age. This decline has mostly occurred in the last twenty-five years. After nearly five hundred years of exposure, accommodation, and resistance to outsiders—Amerindian raiders from other tribes, *conquistadores*, missionaries, government agents, tourists, mining companies, promoters, and academics—only in the past few decades has there been a significant shift from subsistence agriculture to a mixed wage-based economy. Road construction, dramatic expansion of the tribal bureaucracy, new federal or tribally sponsored development projects, and growing national and international markets for native arts and crafts have all created new income opportunities for some Hopi people.[4] While most remain impoverished, the influx of cash, access to commercial food, development on agricultural lands, as well as increased erosion, arroyo-cutting, and persistent regional drought have all served to discourage most families from more than minimal cultivation activities.[5] Monetization of the food supply at Hopi has increased local dependency on the mainstream consumerist, commodity-oriented lifestyle. It has also contributed to the degradation of the cultural and environmental commons that represent concrete "alternatives to the level of consumerism that undermine(s) community-centered traditions of self-reliance."[6]

Terrace Garden Restoration and Sustainability

As agricultural activities declined in the larger fields over the last several decades, another local production strategy also fell into disuse. Sheltered in canyons beneath each village the Hopi have long maintained rock-walled terraces and cultivated spring-fed irrigated gardens.[7] The springs originally permitted human habitation, as well as the possibility of reliable irrigation, in an otherwise arid landscape. In contrast to larger fields around the village and in the washes below, where corn, beans, melons, and squash are dry-farmed by the men, garden work in the one and a half acres which comprise the terraces has been done largely by women and children.

Irrigated terraced gardening was sustained over the long term because Hopi cosmology and custom could not conscionably permit the waste of scarce, life-giving water. Even today the average Hopi household uses ten to fifty gallons of water daily in comparison with the 250 to 400 gallon average for the Southwest as a whole.[8] Archaeological evidence suggests as much as 700 years of continuous terrace gardening, and Spanish chronicles and early Anglo accounts make mention of lush and abundant Hopi gardens. The gardens were an integral part of local subsistence until the completion of potable water systems in the early 1970s. If village residents continued to cultivate gardens at all, they found it much more convenient to water small plots next to their homes from a tap. Most people stopped gardening altogether and the terraces were nearly abandoned for fifteen years.

In 1991 community leaders from Hopi approached anthropologists at NAU for assistance in restoring the terrace gardens to safe and productive use in the community of Paaqavi.[9] Over seven growing seasons, a community-directed, collaborative program of physical restoration, documentation of local practices, and extension for local children developed, which increased local involvement in gardening from two to twenty-five families. Since then, several other communities have initiated terrace restoration projects in collaboration with NAU faculty and students.[10] Although the landscape elements no longer fulfill the "once productive social, cultural, and economic role" that they did previously,[11] they still provide a template for historical trajectories in a larger social project of reconciliation of indigenous and modern lifestyles. While few Hopis can realistically combine contemporary work schedules with full-scale traditional agricultural activities, the gardens provide children and their extended families with an opportunity to connect with the landscape in a manner resonant with Hopi values.

Community leaders across the reservation recognized terraced gardens as a shared feature of the landscape, and their cultural relationship with it as a potential source for motivating young people to hands-on engagement in their communities. Elders were motivated by concern over the loss of language and cultural fluency among young people and of social bonds between generations as a result of distractions of mainstream culture and consumerism—seen as the cause of much of the social dysfunction now all-too-common in Hopi communities.

Garden restoration was also, in part, a reaction to external threats. Concerns over local water supplies due to the rapid draw-down of local aquifers used for coal-slurrying from Black Mesa, the loss of these profitable but controversial mining contracts with imminent power-plant closings, and the contentious use of reclaimed sewer water for artificial snow-making on the sacred San Francisco Peaks, have all led to a sense of disquiet and suspicion locally.[12] By linking ecological restoration of the gardens—a commonly shared landscape feature—with cultural reproduction at the community level, leaders sought to actively re-interpret history and reinforce Hopi cultural identity as a means for, literally, 're-grounding' young people in the 'moral ecology' of the Hopi.

Revitalizing Cultural and Ecosystem: The Benefits of Renaturing

> For the Hopi, growing and caring for plants and working in nature are seen not only as providing food and sustenance, but also as the means to develop both the inner soul and the social faculties. The feeling of working with nature, rather than against her; the encompassing human metaphor—that one cannot harvest anything without first planting and caring for it; the readiness to do whatever is necessary without complaint; a lifestyle oriented to what nature gives, to family collaboration, and to the rhythms of the sun and the seasons—these are ways born of work and service on one's family's lands.[13]

Pueblo scholar Gregory Cajete explains that there is a deep psychological relationship here, "rooted in the inherent focus of Native cultures on participation with nature as the core thought and central dynamic of Native philosophy."[14] "Plants," he says, "present an internalized image of natural life and energy that helps to form our perception of the living Earth. In reality, plants and humans have been biologically and energetically intertwined since the beginning of the human species . . . part of our body memory, conditioned by the oldest survival instinct of humans." Hopi language reflects this interconnection constantly, in a myriad of linguistic root metaphors exemplifying human conditions through floral and botanical images. Parents who participated in the various projects repeatedly expressed fears that in their search for meaning and identity in contemporary reality, their children would look beyond Hopi to the dominant society and that community memory, and ultimately, Hopi culture would gradually erode.[15]

Jordan articulates six essential elements of relations with the natural landscape: the natural landscape itself—natural or historic ecosystems; a reciprocal ecological relationship with these systems with "a genuine exchange of goods and services between ourselves and the natural community"; the engagement of "our physical, mental, emotional, and spiritual capacities"; a "sense of history . . . of our interaction with a particular landscape"; a relationship that is "flexible and capable of a creative expansion and development"; and a means " . . . to articulate and celebrate that relationship in a personally and socially satisfying way."[16]

Hopi terrace renaturing addresses each of these essential elements. First, Hopi terraces have been an integral part of the surrounding natural landscape for centuries. Second, the terrace gardens have provided water, food, medicinal plants, and most recently, a focus for tribal social mobilization. Third, the gardens and the surrounding landscape provide a way to engage physical, mental, emotional, and spiritual capacities through the actual physical work of restoration and maintenance of the site, horticultural planning and implementation, the continual reinforcement of intergenerational and community bonds, and a tangible locus for nature-centered spirituality—the *theology of place* that Cajete refers to.[17] Fourth, restoration of the sites provides old and young alike with memories and oral histories of long interaction with this particular landscape and a focus for youngsters to envision the larger engagement of Hopis, humans, and the natural world. Fifth, use of the terraces for the fusion of cultural and ecological restoration has necessitated "flexible and . . . creative expansion and development" in social mobilization, cultural revitalization, and horticultural innovation. Sixth, activities at the terraces, according to participants, provided family members, neighbors, and visitors, other villages, other tribes and communities, and NAU students and faculty opportunities to collaborate in productive, personally, and socially satisfying ways. Among the benefits of the project that participants and community leaders cited were: ecosystem health; food and nutritional gains; physical health; intergenerational learning for social cohesion and traditional knowledge; self sufficiency; spiritual health; language revitalization; new vocational and professional prospects; and, possibly at some point in the near future, an economic multiplier effect.[18] By reengaging the cultural with the ecological commons, Hopi garden restorations not only restore an important historical landscape, they provide a continuing link for a re-interpretation and renewal of ethnic identity, leading at the same time to other areas of individual motivation and vocational development as well as intergenerational and community mobilization.

WEST RIVER, NEW HAVEN

In colonial New England, coastal salt marshes were harvested for hay and productively managed as common property. As the industrial revolution progressed, the marshes were vilified, privatized, and 'reclaimed.' In 1924, West River Memorial Park was established in New Haven, Connecticut, and included a 54-hectare salt marsh. Tidal flow in the park was intentionally restricted to drain the marsh to eliminate mosquitoes and create upland recreation sites. This process failed, and instead the marsh was transformed into dense stands of *Phragmites australis* as water salinity dropped.[19] This reduced salt marsh grasses, bird habitat, water quality, and the movement of nutrients and marine organisms to and from the estuary.[20] In the early 1990s, the Connecticut Department of Environmental Protection, the city of New Haven, and several universities in and around New Haven developed a plan to restore tidal flow to West River Memorial Park. This plan faltered when municipal politicians became concerned about flooding. The low-income, mostly African American West River neighborhood borders the park to the east. It is separated visually from the river by stands of *Phragmites australis* and access to the park is cut off by a busy four-lane state highway. In 1996, many residents did not know that a river existed only several hundred meters from their homes, and many residents expressed fear of local nature. Only a handful of residents used the park for fishing. As opposed to the Hopi case, the gap between humans and nature in the West River Neighborhood was wide.

Originally designed by Frederick Law Olmsted the park nevertheless had potential to serve as a template for an historical trajectory. While Olmsted envisioned parks as a tool for breaking down social class divisions, the situation along the West River had become the opposite, with dense stands of *Phragmites australis* serving as a racial and social barrier between neighborhoods.

Here, 'how to live life correctly,' was historically linked to the industrial ideal of hard work and pursuit of the 'American Dream,' requiring education, role models, and the accumulation of capital. Industrial collapse in New Haven from the 1960s through 1990s led to capital flight, job loss, high crime, and deterioration of infrastructure and schools. The West River Neighborhood suffered from severe intergenerational breakdown. As was typical of American inner cities in the 1990s, many children were being raised by grandparents because parents were either incarcerated or were victims of gang violence or the crack-cocaine and AIDS epidemics. A recurring theme in conversations was how generations had little in common with each other. Kinship ties and typical American economic household organization were severely compromised so that children had limited access to the intellectual or social skills needed to participate in the legitimate economy.

The reallocation of school funding to construction of new suburban schools has led to deterioration of many urban schools throughout the United States. Also, environmental information, which often feeds back to individuals or small communities in other societies, can become highly institutionalized in urban Western culture. Myriad health, education, economic development, environmental and infrastructural agencies had a stake in the neighborhood, but failed to communicate with each other. These typical processes left West River neighborhood residents economically marginalized, socially disrupted, and de-coupled from natural elements in their landscape.

In 1995, Yale's Center for Coastal and Watershed Systems initiated meetings with leaders of the West River Neighborhood Association to explore opportunities for reciprocal relationships between the restoration project and local needs. Neighborhood leaders were focused on poor school funding and environmental injustice issues, citing suburban growth and diversion of resources as a primary external threat. The deterioration of Barnard Elementary School—immediately across the state highway from the park—served as a powerful symbol of social injustice.

During meetings with community leaders, Yale researchers presented the history of the park, including Olmsted's original vision, and information on salt marsh ecology and hydrology. Neighborhood leaders were intrigued by metaphorical similarities between Barnard School's history of degradation and West River Memorial Park vis-à-vis the demise of urban education and wetlands in general. Degradation of water quality and habitat became a symbol of environmental injustice as the neighborhood adopted the salt marsh restoration. As one resident expressed it: "Our children, like the river, have been forgotten." Residents recognized potential restored biodiversity as something other low-income neighborhoods lacked; thus integrating ecosystem health with their neighborhood identity. As with the Hopi, they saw opportunities in the landscape, but here these were tied to historical narratives of urban decay and the history of environmental degradation, and the park and the salt marsh were seen as a means for motivating and mobilizing the community. The perception of external threat in the form of taxes and reallocated school funding and an identity based on "us against them" also encouraged participation.

Yale researchers and neighborhood leaders collaborated to develop a proposal to turn Barnard School into an environmental education magnet school to leverage state funding for renovations and to use the restoration as a living classroom. Plans included ecotourism with canoe rentals, community gardens, and increased access for fishing and other recreational activities. Barnard School re-opened as a successful environmental education magnet school when renovations were completed in 2006. As interest grew at the state level, funding was added for a grass roof and an 80-kilowatt solar power installation—the largest solar project in New England. A pedestrian skywalk across the state highway was constructed to reconnect the neighborhood with West River Park. Eradication of *Phragmites australis* has opened vistas to the river, and public access points have been established.

As at Hopi, community leaders recognized the motivational potential of linking the re-interpretation of history to a commonly shared landscape feature. Restoration here includes linking ecosystem services like flood protection, water quality, estuarine productivity, and wildlife habitat to cultural reproduction through children learning stewardship of a neighborhood landscape feature and will also lead to greater ethnic integration as magnet students come from other parts of the city. Hands-on engagement with the landscape will provide students with a restorative, embodied nature-based praxis. It also includes essential relational elements with the landscape as proposed by Jordan, engaging the physical, mental, and emotional capacities of neighborhood residents with the historic salt marsh ecosystem.[21] Through ongoing education, the restoration reinstitutes a sense of history of interaction with the landscape and the deeper history of the general relationship of our species with the rest of nature. Cultural reproduction, including identity and cultural transmission of ecological knowledge, has been tied to ecological processes by combining narratives of ecological and social degradation to the marsh and bringing together diverse local actors, educators, scientists and activists. This relationship is flexible, expansive, and allows for ongoing intellectual advances and local cultural evolution.

Unlike the Hopi case, however, the West River project lacks a clear spiritual dimension. It is also more of a challenge to create economic transactions that entail a genuine exchange of goods and services between the people and the ecosystem. Neighborhood use of the park has increased, including participation in fishing. The West River's annual "Canoe Day," which was created as part of the restoration in 1996, draws participants from throughout the area and provides prospects for ecotourism development. But long-term sustainability requires prolonged engagement with the park, leadership, intergenerational bonding, and economic reproduction. City-wide perceptions of the neighborhood are more favorable due to views of the river, which appears to be increasing property values for neighborhood homeowners. As property values rise, however, the neighborhood may experience gentrification. Also unlike the Hopi project, the New Haven project does not integrate natural resources into reciprocal relationships among people and lacks links to kinship. Intergenerational relations are weak because adults are not all directly engaged in the school curriculum. The project facilitates reciprocity at the institutional level, but not necessarily within the community.

LESSONS LEARNED FROM THESE CASE STUDY COMMONALITIES

We had no preconceived notions about commonalities in our case studies. Several salient similarities emerged as we related our stories, especially the observation that commonly shared landscapes were recognized as potential resources in mobilizing people against external threats.

The Importance of History

Humans evaluate the present based on past experience and project these interpretations into the future.[22] This is as true for communities as it is for individuals, where discourse explaining the present must reconcile interpretations of the long- and near-term past. Thus, personal and cultural identities are ongoing, negotiated re-interpretations of history. Metaphors using tangible, non-human entities often play a vital role in elucidating abstractions such as social relationships, historical explanations, and future possibilities. In New Haven, *Phragmites* stands were a visual metaphor for social segregation and economic marginality. Local activists wove the Olmsted promise of parks for social healing into the decay of the urban school system. Hopi narratives admonish living life correctly and are linked with rituals aimed at keeping the cosmos in order. The community-based terrace garden restoration re-grounds young people by reinforcing Hopi cultural identity and moral order. Because this is a collaborative effort in a commonly shared area in which ceremonial plants are grown for exchange, the gardens also reinforce social reciprocity. Both case studies suggest historical reinterpretation can be a powerful motivational force when attached to landscape features. Renaturing can contribute to sustainability because we can learn about ourselves as we learn about the ecosystems in which we live. As a Hopi farmer planting his cornfield explained, "If we gonna stay in a straight line, we gotta look back every now and then to see where we came from." The power of restoration in these cases lies in tying this process to how people think about past uses of landscapes.

Keeping it Local

Both these restoration efforts were initiated by local communities in direct, politically motivated relationships with their landscapes. Through what Clewell called "place-based restoration," the democratic potential of ecological restoration is realized with small, local organizations rather than larger, bureaucratic organizations.[23] This follows Tainter's argument that marginal returns of additional information decrease with increasing social complexity.[24] Since cultural reproduction of industrial societies is highly specialized and stratified, such systems usually respond to an expanding suite of environmental challenges by increasing social complexity, often with disastrous results. As increasingly complex human systems become informationally decoupled from non-human systems, they are less able to respond to environmental change.[25] This is exacerbated by the West's expanding nature–human dichotomy and technological hubris. Restoration, if it is localized or community-based, can reduce complexity, circumventing its negative effects through reiterative interpretation of local history.

Community leaders provide face-to-face negotiation with less hierarchical political complexity, and followers are more likely to trust those with whom they share experiences. Restoration projects that lack local leadership often fail to build enduring public support or cultivate public interest in the ecosystem. Such efforts require time, money, and energy, and may not be sustainable because the target population does not share a cultural identity with the external restoration leadership. In many cases, restoration ecologists themselves may be viewed as a threat.

The Importance of the Commons

Our evolved capacity for cooperation determines that few of us will voluntarily participate in a project where benefits accrue disproportionately to someone else, including ecological restoration. Restoration may be appropriate for private property, but renaturing is probably more feasible when cultural identities are based on common experiences with commonly held resources.[26]

The Hopi case shows the importance of common property for social healing and reciprocity. The Hopi terrace gardens are individual plots passed down matrilineally, but they are constructed in a common area. Children learn gardening from their grandmothers, mothers, and aunties, as well as proper moral behavior, kinship, and gender roles, at the same time that they are learning hydrological and ecological principles. The pools, stairs, and rock walls are maintained communally, and 'how to live correctly,' as an individual Hopi and in cooperation with other Hopi, is symbolically encoded there.

Likewise, harvesting of salt marsh hay by early colonists in Connecticut was based on management of the resource as a commons, and the salt marsh in New Haven remains in a public park. The history of this landscape as common property underlies the ability for leaders to tie urban morality and cultural reproduction to Olmsted's original vision of landscapes ameliorating social injustice.

These cases point out the importance of tying perceptions of social organization and cooperation to shared elements of the landscape, thus creating positive historical trajectories (Figure 18.1). This is less likely to happen if the landscape feature is privatized and potential participants see that specific people will experience a disproportionate benefit.

The Nature-embodied Community

Kidner in Chapter 23 states that sustainability is not so much conscious as it is embodied in a person's entirety. We emphasize the social component of this notion. Humans evolved not as individuals, but as socially interacting groups.[27] Many of our most advanced mental faculties deal more with relations with other people than an ability to solve problems as individuals. "Natural cooperation" might be considered a "third fundamental principle of evolution beside mutation and natural selection," implying that sustainability requires nature to be embodied in social relationships as well as individual minds. Human embodiment of nature also requires active engagement with the landscape.[28] All humans share a basic propensity for

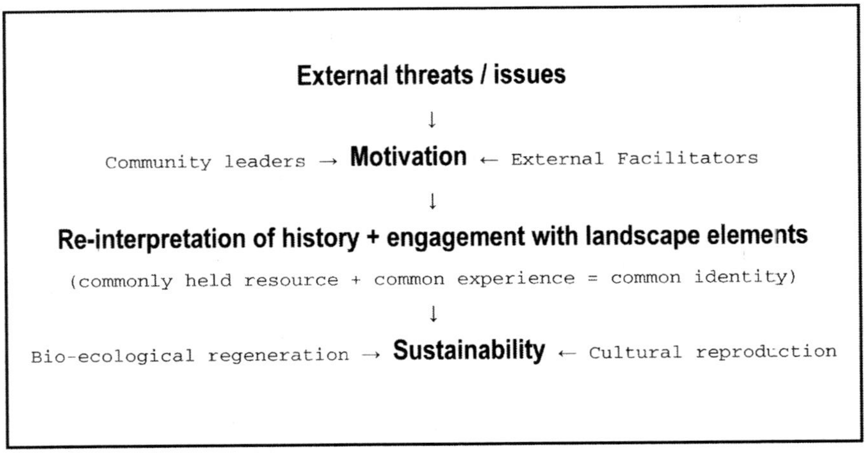

External threats / issues

↓

Community leaders → **Motivation** ← External Facilitators

↓

Re-interpretation of history + engagement with landscape elements

(commonly held resource + common experience = common identity)

↓

Bio-ecological regeneration → **Sustainability** ← Cultural reproduction

---

Figure 18.1   A Cultural-Ecological Restoration Model derived from the Hopi and New Haven Case Studies.

nature-based ritual, and research supports the psychological and physiological restorative power of interacting with nature.[29] Therefore, cultural-ecological sustainability requires both social cooperation and engaged interactions with the landscape. Renaturing contributes to sustainability by creating 'nature-embodied communities.'

External Facilitators

Since marginalized people often lack resources to implement these processes and local communities are often highly resistant to engage external organizations, researchers, government agents, consultants, or others can become crucial facilitators for communicating and finding external resources. This works best when practitioners engage in clear reciprocal relationships with the local population.[30]

CONCLUSION

Social processes, and their linkages to bio-ecological processes, should be considered an important component of restoration. Potential impediments to sustainable renaturing include hierarchical political complexity, privatization of restoration, and the negation of history. Means to overcome these include focusing on common property, promoting local leadership, using local history to respond to threats to cultural reproduction and identity, engaging the public with landscapes, and developing reciprocal relationships between local leaders and external facilitators.

Many opportunities await restorationists interested in long-term, cultural-ecological sustainability. We are proposing a model of environmental information feedback that includes the continuous, localized re-interpretation of history and cultural reproduction. This emphasizes not so much a particular habitat or species, but a coupled human–nonhuman system of information flow from particular habitats. The examples here are not panaceas for the global sustainability crisis we may be facing. Nor is this the only (or best) way to conduct ecological restoration. Our approach is meant to stimulate discussion rather than prescribe. These projects draw on work done in marginalized communities and our model may not apply to more affluent communities unlikely to be sensing threats to their cultural reproduction. Also, our analysis captures only a small part of cultural reproduction for these two groups. However, some aspects, like the focus on common property, engaged interaction with nature, and local leadership, are likely to be applicable on a broad scale. This comparative exercise demonstrates how restoration research could include sustainability in evaluation of projects by considering the following criteria: how well a project makes ties between ecological and cultural processes explicit to participants; whether it strengthens those ties; and whether it contributes to cultural reproduction as well as enhancing ecological processes. Important innovations come from marginalized populations if others are willing to listen. These communities are explicitly linking ecological processes with the reproduction of their cultures. We might consider their experiences in our search for new ideas about renaturing for cultural-ecological sustainability on a global basis.

NOTES

1. William Jordan III, "'Sunflower Forest': Ecological Restoration as the Basis for a New Environmental Paradigm," in D. Baldwin, J. DeLuce, and C. Pletsch, eds., *Beyond Preservation: Restoring and Inventing Landscapes* (Minneapolis: University of Minnesota Press, 1993), 17–34; David G. Casagrande, "The Human Component of Urban Wetland Restoration," in D. G. Casagrande, ed., *Restoration of an Urban Salt Marsh: An Interdisciplinary Approach*, Bulletin No. 100 (New Haven, CT: Yale School of Forestry and Environmental Studies, 1997), 254–70; Miguel Vasquez and Leigh Jenkins, "Reciprocity and Sustainability: Terrace Restoration on Third Mesa," *Practicing Anthropology* 16:2 (1994): 14–17.
2. Charles L. Redman, *Human Impact on Ancient Environments* (Tucson: University of Arizona Press. 1999); Joseph A. Tainter, *The Collapse of Complex Societies* (Cambridge, UK: Cambridge University Press, 1988).
3. Gregory Cajete, "Look to the Mountain: Reflections on Indigenous Ecology," in G. Cajete, ed., *A People's Ecology: Explorations in Sustainable Living* (Santa Fe, NM: Clear Light Books, 1999), 3–20.
4. Miguel Vasquez, "The Hopi of Arizona," in Tom Greaves, ed., *Endangered Peoples of North America: Struggles to Survive and Thrive* (Westport, CT: Greenwood Press, 1998), 79–95.

5. Vasquez, "Hopi of Arizona;" Shawn Kelley, "The Role of Applied Anthropologists in Sustaining (Hopi) Agriculture," (M.A. Internship Thesis, Department of Anthropology, Northern Arizona University, Flagstaff, AZ, 2006); Gregory Glassco, "Feasibility Study of Hopi Terrace Garden Construction and Operation Below Suvipa Spring," (MA Thesis, Department of Anthropology, Northern Arizona University, Flagstaff, AZ, 1999).
6. C. A. Bowers, "Educating for a Sustainable Future: Mediating Between the Commons and Economic Globalization," http://cabowers.net/CAbookarticle.php, 2005.
7. Vasquez and Jenkins, "Reciprocity and Sustainability."
8. Masayesva, personal communication.
9. Vasquez and Jenkins, "Reciprocity and Sustainability."
10. Glassco, "Feasibility Study;" Kelley, "Role of Applied Anthropologists."
11. Vasquez and Jenkins, "Reciprocity and Sustainability."
12. Vasquez, "Hopi of Arizona."
13. Vasquez and Jenkins, "Reciprocity and Sustainability."
14. Gregory Cajete, *Native Science: Natural Laws of Interdependence* (Santa Fe: Clear Light Books, 2000), 108.
15. Vasquez and Jenkins, "Reciprocity and Sustainability."
16. Jordan, "Sunflower Forest."
17. Cajete, "Look to the Mountain."
18. Kelley, "Role of Applied Anthropologists."
19. *Phragmites* spp. are reed-like wetland plants common to both case studies. Cultural values attributed to this genus vary radically. The Hopi cultivate *Phragmites* in their terraced gardens for utilitarian and ceremonial purposes. Paaqavi, the Hopi name of one of the villages in this study, means 'place of reeds.' In Europe, large, dense stands of *Phragmites australis* are considered important habitat for species like the reed warbler. *Phragmites australis* is native to eastern North America, but ecologists generally consider large, dense stands of *Phragmites australis* to be biologically impoverished, anthropogenic habitats. Many coastal restoration projects are attempts to eradicate *Phragmites australis* to enhance wildlife habitat.
20. David G. Casagrande, "The Full Circle: A Historical Context for Urban Salt Marsh Restoration," in D. G. Casagrande, ed., *Restoration of an Urban Salt Marsh: An Interdisciplinary Approach*, Bulletin No. 100 (New Haven, CT: Yale School of Forestry and Environmental Studies, 1997), 13–40.
21. Jordan, "Sunflower Forest."
22. L. W. Barsalou, "Deriving Categories to Achieve Goals," in *The Psychology of Learning and Motivation*, vol. 27 (New York: Academic Press, 1991), 1–64; W. Kempton, "Lay Perspectives on Global Climate Change," *Global Environmental Change* 1 (1991): 183–209.
23. A. Clewell, "Downshifting," *Restoration and Management Notes* 13 (1995): 171–75; A. Light, "Hegemony and Democracy: How politics in Restoration Informs the Politics of Restoration," *Restoration and Management Notes* 12 (1994): 140–44.
24. Tainter, "Collapse of Complex Societies."
25. J. R. Stepp, E. C. Jones, M. Pavao-Zuckerman, D. Casagrande, and R. K. Zarger, "Remarkable Properties of Human Ecosystems," *Conservation Ecology* 7 (2003): 11.
26. M. A. Nowak, "Five Rules for the Evolution of Cooperation," *Science* 314 (2006):1560–1563; L. Cosmides and J. Tooby, "Cognitive Adaptations for Social Exchange," in J. H. Barkow, L. Cosmides, and J. Tooby, eds., *The Adapted Mind: Evolutionary Psychology and the Generation of Culture* (Oxford: Oxford University Press, 1992), 163–228.

27. R. I. M Dunbar, "The Social Brain Hypothesis," *Evolutionary Anthropology* 6 (1998): 178–90.
28. Nowak, "Five Rules" 1563; Jordan, "Sunflower Forest."
29. T. Hartig, P. Bowler, and A. Wolf, "Psychological Ecology: Restorative-environments research offers important conceptual parallels to ecological restoration," *Restoration & Management Notes* 12 (1994): 133–37.
30. Vasquez and Jenkins, "Reciprocity and Sustainability."

19 Models for Renaturing Brownfield Areas

Lynne M. Westphal, Paul H. Gobster,
and Matthias Gross

While the term "restoration" is widely used in the United States and Europe, many projects and activities falling under this rubric might more appropriately be labeled "renaturing." *Restoration* often aims to recreate presettlement conditions (in the United States) or some other chosen point in the past. We are not alone in questioning this focus; many of the authors in this volume challenge and evaluate the use of a single historical point to frame restoration activities. We find that "restoration" is especially problematic in urban situations, where the settlement activity impacts soils and nutrients, fragments land cover, alters hydrology, and can change human values for the land and thereby seriously restrict hopes of returning a site to historic conditions with any degree of authenticity. For example, a prairie restoration at the scale of a nature garden or even one 50 acres (20 hectares) in extent will never be home to a bison, the prairie's keystone species. Instead, most projects focus on recovering or reintroducing the key flora of a target community and hope to attract smaller fauna such as butterflies and reptiles. A dune restoration cannot be given the freedom to shift across a park road or into a neighbor's backyard. Instead, plant communities in the urban world are necessarily fixed in space and any movement of elements in the community must take place within site boundaries. And while prescribed burning may be used to manage the understory of an open oak woodland or savannah restoration, setting back succession with a stand-consuming crown fire is not in the urban restorationist's playbook. While the historical context can provide ideas of what plants to use or stream shapes to reconfigure, we also need creative, contemporary strategies to deal with the new realities of climate, species arrivals (be they invasives or endangered species finding new habitat), fragmentation, changes in human population and settlement patterns, industrial legacies, altered hydrology, and other unknowns involved in bringing back ecological structure and function.

In coming to grips with such realities, a number of emerging models have been posed as alternatives to traditional restoration concepts and are premised on what functions best in a landscape for different objectives, such as the provision of habitat, recreation, or ecosystem services. These renaturing models include "reconciliation ecology," "ecological rehabilitation," "designer ecosystems," "new wilderness," and "invented nature."[1] Common to each of these renaturing models is the acknowledgment that the issues are more complex than returning a landscape back to a single desired ecological reference point. Renaturing activities in Europe and increasingly in the United States often must grapple with accommodating the multiple layers of history present at a site, histories that are difficult to separate into discrete categories of "natural" and "cultural."

In this chapter we examine how urban renaturing activities in the United States and Europe are attempting to incorporate the natural and historical legacies of sites into plans for current use. Two research settings are highlighted, both focused on renaturing in an industrial setting: the Calumet Initiative, a project to revitalize the rustbelt landscape of the southeast side of Chicago and northwest Indiana, USA; and the renaturing of open cast mining operations around Leipzig, Germany. These two projects make it clear that we value in some way what was lost and/or what might be gained by increasing the number or viability of the plant and animal communities and the biogeophysical processes upon which they depend. This is fundamental: restoration and renaturing are inherently human activities based on individual and group attitudes, beliefs, and values.

RESEARCH SETTINGS

The Calumet Initiative

The Calumet region is extensive, covering 160 square miles (414 square kilometers) of northeast Illinois and northwest Indiana in the United States. It includes 10 percent of the city of Chicago, many of its southern suburbs, Indiana communities including Gary and Hammond, and the Indiana Dunes National Lakeshore. The Calumet Initiative is a partnership of government, non-profit groups, business, and local residents aimed at revitalizing the local economy and environment. The Calumet region was the largest area of heavy industry in the world after World War II. It is still an industrial region, but one with many brownfields and contaminated sites. It is also home to many wetlands, prairie, dune and swale, savannah, and woodland patches. These range in size from the very small to the 15,000 acres (6070 hectares) of the Indiana Dunes National Lakeshore. They also range in health from severely degraded to robust natural systems.

At the urging of local residents, funding was provided to study the Calumet region for inclusion in the U.S. National Park system. The study indicated that much of the region fit the criteria for a national heritage area, having sufficient cultural and natural assets to warrant protection. Some of the cultural resources include the former company towns of Pullman in Chicago and Marktown in East Chicago, Indiana. These were built at the turn of the last century and are still distinct communities today. The Underground Railroad is believed to have had safe houses in the region, and efforts are underway to identify their locations and preserve the sites. While there was significant support for designating sections of the Calumet region a Heritage Area, there was also opposition. Today, the debate continues on this and other projects and developments.

Calumet also has a rich natural history. In the late 1800s and early 1900s, University of Chicago botanist Henry Cowles conducted his investigations of Calumet's dunes and bogs, leading to his contributions to early ecological theory in plant community succession. In the early 1900s, another researcher from University of Chicago, Norma Pfeiffer, found *thismia americana* in Calumet, the only known location of this plant in North America. The plant was last seen in 1916. For several years in the mid 1990s, annual *thismia americana* hunts were conducted. Volunteers came from miles around in hopes of rediscovering this tiny plant. Their efforts did not yield the *thismia*, but other plant species were identified and added to Calumet's list. In 2002, the Calumet Bioblitz amplified this species hunt, and in a 24-hour collection period, over 130 scientists and scores of local volunteers scoured two Calumet sites for as many species as possible. The results: Over 2200 species of plants and animals were identified, some newly recorded for the region. Together, this pastiche of industry, nature, and the history of both, is Calumet.[2]

Renaturing of Open Cast Mining Operations Around Leipzig, Germany

Around Leipzig, open cast mining of coal has utterly transformed the landscape within the span of a few decades drastically disturbing centuries-long cultural patterns of farms and woodlands. Now eastern Germany's economic shifts and a shrinking population have opened the way for the renaturing of mining sites, creating the *Bergbaufolgelandschaft*, or post-mining landscape of lakes, woodlands, and other "natural" places for the area's 150,000 residents and anticipated visitors and tourists.

The approach to creating mines in the former German Democratic Republic was to destroy villages to expand the mines, often relocating an entire village's population within a single apartment block. Altogether 25,000 people were resettled between the 1950s and the late 1980s, and seventy villages and small towns were wholly or partially destroyed.[3]

The end of mining was welcomed by many. As one resident stated in an interview: "In 1990 it did not matter what was suggested as an alternative for mining, as long as the mining, the pollution, and the destruction of the land was stopped."[4] Not everyone felt this way, especially some of the former mining employees who lost their jobs, but the closing of the mines created a new beginning in the hearts and minds of Leipziger residents.

Beginning in the early 1990s, many of the open cast pits were flooded, some by naturally rising water tables, others filled with industrial waters from neighboring active mining pits, and still others filled with water diverted from adjacent rivers. These changes have been taking place rapidly, with seven lakes to be created by 2012 in the immediate vicinity of Leipzig, and more throughout the Freestate of Saxony. Several lakes have been completed already, including Lake Cospuden, offering boating, swimming, hiking, and other recreational activities. The lakes range from 69 to 2258 acres (28 to 914 ha). Along with the increased surface water, woodlands are also expected to increase. In 1996 only 17 percent of the mining-counties' landscapes were woodland, but by 2050 it is expected that some 40 percent of these landscapes will be covered with woodland and new forests. By contrast, in 1996, 31 percent of the landscape was farmland, but by 2050 it is expected that only 20 percent will be farmed. These changes also reflect the expected patterns of human land use in the future.

Certainly this renatured landscape is substantially different from that which existed before mining. The new nature replacing the closed mines is created by using pre-mining landscapes as a guide, and emulating lakes formed through natural processes. While this large-scale effort is not restoration in the sense of returning an area to some earlier condition, the past can nonetheless guide current projects, as by suggesting plant species (including rare and native species) and approaches to managing wildlife as it returns. This approach is taken in some areas of the former mining landscape south of Leipzig, where natural lakes of the region are used as a design template for the now-flooded pits in an attempt to fit them into the landscape as if they had existed in the nineteenth century, that is, as if nature, itself, had built the lakes.

The end of mining has opened space for hopes and dreams for the future, something that was missing during the mining era. And the ways in which these aspirations are expressed indicate the strong connection between the land and the expression and creation of culture. One example is found around the new Lake Störmthal. This lake will be fully flooded by 2012, but eight annual Störmthaler Seefests, or lake festivals, have been held already, celebrating a yet-to-exist lake. Other indications of the new possibilities created by the new lakes are residents opening wind surfing shops and other businesses counting on the recreational and tourist business opportunities created by the new lake district.

RENATURING MODELS

These settings and the projects taking place across the Calumet and Leipzig landscapes vary considerably, underscoring the fact that not all restoration or renaturing projects are the same. The presence of native and exotic species and the type and extent of human presence in the landscape differ considerably even within these two settings. Gobster articulated several types of renaturing projects he found in the urban parks he has studied.[5] These include (1) the "classical" model, (2) the sensitive species model, and (3) the habitat model. Westphal and Gross find examples of these three in the industrial legacy sites they work in, but add two additional models, (4) the cultural landscape restoration model, and (5) the rehabilitation model. These models are discussed here, and while they are not meant to be an exhaustive list, they illustrate the different types of approaches, questions, values, and needs being addressed by renaturing work worldwide. Each model is described, followed by an example from Leipzig or Calumet.

1. "Classical" model: This type of renaturing builds on significant remnant indigenous flora that are left because the land has been protected from development for a variety of reasons (e.g., steep hillsides, wetlands). Renaturing of these urban sites conforms most closely to a "classical" model of ecological restoration, where native plant diversity is maintained and enhanced through invasive species control and replanting, though these activities are sometimes accomplished in uncommon ways to deal with structural and social constraints.

Example: Powderhorn Lake, Marsh, and Prairie is a 175 acre (~71 ha) protected Illinois Natural Area with intact oak savannah and dune and swale structure. The site provides habitat for threatened plants and animals, while also providing fishing and other recreational activities. It is one of the rare places where one can stand on native soils in the industrialized region of Chicago's Calumet. Restoration on the site has focused on preserving the rare ecosystem, and has been done through many volunteer hours of brush cutting, prescribed burns, and other "boutique" methods for removing invasive plant species. For several years, volunteer restoration activities were halted due to controversy over the use of pesticides.[6] The restoration moratorium has been lifted, and restoration work is again underway. Powderhorn was one of the sites for the Calumet Bioblitz mentioned earlier.

2. Sensitive species model: Some urban sites harbor plant or animal species that have been identified as rare, threatened, or present some other reason of conservation interest to the region. In contrast to the plant community focus of the classical model, restoration of these sites focuses in significant part on protecting and enhancing the populations of these sensitive individual species. The weight these species are given in site management invokes a kind of "ecological primacy" that makes the existence of incompatible exotics and access by incompatible uses much less negotiable. This primacy is sometimes controversial when sensitive species are re-introduced into a restoration area where they have been extirpated, and may be seen by some critics as a move by restorationists to close off public park space to a special interest.

Example: Calumet's Indian Ridge Marsh and the wetlands immediately around it are managed for the state endangered Black-crowned night heron.

The marsh hosts one of Illinois' most important rookeries for the herons, where the birds use the structure provided by phragmities (an invasive with both native and exotic varieties) for their nests. The need for a viable rookery takes precedent over other considerations, including other potential uses for the site, such as recreational trails. And because the herons use the phragmities, typical restoration techniques like the removal of invasive species are also precluded, at least until the site has alternative nesting structure for the rookery.

3. Habitat model: More broadly conceived than the sensitive species model, the habitat model of restoration aims at providing the appropriate set of conditions for a range of species of interest such as neotropical song birds or wetland dependent birds. Non-native species that provide food and cover are often tolerated or even added in habitat projects, though there is increasing concern in avoiding plants that may produce invasive monocultures.

Example: One of the prominent features of the landscape south of Leipzig is a large hill, called Trages heap. This hill does not date from the glacial age, but was made from the overburden removed from the local coal mines. It was planted with trees and now is heavily vegetated. A viewing tower was built on top and hiking is permitted on some trails and the one road. The trees are now mature enough for timber harvest. Once-familiar species of plants and

Figure 19.1 The Calumet Region is a pastiche of industry, nature, and the history of both. (Photograph by Lynne M. Westphal)

animals (including deer, hare, and boar) have re-colonized the hill, increasing the local biodiversity. The tree planting renatured the overburden pile creating habitat, as well as economic and recreation opportunities.

4. **Cultural Landscape Restoration model:** In this type, the role of humans in shaping the landscape is more widespread than recognized in other renaturing models. The Cultural Landscape Restoration model recognizes the importance of human endeavors in a region, the values people have for the land and their history as reflected on the land, as well as for the plant and animal assemblages that make use of the land. The European Landscape Convention from 2000 defines landscape as "an area, as perceived by people, whose character is the result of the action and interaction of natural and/or human factors."[7] This definition is valuable since it not only addresses the negative impacts of people on the "natural" landscape, but also the potential benefits, the values attached to nature, and the ways people contribute to the distinctiveness of the landscape. In Chapter 17 in this volume, Tomblin discusses cultural restoration, and the use of American Indian restoration activities as an act of resistance against threats to a culture or subculture. This can also be the case in the Cultural Landscape Restoration model, when organizing around natural areas is also an expression of identity and/or used as a tool in negotiating with external entities.

Example: Leipzig's new lake district is creating new cultures as well as new landscapes. The lake festivals and restaurants located along lakes that don't yet exist indicate the power that even potential positive change in the landscape—the creation of new natures—can have for people, particularly when local cultures have been stressed and dramatically changed by powers outside their control. The creation of new cultures, however, does not wipe the old cultures from the landscape. An Association of Lost Villages has formed, holding village reunions and safeguarding artifacts from lost villages. In one case, residents saved the town's church bell. When the lake is completed, they plan to build a replica of the church steeple on an island where the church stood, and reinstall the bell in the steeple. Tours of completed lakes include what had existed before: "here was the school, here was the church." This landscape carries with it the old and the new, allowing for new futures and new cultures while also carrying memories of the past.

5. **Rehabilitation model:** In this model, the landscape has been so degraded that restorationists are working almost from scratch to reestablish natural systems and assemblages. Brownfields in the form of mining areas, gas stations, air strips, and landfills have all been reclaimed with native vegetation, (re)introduced streams and lakes, and other "natural" features. Sites in this category often have contamination that must be addressed. Planting and renaturing can help, as in the case of phytoremediation and tree planting to address mine tailings. In other cases, renaturing might need to be restricted so as not to introduce contaminants into the food chain.

Figure 19.2 This is a Leipzig restoration site, Germany: *Cultural Landscape Model.* (Photograph by Matthias Gross)

Example: Calumet's Indian Creek did not exist as a creek or stream before European arrival. It was formed when the existing wetlands were filled over the decades with slag and cinder from local industry, leaving a ditch that was optimistically named Indian Creek. There is some historical evidence that it was maintained as a channel through the wetland by the indigenous Americans, the Potawatomi, but it was never a creek. At the beginning of the Calumet Initiative, Indian Creek was an aquatic dead zone. Research determined how to improve the aquatic habitat by making it more stream-like. This meant changing its shape to increase flow speed so that calcite that leached from the slag would flush from the stream bed, and adding pools and riffles to provide resting and feeding places for fish. This transformation has taken place on a part of Indian Creek, with some promising early results.

CONCLUSIONS

These models suggest the variation that is possible among restoration and renaturing projects, especially those done in urban, industrial legacy areas.

216   *Lynne M. Westphal, Paul H. Gobster, and Matthias Gross*

These five models by no means describe the wide range of renaturing projects. But these models do make clear that humans undertake restoration and renaturing projects because they meet human needs and values. Even those who focus on restoration and renaturing for the sake of non-human nature do so because there is something about non-human nature they value. The examples, provided with the different models, illustrate some of the underlying values that motivate the restoration and renaturing projects. One value that can vary across these models is the acceptability of evidence of human beings. This is generally less acceptable in the classical model, but is potentially acceptable in the sensitive species model, and is a fundamental element in the cultural landscape restoration model. Another range of values is the primacy of a single species or the primacy of a community or assemblage. The sensitive species model focuses on a given species, while the habitat model reflects values of a broader nature. The Indian Creek example from Calumet is indicative of a value set that accepts limitations and does not seek an "ideal" landscape, while the creation of a new lake district around Leipzig integrates past landscapes and local culture with new possibilities.

These models, and their underlying human values, matter for restoration projects. The aims of a project and the clarity of its goals—as shown by identifying the model a project best fits—can help in the decision process for a site. Are all non-indigenous invasives to be removed? Yes, in the classical model; not necessarily, in the sensitive species or rehabilitation models. Restoration and renaturing activities can be contentious, both within a project and between project proponents and others.[8] Understanding the different models can make it easier to see the root of a conflict, and therefore to create more lasting resolutions. Listen closely and you may hear one person speaking on behalf of idealized past landscapes—the classical restoration model—while another espouses the rich cultural legacy present in the landscape, or the cultural landscape restoration model. Whichever the case, restoration and renaturing are done because people value the outcome, and because people are intrinsically connected to the land.

NOTES

1. R. R. Britt, "The New Nature: Cities as Designer Ecosystems," *LiveScience* (January 10, 2004). City of Chicago Department of Environment, *Calumet Area Ecological Management Strategy* (Chicago: City of Chicago, Department of Environment, 2002); M. Gross, *Inventing Nature: Ecological Restoration by Public Experiments* (Lanham, MD: Lexington Books, 2003); M. L. Rosenzweig, "Reconciliation Ecology and the Future of Species Diversity," *Oryx* 37:2 (2003): 194–205; I. Kowaik and S. Körner, eds., *Wild Urban Woodlands: New Perspectives in Urban Forestry* (Berlin: Springer, 2005); F. Turner, "The Invented Landscape," in A. A. Baldwin, J. DeLuce, and C.

Pletsch, eds., *Beyond Preservation: Restoring and Inventing Landscapes* (Minneapolis: University of Minnesota Press, 1994).
2. For more about Calumet go to http://www.nrs.fs.fed.us/4902/focus/calumet/; http://www.fieldmuseum.org/bioblitz/; http://www.fieldmuseum.org/calumet.
3. Sigrun Kabisch and Sabine Linke, *Revitalisierung von Gemeinden in der Bergbaufolgelandschaft* (Opladen: Leske + Budrich, 2000).
4. Interview with Bernd Walther, June 2005.
5. P. H. Gobster, "Models for Urban Forest Restoration: Human and Environmental Values," in J. Stanturf, ed., *Proceedings of the International Conference on Forest Landscape Restoration, 14–19 May 2007, Seoul, Korea* (Seoul: International Union of Forestry Research Organizations and Korea Forest Research Institute, 2007): 10–13.
6. P. H. Gobster and R. B. Hull, *Restoring Nature: Perspectives from the Social Sciences and Humanities* (Washington DC: Island Press, 2000).
7. Council of Europe. Landscape Convention, 2000: http://conventions.coe.int/Treaty/en/Treaties/Html/176.htm. Accessed November 2, 2006.
8. Gobster and Hull, *Restoring Nature.*

20 Sidebar

Conflicting Restoration Goals in the San Francisco Bay

Laura A. Watt

The South Bay Salt Ponds Restoration Project aims to restore up to 15,000 acres of solar salt production ponds along the edges of the San Francisco Bay to a mix of tidal and wetland habitats. It is a project primarily dominated by biological, hydrological, and recreation variables, but it is taking place in a landscape that has been part of the Bay Area's industrial history for over 150 years. Early developers of these lands were interested in a form of industrial-scale gardening by modifying the natural landscape to be productive and profitable: wetlands were considered wastelands until they could be "improved" for human utilization. Some improvers aimed to reclaim the marshes for agricultural use; others carved ponds out of the landscape by installing levees and dikes for producing crystallized salt from the natural evaporation of bay water. A single company, Leslie Salt Company, gradually came to control 47,000 acres around the Bay for the ever-expanding salt market through the 1970s, while also considering possible future "highest and best uses" such as factories, shipping facilities, and then residential developments like Foster City and Redwood Shores.

With the creation of the Don Edwards National Wildlife Refuge in 1972, the trajectory of marsh development began to point along a different course, leading eventually to the current salt ponds restoration project, which implies that the ultimate "highest and best use" of the Bay marshlands is to return them to a more natural form and function. But as the restoration project moves forward, it must accommodate several differing goals: rewilding some ponds by breaching levees and allowing tidal flow to return, creating habitat for several endangered marsh species; continued management of additional ponds to provide habitat for shorebirds and other wading species; and the very human-centered goals of flood protection while providing recreational spaces around the Bay. The resulting restored landscape will likely be an eclectic mix of wild and managed space, yet the concept of "restoration" does not easily acknowledge the continuing role of human intervention, instead implying a return to a more natural state.

But identifying this "more natural" condition is a complicated issue. Most public portrayals of the project suggest that it will return the Bay's salt marshes to some original state—even to the extent of superimposing maps of the current-day ponds on an 1880s-era map of historic slough channels, thereby creating an impression (if not an intention) that the end result will look and function just like it did before the land was reclaimed for salt production. Even the main restorative method proposed, which will rely on breaching selected levee walls to reintroduce tidal flow to the ponds, implies that all one needs to do is to "liberate" natural processes, and then stand back and watch nature heal itself.

In reality, the salt ponds landscape is riddled with a legacy of uses and changes. It is overlain with an elaborate network of levees, some of which will be breached for the project but not removed, as complete removal is prohibitively expensive. Channelization of many of the old sloughs may have altered their topography in permanent ways. A long history of groundwater pumping in the South Bay has caused subsidence of the floor of many salt ponds, leaving them too deep for salt marsh to develop. Questions also remain about whether there is sufficient sediment in the Bay's waters to rebuild the extent of salt marshes planned by the project; it is possible that in places, levees will be breached and the result will only be more open water. Even where the project succeeds in producing healthy salt marsh ecosystems, these will not be exact replicas of salt marshes that existed before salt extraction.

Beyond these aims to recover habitat and protect biodiversity, the other core goals of controlling floods and providing recreational space are not necessarily compatible with a restored wildness; there are already debates about placement of trails or boardwalks and their effects on sensitive species, and concerns that increasing wetlands around the Bay will promote mosquito populations and the diseases they carry, such as West Nile virus, perhaps necessitating the use of chemical controls. There is even disagreement amongst ecologists about how much land should be converted to tidal salt marsh, which benefits such endangered species as the clapper rail and the salt marsh harvest mouse—and how much should remain as managed ponds, which provide habitat for shorebirds and other waterfowl, including the threatened snowy plover. Hence much of this project is explicitly *not* a "return to nature," but will remain a highly managed and manipulated landscape, surrounded by intense urbanization and layered with a variety of human goals and values, many of which may have little to do with ecological restoration or anything very wild. Yet the word *restoration* allows us to overlook much of these more complex nuances, by appealing to our nostalgia for an idyllic ecological past—even though this kind of re-creation will not actually be taking place. Like preservation, the concept of restoration masks continuing human involvement in the landscape, and the very human-centered goals and values that the revised landscape will serve.[1]

Figure 20.1   This former salt pond was owned by Leslie Salt Company, southern San Francisco Bay. (Photograph by Laura Watt)

NOTE

1. For more detail of salt ponds project, see Laura A. Watt, *South Bay Salt Pond Restoration Project Historic Context Report and Final Cultural Resources Assessment Strategy Memorandum* (Prepared by EDAW, Inc. for the CSCC, USWS, and CDFG, 2005); see also Laura A. Watt, "The Trouble with Preservation, or, Getting Back to the Wrong Term for Wilderness Protection: A Case Study at Point Reyes National Seashore," *Association of Pacific Coast Geographers Yearbook*, Volume 64 (2002): 55–72.

Part V

Changing Concepts in Restoration

21 Nature Without Nurture?

Kathy Hodder and James Bullock

BACKGROUND: "SOMETIMES THE BEST MANAGEMENT IS NO MANAGEMENT"

In a passionate diatribe against excessive pesticide use on species invading North American ecosystems, Herman proposes that "sometimes the best management is no management." This sentiment is also surprisingly common in Western Europe, perhaps as a backlash against decades of prescriptive conservation management. Although focus on virgin wilderness (and stupendous scenery) has been described as one of the true idiosyncrasies in the American character, the emotional pull of "self-willed land" has extended across the Atlantic, despite the fact that nearly all European ecosystems are certainly not wilderness in the untrammelled sense.[1]

The possibility of stepping back from intensive conservation management and "allowing nature to take its course," has generated considerable support in the United Kingdom and there has been increasing interest in landscape scale conservation including the prospect of creating "new wilderness" areas.[2] For instance, in the early 1990s, the National Parks Review Panel recommended that:

> A number of experimental schemes on a limited scale should be set up in the [upland] National Parks, where farming is withdrawn entirely and the natural succession of vegetation is allowed to take its course.[3]

More recently, rewilding has been advocated as the optimal conservation strategy for the maintenance and restoration of biodiversity in Europe. Specifically, this includes the restoration of wild large herbivores, i.e., "naturalistic" grazing. The level of interest in this discourse is evidenced by a consortium of thirty-eight ecologists and policy makers who recently placed the consequences of rewilding as one of the top 100 ecological questions of high policy relevance in the United Kingdom.[4]

This chapter explores the well-travelled but poorly mapped terrain of how far nature needs to be nurtured, with specific focus on the United Kingdom. Whether involved in conservation ecology, ecological restoration, biodiversity management, rewilding or related pursuits, the ultimate purpose is surely to work toward sustaining the quality of life on earth. With this in mind, our discussion is based on real case study sites in which a more extensive approach is being considered. Talking to the people involved in the day-to-day running of sites keeps the focus on real practical issues. What might arise *if* rewilding or naturalistic grazing schemes were considered in the English landscape?

ORIENTATION

Yet another reflection on the meaning of "naturalness" or "wilderness" may not be well received by hands-on conservationists; we risk being accused of academic displacement activity. It would be convenient if instead we could follow the advice of Cato the elder by "grasp[ing] the subject" and assuming that the "words will follow." Unfortunately, a brief look through the academic and popular literature reveals startling differences in interpretation both within and between these categories. And perhaps this is not surprising—as the subject reflects on nothing less than the role of people in the natural world. Hence, brushing the surface of the huge field of environmental philosophy, we have included brief notes on the implications of some key terms in the context of conservation in the United Kingdom, which will hopefully orient the reader, and perhaps make the text more useful.[5]

Natural(ness)

Attempting to move beyond the philosophical quagmire surrounding the concept of "naturalness" George Peterken suggested describing different states of naturalness, paraphrased as:

(i) *Original-natural*: the state that existed before people became a significant ecological factor. (In Britain this was commonly recognised as the pre-Neolithic period, prior to 6000 years BP, but the impacts of Mesolithic cultures are increasingly recognised[25]).

(ii) *Present-natural*: the state that would prevail if people had not become ecologically significant.

(iii) *Future-natural*: a hypothetical state that would develop if human influence could be removed.[6]

These ideas are particularly relevant in the United Kingdom where human influence is so pervasive. In this cultural landscape *future-natural* is, of course, an emphatically hypothetical state: human impacts are clearly unavoidable. Instead of encouraging what William Cronon cautions to be the "dualistic vision in which the human is entirely outside the natural," the idea of a *future-natural* state may help us to focus on how our impacts can be managed and designed to allow people to coexist, in Eric Higgs's words, "more generously with other living things."[7]

Wilderness

Although in common parlance the word *wilderness* is easily defined as an uninhabited, uncultivated place—the meaning of the word, its cultural context, and the concomittant implications for conservationists and restorationists, have been hotly debated, especially in North America. "Post-modern-deconstructionist academics" stand accused of deliberately or ignorantly fogging the meaning of "wilderness," especially by suggesting that it must be pristine. In fact the 1964 U.S. Wilderness Act only requires an area to be *primarily* affected by the forces of nature.[8] More recently, writing from the Aldo Leopold Wilderness Research Institute, David Cole is clear that the ubiquity of human disturbance forces us to "confront the fact that we cannot have wilderness that is truly wild or natural—let alone wilderness that is simultaneously wild *and* natural."[9] Nevertheless, people cling tightly to the perception that they are experiencing untouched nature. For instance, in the Val Grande National Park (an "alpine wilderness area" in Italy) 62 percent of polled visitors came to experience "untouched nature" despite the fact that large areas had been cultivated for centuries and only abandoned a few decades ago.[10] In the cultural landscape of the United Kingdom specific wilderness legislation would surely be an irrelevance.

Rewilding in the United Kingdom

The idea of rewilding parts of the United Kingdom has recently received much attention but the word is often used in different contexts and with different meanings. For some, it means allowing areas to develop with "undefined outcomes" for habitat and species composition with the hope that "wild nature can do better" than interventionist management.[11] In other cases the term is used with a quite different meaning. Some texts reflect a North American wildland ethic with emphasis on providing suitable habitat for large charismatic animals, including carnivores. In the UK literature though, as reflected in this volume by Rotherham, Harrison, and Smout, rewilding may also denote landscape scale conservation management, or even describe the planned management of relatively small and seasonally grazed sites. The term is somewhat in danger of becoming a panchreston. A Wildland Network for the United Kingdom was founded in 2005: in summary their web pages state objectives to (1) promote the recognition and appreciation of wild land, (2) protect and conserve qualities of wildness, and (3) promote the establishment of complete ecosystems on a large scale. Their web pages also acknowledge the difficulty of defining "wild land."[12]

Naturalistic Grazing

Grazing management is increasingly used for biodiversity conservation in the United Kingdom, and a large body of research has developed on the science and practice of conservation grazing. This recognizes the key importance of large herbivores and their strong direct and indirect influences on ecosystem dynamics.[13] Indeed, most countries in western Europe have grazed reserves that are outstanding in terms of biological diversity: the Camargue in France, the New Forest in England, the Borkener Paradies in Germany, and the Junner Koeland in The Netherlands. Even grazing by sheep (a non-native in the United Kingdom) can be beneficial when numbers are not kept artificially high. Soay sheep in the St. Kilda archipelago, for example, whose numbers are naturally regulated by food availability, appear to have a positive effect on plant biodiversity.[14] So the utility of *extensive* grazing for conservation of unenclosed habitats is well established, but in our case study sites and elsewhere in the United Kingdom, conservation managers have been considering adopting *naturalistic* grazing methods as pioneered in the Oostvaardersplassen, Netherlands.[15]

Described as "new nature below sea level" the Oostvaardersplassen was created in 1968 as a polder (or track of low-lying land), but was not developed for industry due to economic recession. A wetland component developed into an important nature reserve, and since the 1980s, 2,000 ha of grassland, which had been partly developed for agriculture, has been added to the reserve and grazed by free-ranging herds of Heck cattle, konik ponies, and red deer.[16] Rather than maintaining a prescribed stocking level, the large herbivore populations are allowed to fluctuate through births and deaths, and animals are culled only when their condition and behavior indicate that they are near death. This approach was reviewed in 2006 by an international committee which recommended earlier culling for welfare reasons, managing to defined targets, and increasing intervention, such as providing shelter for grazing animals.[17]

How does *naturalistic* grazing differ from other forms of extensive grazing? In naturalistic grazing there is:

- *no specified herbivore density*; instead populations are resource limited so numbers fluctuate according to factors such as food availability, climate, pathogens and parasites.
- *an assumption that grazing animals are key ecosystem drivers*, and natural processes are allowed free rein, rather than aiming to achieve targets for habitat and species composition.
- a reduction of direct management to a minimum, so that the *natural process is seen as an aim in itself.*

There is therefore considerable contrast with most extensive grazing management, which seeks to achieve management targets (such as species composition) through application of specific grazing pressure.

Reference Ecosystems, Templates, and Shifting Baselines

The merits and limits of applying knowledge of past landscapes in ecological restoration deserve, and have received, significant attention in the literature.[18] Certain conceptual and practical issues emerge repeatedly:

- Are we aiming for the right target? There is an inevitable uncertainty about whether the reference period is "the best" to aim for in terms of biodiversity or natural capital. If the reference system is itself degraded, does this lead (as Vera asserts in this volume) to conservation goals that are not ambitious enough? In other words the baseline has shifted from a supposed pristine ideal, that is, the "original natural."
- Moving targets. How meaningful is restoration in a world where biophysical conditions constantly change? Pragmatism seems prudent in this case. Although restorationists may be accused of trying to replicate the "real thing," no one really believes that past landscapes can be restored exactly, but instead that invaluable practical lessons are learned by looking backward and that we can strive toward, but never reach, a future natural state. Restoration targets can be flexible and include a range of variability based on historical data. Goals will also necessarily adapt to changing conditions such as climactic perturbations.[19]

CASE STUDIES: REWILDING ENGLAND?

In the United Kingdom, and much of Europe, cultural systems have had a dominant influence on vegetation, wildlife, and nutrient, energy, and water flows for millennia. Despite this, advocacy of rewilding, and particularly naturalistic grazing, has been influential in recent years. So in 2003, a UK government agency (English Nature, or "Natural England" since 2006) commissioned us to investigate the ecological, cultural, and welfare implications of rewilding in England.[20] This was very much a "what if" study, because *resource limitation* of large herbivores is central to the naturalistic approach, but animal welfare legislation in the United Kingdom would prohibit Oostvaardersplassen-type food limitation of cattle and ponies

To focus on practical issues, our review covered management discourse by questionnaires and interviews with site managers, owners, and advisors from three contrasting landscapes (3000–5500 ha) which were chosen to give maximum coverage of possible ecological, economic, and cultural scenarios: an upland area, a fertile agricultural/forestry site in lowland England, and a coastal site of varied habitats with high conservation value. Opinions differed within and between sites on priorities and methodologies (not surprisingly), but three common themes emerged: *bigger is better, aiming without a target,* and *wilderness views.*

Bigger is Better

There was no disagreement that scaling-up of management had potential for ecological benefit in reducing isolation, in addition to potential economic savings. However, landscape-scale management was seen as a separate matter from the prospect of managing with minimum intervention, and the move toward the use of naturalistic or resource limited grazing animals, was a distinct issue.[21]

Aiming Without a Target

The practical difficulty of attempting to reconcile "naturalistic" ideology with the day-to-day issues of site management was a major theme emerging from our case studies. Although the intention to step back from management, and allow "natural processes" to shape the landscape, was a stated general aim for two of the three sites, it was more difficult to apply this objective in detail. Many overall aims were concerned with creating wilderness areas, removing unnatural boundaries, and allowing room for natural processes; but on closer deliberation these aspirations were not always compatible with more specific objectives. None of the managers expressed an intention to give natural processes entirely "free rein," as once suggested for upland National Parks and since advocated. Even when a general ambition to "allow nature to take its course" was expressed, managers were understandably reluctant to accept losses when pressed about individual species or valued habitats.[22]

Where conservation of traditional or cultural landscapes is favored, the unknown outcomes implicit in a naturalistic approach are incompatible. This situation applies to much of lowland Britain. One respondent, writing from the populous south of England, noted that "Much of our woodland interest has been lost through lack of traditional management [and] much of our woodland interest in the region is restricted to a few sites. The immediate need is to secure these remaining populations and expand them into neighbouring areas. This requires direct management intervention." There is good evidence that many species and communities have come to depend on maintenance of semi-natural landscapes by people and cessation of management could result in permanent losses of biotopes and biodiversity.[23]

Even in reserves larger than our case studies, such as the envisaged wetland expansion around Wicken Fen in Cambridgeshire, active management would be required. The target for the expanded area is a habitat mosaic; maintenance of the correct density of grazing animals will be vital to prevent succession from open fen to fen woodland and "rewetting" of once-arable fields will require hydrological management using sluices.[24]

In seeking workable alternatives to prescriptive management targets there was much discussion of the use of more flexible limits to acceptable change in the landscape. At sites where biodiversity loss could be significant, limits to acceptable ecological change are likely to differ very little from targets already used for conservation management. In sites with less conservation interest at their starting point, relatively flexible limits could be applied to allow more scope for "natural processes" to shape the landscape. On specific sites these could be determined by ecological factors as well as cultural sensitivities, such as landscape preferences. At all sites "limits to acceptable change" were identified that would need to be adhered to; these could be grouped into three categories.

(i) *Protection of special features.* This category included cultural artefacts or rare species and habitats that required absolute protection, which might include fencing them off from grazing animals.

(ii) *Control of unwanted species.* The second category was concerned with the control of unwanted species such as invasive aliens (e.g., *Rhododendron ponticum*), or pernicious weeds that might cause problems on adjacent landholdings. Direct management intervention was suggested for control of these species.

(iii) *Change in habitat extent or species population size.* The third category was more flexible and respondents suggested that change in the proportion of habitat types (e.g., grassland and scrub) could be monitored and action taken if the relative proportions exceeded agreed limits. The preferred method was manipulation of grazing; however, this may not be easy for free-ranging animals that have formed social groups. Acceptance of this approach was not unanimous, and some advisors and respondents felt that more direct intervention might be necessary, particularly in the early stages of reserve development.

Flexibility of limits to change was strongly affected by the initial biodiversity interest at the site. Where important species and habitats were already present, there was understandable concern that any change in management could potentially be detrimental. Hence in such cases, there could in practice be little, if any, difference from more conventional target-led conservation management. Even on the sites where initial biodiversity interest was relatively low, managers sought to find a means of planning that would guide the development of the area toward greater ecological diversity and interest.

The open wood-pasture type landscape that has been envisioned for naturalistic lowland areas depends on the development of a shifting mosaic of vegetation types including open grassland and woodland glades.[25] Some case study respondents hoped that such shifting mosaics would develop as a result of "natural processes," and particularly through naturalized grazing. However, the scope for shifting mosaics to operate, if stock levels are manipulated to maintain proportions of habitat within pre-determined limits, must surely be low. If management aims for a reserve include grazing at sufficient density to maintain short sward in grasslands, this would preclude the woodland regeneration phase of the shifting mosaic. Herbivore population crashes would be required to provide windows of opportunity for scrub and tree regeneration. This could potentially be managed by simulating population crashes by periodically reducing stock density, but of course, this would not be "naturalistic grazing." Timescales for this sort of simulation would need to be long, at least decades, because temperate woodland regeneration may take at least thirty years.[26]

In the Oostvaardersplassen, more than twenty years after the start of grazing by cattle, ponies, and deer, the fertile soil supports a high density of the herbivores on a close cropped turf. There are patches of scrub (mainly willow and elder) that colonized the marginal area prior to its addition to the grazing reserve, but since then, bark stripping by the herbivores has killed most of these trees. There is virtually no sign of tree or scrub regeneration, and it seems likely that a major population crash would be required to start this process. There is no way of accurately predicting the temporal or spatial patterns that might emerge.

Wilderness Views

The importance of management to create an *appearance* of wilderness was not underestimated by our case study respondents; particularly the need to provide unobstructed views, and to remove unsightly artificial boundaries. This was reflected in the visions, or overall aims, of the sites designated for creating wilderness areas and allowing room for natural processes. In some cases, however, conflation of the wilderness experience (which often requires direct management) with increased scope for natural processes (deliberate removal of management) resulted in impasse.

Particularly in upland areas, the creation or preservation of a sense of remoteness may be a significant factor guiding reserve design and management. Visitors to Ennerdale in Cumbria, for example, enjoy views of spectacular craggy mountains. Unimpeded regeneration of conifers could block these views, significantly detracting from the sense of wildness. Landscape management goals are therefore very important in the design and planning of such upland sites. These landscape management aims should not be confused with an intention to allow unchecked natural processes to act in an area. The "Wild Ennerdale" scheme cites the preservation of a "sense of wildness" as a key aim and provides an excellent example of a large-scale and extensively managed initiative where great care is being taken to disentangle the various distinct goals (landscape and ecological) in order to explicitly state them and effectively manage toward them.[27]

CONCLUSIONS: MAKING A DIFFERENCE

The aspirant language of "conservation politics" is vital for rallying support, but can also cause confusion when vision statements are converted into practice. Site planners and managers then have to try to disentangle the various motivations and global aims to produce something that can be put into practice in the context of the landscapes they manage. Confounding factors which became increasingly evident in the course of our review included semantic issues, and more crucially, disparities between global ideologies and the practicalities of managing for biodiversity.

There is a strong emotional driver for rewilding, linked to a tendency to view semi-natural habitats as in some way *deficient* because they result from human activity. Is British nature conservation a form of gardening? Are we "just rearranging nature to satisfy the human ego," and should we lament a loss of "respect" for wildlife and a disappearing sense of mystery when nature management is driven by action plans and their targets?[28] In this context the appeal of a minimum intervention, target-free, nearly natural alternative is hardly surprising. Add to this an Arcadian image of a "natural" parkland landscape resulting from naturalistic grazing in the lowlands, as well as reduced outlay on practical management (compared to hands-on regardening) and rewilding appears something of a panacea.[29] However, there is no evidence that parkland or wood-pasture landscapes would necessarily dominate a "naturalistic" landscape, so there is a need for caution: the actual landscapes that might result from naturalistic grazing regimes cannot, by definition, be predicted.

Proponents of naturalistic grazing may in some cases seek reduction in management intervention *because* the "natural process" is an aim in itself. More often, when actual landscapes, habitats, and species are discussed, it becomes evident that the real aim is for extensification of management, and a change to practices that are less *conspicuously* artificial. In other words, the aim is for providing idealistic or cultural ecosystem services (non-material benefits such as recreation and spiritual enrichment), which incidentally comprised one of four main categories adopted by the Millennium Ecosystem Assessment. Although an abstract desire to move away from highly prescriptive target-led management was often mooted, the potential loss of biodiversity or of highly valued habitats was rarely seen as acceptable. This extensive management is *not* the same thing as the naturalistic approach used in the Netherlands.[30]

Enlarging and linking nature reserves and planning on a landscape scale has numerous advantages over conservation in small fragmented reserves, and if we are to conserve and enhance biodiversity in Europe—or in any deeply humanized landscape—then this approach is likely to be essential. Currently there is increasing effort being made to extend agri-environment schemes and move toward large-scale habitat restoration in the wider countryside.[31]

Although extensively managed herds of large herbivores would play a key role, these networks and large reserves do not require the application of *naturalistic* grazing. Despite copious evidence detailing the benefits of grazing for conservation, the impacts of naturalistic grazing cannot be extrapolated from results obtained in more controlled systems.[32]

Where there is little to lose in biodiversity interest, there may be the greatest scope for developing naturalistic regimes, but in areas currently of high conservation interest, more management intervention will be required. The principal issue raised is the unpredictability of outcomes: if the conservation of natural processes becomes the *goal*, it is then difficult to define targets and evaluate development of wild areas. Natural processes are as slippery to define as "naturalness" and could be said to occur in any circumstance. Most ecologists in recent decades embrace a "flux-of-nature paradigm" with many possible steady states and multiple trajectories. That nature reserves must allow for dynamic processes while maintaining mosaics of habitat patches in disturbance and succession cycles has long been recognized.[33] This does not, however, equate with replacing definable targets with managing *for* process. How do you know whether this process is working?

In practice, managers may opt for allowing natural change to occur up to predefined limits in landscape and biodiversity, and close monitoring could enable adaptive management intervention. If such trials were attempted, managers should be aware that the practical implications of either stepping back from active management or aiming for a site that *appears* to be in a state of wilderness, often conflict. Value-orientated or spiritual motivations for ecological restoration often complement pragmatic and scientific justifications. However, confusing these separate objectives may make it more difficult to grasp the exciting opportunities we have for developing large interconnected nature reserves for the benefit of plants, animals, and people.[34]

Although there is undoubtedly a large political or social element in decision making on the management of tracts of land, ecologists have a crucial role—centered on improving and communicating knowledge of the potential impacts of management strategies on biodiversity, soil processes, etc. Perhaps there are parallels with gardening, which may be less palatable to some than aspiring to re-create the wild, but it is likely that both rewilders and regardeners ultimately aim to make a positive difference in the management of human impact. As renowned British ecologist, Charles Elton, mentioned over fifty years ago, we need to seek "some wise principle of co-existence between man and nature, even if it has to be a modified kind of man and a modified kind of nature."[35]

ACKNOWLEDGMENTS

Many thanks are due to the respondents to our questionnaires for giving their time to provide comprehensive responses (names not included for confidentiality). Thanks also to Jan Bakker, Fred Baerselman, Hans Kampf, Feiko Prins, Henk Siebel, Frans Vera, Michiel WallisDeVries, and Saskia Wessels in the Netherlands for generously guiding our study tour. Helen

Armstrong, John Bacon, Jan Bokdam, David Bullock, Matthew Oates, Neil Sanderson, Jonathan Spencer, Peter Taylor, and Sandie Tolhurst for valuable constructive discussion.

NOTES

1. S. G. Herman, "Wildlife biology and natural history: time for a reunion," *Journal of Wildlife Management* 66, (2002): 933–46; P. Taylor, *Beyond Conservation: A Wildland Strategy* (London: Earthscan, 2005); P. Marren, *Nature Conservation—a review of the conservation of wildlife in Britain, 1950–2001* (London: Harper Collins, 2002); P. Shepard, *Man in the landscape: a historic view of the esthetics of nature* (Athens, Georgia: University of Georgia Press, 2002); J. Aronson and J. v. Andel in *Restoration Ecology: the New Frontier,* ed. by J. v. Andel and J. Aronson (Blackwell Science Ltd, London, 2006).
2. See, for example, Taylor, *Beyond Conservation*; J. Fenton, "Wild thoughts ... A new paradigm for the uplands," *Ecos* 25 (2004): 2–5; P. Taylor, "Whole ecosystem restoration: re-creating wilderness?" *Ecos* 16 (1995): 22–28; T. Aykroyd, "Wild Britain—a partnership between conservation, community and commerce," *Ecos* 25 (2004): 78–83; A. Whitbread and W. Jenman, "A natural method of conserving biodiversity in Britain," *British Wildlife* 7 (1995): 84–93.
3. R. Edwards (Cheltnham: Countryside Commission, 1991).
4. F. Vera, *Grazing Ecology and Forest History* (Wallingford: CABI International, 2000); W. J. Sutherland et al., "The identification of 100 ecological questions of high policy relevance in the UK," *Journal of Applied Ecology* 43: 617–27.
5. E. Knowles, ed., *The Oxford Dictionary of Quotations* (Oxford: Oxford University Press, 1999); Aronson and Vallejo in *Restoration Ecology: the New Frontier*; see, for example, W. M. Adams, *Future Nature: a Vision for Conservation* (London: Earthscan, 2003).
6. G. F. Peterken, *Natural Woodland: Ecology and Conservation in Northern Temperate Regions* (Cambridge: Cambridge University Press, 1996); G. F. Peterken, *Woodland Conservation and Management* (London: Chapman and Hall, 1981).
7. W. Cronon, ed., *Uncommon Ground Rethinking the Human Place in Nature* (W. W. Norton & Co., 1996); E. Higgs, *Nature by Design: People, Natural Process, and Ecological Restoration* (Cambridge, MA: MIT Press, 2003).
8. D. Foreman, *Rewilding North America: A vision for conservation in the 21st century* (Washington, DC: Island Press, 2004).
9. D. N. Cole, "Management dilemmas that will shape wilderness in the 21st century," *Journal of Forestry* 99 (2001): 4–8.
10. F. Hochtl et al., "'Wilderness': what it means when it becomes a reality—a case study from the southwestern Alps," *Landscape and Urban Planning* 70 (2005): 85–95.
11. J. Fenton, "Wild thoughts;" M. Fisher, "Self-willed land: Can nature ever be free?," *Ecos* 25 (2004): 6–11.
12. Foreman, *Rewilding North America*; Taylor, *Beyond Conservation*; R. Neale, "Wilder slopes of Snowdon," *Ecos* 25 (2004); Wildlands Network: http://www.wildland-network.org.uk/index.htm visited on March 1, 2009.
13. S. C. Palmer et al., "Introducing spatial grazing impacts into the prediction of moorland vegetation dynamics," *Landscape Ecology* 19 (2004): 817–27; B. A. Woodcock et al., "Grazing management of calcareous grasslands and

its implications for the conservation of beetle communities," *Biological Conservation* 125 (2005): 193–202; D. J. Bullock and H. M. Armstrong in *Grazing management: British Grassland Society Occasional Symposium* 34 ed. by A. J. Rook and P. D. Penning (Reading, 2000): 191–200.

14. M. F. WallisDeVries in *Grazing and Conservation Management*, ed. by M. F. WallisDeVries et al. (Kluwer Academic Publishers, Dordrecht, 1998): 1–20; T. H. Clutton-Brock and J. M. Pemberton, eds., Soay Sheep: *Dynamics and Selection in an Island Population* (Cambridge: Cambridge University Press, 2004); M. J. Crawley et al. in *Soay Sheep: Dynamics and Selection in an Island Population*, ed. by T. H. Clutton-Brock and J. M. Pemberton (Cambridge: Cambridge University Press, 2004).

15. V. Wigbels, *Oostvaardersplassen: new nature below sea level* (Staatsbosbeheer, Flevoland-Overijssel, 2001); ICMO, *Reconciling Nature and human interests*, Report of the International Committee on the Management of large herbivores in the Oostvardersplassen (ICMO), Wageningen UR-WING rapport 018 June 2006 (The Hague/Wageningen, Netherlands: 2006).

16. Wigbels, *Oostvaardersplassen*; J. Vulink and M. Van Eerden in *Grazing and conservation management*, ed. by WallisDeVries et al. (Dordrecht: Kluwer Academic Publishers, 1998).

17. R. Tramper, "Ethical Guidelines," (Utrecht: Centre for Bioethics and Health Law, University of Utrecht, 1999): http://www.grazingnetworks.nl/userImages/File/ethical%20guidelines%20tramper.doc, visited on March 1, 2009; ICMO, *Reconciling Nature and human interests*.

18. See, for example, D. Egan and E. A. Howell, eds., *The Historical Ecology Handbook: a Restorationists' Guide to Reference Ecosystems* (Washington. D.C.: Island Press, 2001); M. Hall, *Earth Repair: A Transatlantic History of Environmental Restoration* (Charlottesville: University of Virginia Press & Center for American Places, 2005).

19. J. A. Harris and R. V. Diggelen in *Restoration Ecology: the New Frontier*, ed. by J. v. Andel and J. Aronson (London: Blackwell Science Ltd, 2006); J. E. Dizard, Chapter 5, this volume; P. S. White and J. L. Walker, "Approximating nature's variation: selecting and using reference systems in restoration ecology," *Restoration Ecology* 5 (1997): 338–49; J. A. Harris et al., "Ecological restoration and global climate change," *Restoration Ecology* 14 (2006): 170–76.

20. Aronson and Andel in *Restoration Ecology*; K. H. Hodder et al., "Large herbivores in the wildwood and modern naturalistic grazing systems," *English Nature Research Reports* 648 (2005) http://www.english-nature.org.uk/pubs/publication/PDF/648.pdf.

21. Wigbels, *Oostvaardersplassen*.

22. Edwards, Countryside Commision; Fenton, "Wild thoughts."

23. J. P. Bakker in *Vegetation Ecology*, ed. by E. van der Maarel (Blackwell Science, Oxford, 2005): 309–31.

24. L. F. Friday and T. P. Moorhouse, *A Report to the National Trust* (Cambridge: 1999); A. Colston, "Wicken fen—realising the vision," *Ecos* 25 (2004): 42–45.

25. Vera, *Grazing Ecology and Forest History*; H. Olff et al., "Shifting mosaics in grazed woodlands driven by the alternation of plant facilitation and competition," *Plant Biology* 1 (1999): 127–37.

26. R. Harmer et al., "Vegetation changes during 100 years of development of two secondary woodlands on abandoned arable land," *Biological Conservation* 101 (2001): 291–304.

27. G. Browning and R. Yanik, "Wild Ennerdale—letting nature loose," *Ecos* 24 (2004): 34–38.

28. M. Toogood, "Semi-natural history," *Ecos* 18 (1997): 62–68; N. Henderson, "Wilderness and the nature conservation ideal: Britain, Canada and the United States contrasted," *Ambio* 21 (1992): 394–99; J. Robertson, "Designed by nature?," *Ecos* 16 (1995): 9–12; Marren, *Nature Conservation*; Foreman, *Rewilding North America.*
29. Vera, *Grazing Ecology and Forest History.*
30. A. F. Clewell and J. Aronson, "Motivations for the restoration of ecosystems," *Conservation Biology* 20 (2006): 420–28; *Ecosystems and Human Well-being: Synthesis Report*, Millennium Ecosystem Assessment (Wash., D.C.: Island Press, 2005); Wigbels, *Oostvaardersplassen.*
31. W. J. Sutherland, "Restoring a sustainable countryside," *Trends in Ecology & Evolution* 17 (2002): 148–50.
32. See, for example, S. C. F. Palmer et al., "Introducing spatial grazing impacts into the prediction of moorland vegetation dynamics," *Landscape Ecology* 19 (2004): 817–27.
33. S. P. Lawler et al. in *The functional consequences of biodiversity: empirical progress and theoretical extensions*, ed. by A. P. Kinzig et al. (Princeton: Princeton University Press, 2001): 294–313; E. C. Stone, "Preserving vegetation in parks and wilderness," *Science* 150 (1965): 1261–1267; S. Pickett and J. Thompson, "Patch dynamics and the design of nature reserves," *Biological Conservation* 13 (1978): 27–37.
34. J. D. C. Linnell et al., "Conservation of biodiversity in Scandinavian boreal forests: large carnivores as flagships, umbrellas, indicators, or keystones?," *Biodiversity and Conservation* 9 (2000): 857–68.
35. C. Elton, *The Ecology of Invasions by Animals and Plants* (London: Methuen and Co Ltd, 1958).

22 Toward a Multiple Vision of Ecological Restoration

Jozef Keulartz

There has recently been growing interest in the role of metaphors in environmentalism and nature conservation. Metaphors not only structure how we perceive and think but also how we should act. The metaphor of nature as a book provokes a different attitude and kind of nature management than the metaphor of nature as a machine, an organism, or a network. This chapter explores four clusters of metaphors that are frequently used in framing ecological restoration: metaphors from the domains of engineering and cybernetics; art and aesthetics; medicine and health care; and geography. It is argued that these metaphors, like all metaphors, are restricted in range and relevance, and that we should adopt a multiple vision on metaphor. The adoption and development of such a multiple vision will facilitate communication and cooperation across the boundaries that separate different kinds of nature management and groups of experts and other stakeholders.

ENGINEERING AND CYBERNETICS

Activities to repair environmental damage were initiated by environmental engineering, a discipline that evolved in the early 1970s from sanitary engineering when biochemistry, microbiology, fluid mechanics, physical and chemical oceanography, meteorology, et cetera, were integrated into traditional drinking water, wastewater, water quality, and air pollution courses.[1] The emerging discipline had its own journal, *The Journal of Environmental Engineering*, published by the American Society of Civil Engineers.

The late 1980s witnessed the birth of a new discipline: ecological engineering. The ecological engineers launched their own journal in 1992, *Ecological Engineering, The Journal of Ecotechnology*. Ecological engineers or "ecotechnologists" filled niches left vacant by environmental engineers and gained a foothold especially in the areas of wetlands creation and ecosystem restoration. Mainly based on the work of Howard Odum, who was named honorary editor of *Ecological Engineering* on the occasion of his seventieth birthday, ecological engineering was defined as "the design of human society with its natural environment for the benefit of both."[2]

Ecological engineering is an important offshoot of the so-called New Ecology, a new approach within the field of landscape planning and conservation biology that can be traced back to cybernetics, which flourished in the United States in the early post-World War II years, in a climate of technocratic optimism. The politicians, having proved unable to cope with the problems of a complex industrial society, were urged to make way for social engineers who would then manage society as a self-regulating machine. One of these technocrats, Evelyn Hutchinson, was to leave an indelible mark on post-war ecology, particularly through the work of Howard Odum and his brother Eugene.[3]

Cybernetics has displayed extraordinary communicative power because it is compatible with two comprehensive root metaphors that have shaped belief systems or "world hypotheses" with a long history in Western tradition: organicism and mechanism.[4] This compatibility can be clearly observed in the work of Norbert Wiener, the founder of cybernetics. When he first coined the word "cybernetics" in 1945, he defined it as "control and communication in the animal and the machine." Wiener brought together two fields of research. On the one hand he elaborated on the engineering-oriented research into the "servomechanical" nature of control and communication in machines, using the ideas of information flow, noise, feedback, and stability. On the other hand he built on what physiologists like Walter Canon had developed under the headings of "homeostasis": a variety of mechanisms in the organism to maintain fixed levels of blood sugar, blood proteins, fat, calcium, as well as adequate supply of oxygen, a constant body temperature, and so on.

Whereas Wiener sought to extend his cybernetic program to social systems, Hutchinson and his students Howard and Eugene Odum applied this program to ecosystems, stressing their tendency to maintain or restore homeostasis through self-regulating, feedback mechanisms.

Because the cybernetic concept of ecosystems is consistent both with organic and mechanic worldviews it could fulfill an intermediary role between ecologists and politicians during the preparatory years (1963–68) of the International Biological Programme, which coincided with a general wave in environmental consciousness.[5] It appealed both to technocrats and environmentalists. To technocrats this concept offered an image of a closed system that could be controlled and manipulated from a position outside or superior to the system and so gave rise to an immense technological optimism. On the other hand the cybernetic concept allowed environmentalists to consider and admire ecosystems as large interdependent wholes that are definitely more than the sum of their parts.

However, from the early 1970s onward, the cybernetic concept of ecosystems as an intermediary started to disintegrate and lose its attraction to environmentalists. At first glance, its holism seemed to counteract the reductionist implications of the mechanistic and materialistic metaphors of nature that emerged during the scientific revolution of the seventeenth century, and that were held responsible for the decline in community spirit and the alienation from nature.[6] However, on further consideration, environmentalists became convinced that the cybernetic metaphors seemed to be little more than sophisticated versions of the same old mechanistic metaphors.[7] In fact, as some argued, systems ecology does not proceed in a less reductionist fashion than classic natural sciences, but its reductionism is of a totally different order. Reductionism in classic natural sciences refers to a reduction to elements, such as atoms or molecules that are identical in a *material* sense. Systems ecology, on the other hand, is concerned with components that are identical in a *functional* sense because they perform the same function within the ecosystem, for example that of producers (green plants), consumers (herbivores, carnivores, and omnivores), or decomposers (bacteria and fungi). Such reductionism allows for a trade-off, as it were, between organisms with a similar function with the aim of optimising the biomass yield.[8]

The terminology of *producers*, *consumers*, and *biomass yield* brings to light another domain of metaphors that is important in shaping the outlook of the New Ecology: modern economy. The mechanistic view of nature supports an economic model of human–nature interactions. As Donald Worster has shown in detail, exponents of the New Ecology view nature as a set of resources with cash value; they have transformed nature into a reflection of the modern corporate state, a chain of factories, an assembly line. "It is not fanciful to attribute to the mechanistic, energy-based bio-economics of the New Ecology a built-in bias toward the management ethos, and even toward a controlled environment serving the best interests of man's economy."[9]

The mechanistic and economic metaphors of nature share a pronounced anthropocentric character. Many environmentalists are deeply concerned about the destructive consequences of such an anthropocentrism. Whenever man sets himself up as the measure of all things—so runs their critique of anthropocentrism—nature, including human nature, ceases to be an independent and inexhaustible source of value and becomes instead a mere resource to be disposed of at will, with all the detrimental consequences for the environment. If we are to prevent the environmental crisis from ending in catastrophe, environmental philosophers agree, we must convert to non-anthropocentrism, judging life forms on their intrinsic value and not on their instrumental value.

Following famous philosopher Martin Heidegger, many environmentalists are convinced that we should no longer approach nature in terms of its utility and availability to our insatiable will to power but instead adopt an attitude of responsiveness and "releasement" (*Gelassenheit*). Man should learn to behave more like a shepherd than as a lord of being. We will get nearer to such an attitude if we turn to another source of metaphors, the field of art and aesthetics, that has inspired the theory and practice of ecological restoration, at least in its formative stages.

ART AND AESTHETICS

In 1981 William R. Jordan III, who coined the term "ecological restoration," founded the oldest journal that deals exclusively with the subject of restoring ecosystems, first as *Restoration & Management Notes* and since 1999, as *Ecological Restoration.* Jordan is also a founding member of the Society for Ecological Restoration International (SER) that was established in 1988. In 1993 SER published the first issue of its flagship journal, *Restoration Ecology.*

Ecological restoration is considered an intentional activity that initiates or accelerates the return of an ecosystem to its historical origin. Ecological restoration differs from ecological engineering with respect to the predictability of outcomes. "Predictability is a primary consideration in all engineering design, whereas restoration recognizes and accepts unpredictable development and addresses goals that reach beyond strict pragmatism and encompass biodiversity and ecosystem integrity and health."[10]

Because restoration attempts to return an ecosystem to its historic trajectory, historic conditions are the ideal starting point for restoration design. Although it may be difficult or impossible to determine the precise historic trajectory of a severely impacted ecosystem, the general direction and boundaries of that trajectory can nevertheless be established through knowledge of the damaged ecosystem's pre-existing structure, composition, and functioning (palaeo-references), and through studies of comparable intact ecosystems elsewhere (actuo-references).

From the outset, ecological restoration's attempt to return degraded ecosystems to their original state has been interpreted in terms of the restoration of works of art. This metaphor was put forward by environmental philosophers in particular. At first, the comparison of nature with art was made in order to discredit ecological restoration. In his famous 1982 paper *Faking Nature*, Australian philosopher Robert Elliot argued that ecological restoration is akin to art forgery. Just as a reproduction or replicate cannot reproduce the value of the original, restored nature cannot reproduce the value of original nature. "What the environmental engineers are proposing is that we accept a fake or forgery instead of the real thing."[11] A Van Meegeren will of course always be inferior to a real Vermeer!

In his 1992 paper *The Big Lie: Human Restoration of Nature*, environmental philosopher Eric Katz further argued that whatever is produced in a restored landscape it certainly cannot count as having the original value of nature, particularly wild nature, and necessarily represents a form of disvalue and domination of nature. "Once we dominate nature, once we restore and redesign nature for our own purposes, then we have destroyed nature—we have created an artifactual reality, in a sense, a false reality, which merely provides us the pleasant illusory appearance of the natural environment."[12]

Other environmental philosophers are less harsh in their judgement. Andrew Light (2003) for instance, thinks that Eliot's and Katz's criticism is only valid with respect to a particularly malicious kind of restoration— restoration that is used to justify the disturbance or destruction of nature, for instance for the benefit of some industrial activity, with the argument that it is possible nowadays to create a piece of nature with the same value in a later stage or at a different place. But this kind of restoration, Light insists, is relatively rare. Most restoration efforts are undertaken to correct a past harm. In these cases ecological restoration is more akin to art restoration than to art reproduction or art forgery.

However, this metaphor is not unproblematic at all because the very idea of restoration of works of art is itself controversial, as Light points out with reference to the early aesthetic theory of Mark Sagoff. In 1978, before he ever turned to environmental questions, Sagoff published an article, "On Restoring and Reproducing Art," in which he argued that not just any art restoration should count as a legitimate restoration. Here Sagoff draws a distinction between integral and purist art restorations. An integral restoration puts new pieces in the place of original fragments that have been lost. The point of integral restoration is to make the whole work look original. Sagoff rejects this kind of restoration as both aesthetically and ethically troublesome and he is obviously in favor of purist restorations. "A purist restoration limits itself to cleaning works of art and to reattaching original pieces that may have fallen. Purists contend that nothing inauthentic— nothing not produced by the original artist—may be shown. If damage obscures the style of the original, a purist may allow a few substitutions, but only in outline or in another color, to avoid any pretense of authenticity." A purist restoration allows viewers to imagine a work of art as complete, yet, at the same time, to see what is authentic and what is not, whereas an integral restoration only succeeds if viewers cannot tell the difference between the original and the restored work. Integral restorations not only diminish or even destroy the original value of an artwork, purists argue, they also deceive the public.[13]

Extending Sagoff's negative judgment of integral restorations to ecological restoration projects, Light claims that any integral restoration of nature, even if it is benevolent, cannot have the same value as original nature. He seems to be somewhat less rigorous than Sagoff, however, in assuming that one can distinguish between different kinds of integral ecological restorations as better or worse. But, more importantly, Light claims that many ecological restorations amount to something more akin to purist than to integral restorations. Light mentions clean-ups as the most obvious cases. These cases would include the bio-activation of existing micro-organisms in soils to allow the land to essentially clean itself up, and cleaning out exotic plants that were introduced at some time into a site and allowing the native plants to reestablish themselves. "Here we have human meddling in nature for the same purposes as the purist restoration of art. No new 'work' is produced, but suppressed elements of nature are allowed to once again perform their functions."[14]

However, the purist view of ecological restoration is less unproblematic than Light seems willing to acknowledge. Purists respect a work of art as the result of a particular process, as the creation of a particular artist working at a certain place and time, and not merely as a bearer of aesthetic properties. But here the art–nature analogy may ultimately break down. Even if we would allow the existence of a master craftsman to whom we may attribute the authorship of nature, his or her products can never be traced back to a particular place and time. Among ecologists and conservationists there is an ongoing discussion over the question of which historical reference one should choose. Should one go back to the last interglacial era when man did not even yet have projectile weapons (such as the bow and arrow) and therefore was not yet capable of submitting his natural enemies? Should one go back to the times before the emergence of agriculture, or should one only have to go back to pre-industrial times and resort to traditional agrarian techniques such as reed and brushwood cultivation, tree planting and felling, mowing, and turf cutting?

The answer to this question depends on how one considers the relationship between humans and nature. This relationship can only in a severely limited way be compared to our aesthetic attitude toward works of art. This attitude is one of sympathetic contemplation of an object for its own sake. According to eighteenth-century Philosopher Immanuel Kant, works of art possess a certain "Zweckmässigkeit ohne Zweck" (purposiveness without a purpose). He describes the way we experience their beauty as "interesseloses Wohlgefallen" (disinterested pleasure). But nature is never solely the object of experiences of beauty or the sublime but has many other functions. Water for instance is of importance for traffic, transportation, food supply, irrigation, recreation, cooling of power stations, and domestic use.

Another major difference between art and nature, as Rolston has argued, is that works of art are entirely passive and, left to themselves, inevitably decay. We restore them; they do not restore themselves. In contrast, left to itself nature flourishes and can restore itself. But this argument is only valid as long as we stay focused on the *visual* arts. If we turn to *performing* arts like theatre, dance, or music, the restoration metaphor acquires quite a different meaning. A ballet, symphony, or play is anything but static but derives its very life from being recreated time and again. Such an artwork obtains its identity only through the multitude of its successive performances. This equation of ecological restorations with artistic performances was made repeatedly by Jordan, who denounces "environmentalism's blindness to the performative or expressive aspect of restoration—to what might be called its ritual value."[15]

Moreover, the performing arts cannot thrive without an audience. Because artistic performances are almost by definition public rituals, this version of the art restoration metaphor is akin to the community metaphor, which has a long tradition in ecology, especially in the land ethic of Aldo Leopold. According to Leopold's famous statement, "a land ethic changes the role of *Homo sapiens* from conqueror of the land community to plain member and citizen of it." Restorationists who have adopted the community metaphor perceive themselves as participants rather than as curators of museum pieces. Participation is supposed to strengthen the ties between humans and between the human community and the larger ecological community.[16]

The analogy between ecological restoration and artistic recreation does more justice to the dynamic interplay of nature and culture that follows from our multifunctional use of natural resources than the parallel of ecological restoration with the restoration of parts or pieces of a museum collection. In this respect there is some resemblance with metaphors from the domain of engineering and cybernetics—metaphors that are also human-inclusive rather than human-exclusive. But they differ with respect to the standard by which ecological restorations should be evaluated. In the first case "authenticity" functions as such a yardstick. An artistic performance should be true to the original score, script, or scenario. Although the players, the props, the scenery, and costumes constantly change, the performance has to remain Swan Lake. Metaphors from the domain of engineering and cybernetics, on the other hand, are not about originality or authenticity but rather about "functionality." In this regard these metaphors have more in common with a cluster of metaphors that more recently emerged from the domain of medicine and health care.

MEDICINE AND HEALTH CARE

Since about 1990, the notion of "health" made an amazing career within environmentalism and ecology. The domain of application of this notion has been extended from the level of the individual (clinical and veterinary medicine) and the population (epidemiology and public health) to the level of ecosystems. An interdisciplinary field of research has developed in which the relations between human activities, social organizations, natural systems and health are being systematically explored. At present, the notion of health functions as a focal point for the integration of three highly overlapping areas of research activity: ecosystem medicine, geographical medicine, and conservation medicine.

"Ecosystem medicine" emerged in the early 1990s and gained momentum with the establishment in 1994 of the International Society for Ecosystem Health (ISEH). Since 1995, ISEH has published the journal *Ecosystem Health*. The society is dedicated to the idea that a healthy ecosystem is one that provides services supportive of the human community, such as food, potable water, clean air, and the capacity for assimilating and recycling wastes. Ecosystem medicine aims to develop "a systemic approach to the preventive, diagnostic, and prognostic aspects of ecosystem management, and to the understanding of relationships between ecosystem health and human health."[17]

This approach is not entirely new. Aldo Leopold already refers to "the art of land doctoring" and "the science of land health." Also worth mentioning is the so-called "landcare movement" that spread all over Australia in the early 1980s. Some scientists go as far in their application of the notion of health to ecosystems as to recommend palliative care for degrading landscapes that should be considered terminally ill.[18]

Ecosystems are regarded as healthy as long as they have the capacity to maintain structure and function in the face of stress. Proponents of this approach talk about the "Ecosystem Distress Syndrome" (EDS). Indicators of this syndrome are: changes in primary productivity and in nutrient cycling, loss of species diversity, and return to early stages of succession.

The second area of research activity that uses a broad concept of health is called "geographical medicine" or "geomedicine." It is a subdiscipline of epidemiology that studies the impact of the environment on the geographical distribution of health and illness. Recently there is growing concern over the influence on the health of human populations by global economic, technological, and environmental changes, including climate change, ozone depletion, loss of biodiversity, land degradation, desertification, deforestation, the increasing gap between the rich and poor, world-wide urbanization and mass migration due to war or natural disasters. From 2000 to 2004 the journal *Global Change and Human Health* provided a platform for scientific research into the health impacts of globalization processes.[19]

The last area of research focusing on a broad health concept is called "conservation medicine." This new discipline combines techniques, facts, and concepts from public health, veterinary medicine, conservation biology, and plant pathology. Conservation medicine evolved out of a crisis: unprecedented levels of disease in many species as a result of the worldwide transformations of the host–parasite relations by climate change, chemical pollution, animal trade, encroachment into wildlife areas, habitat fragmentation, and loss of biodiversity.

With the launching of the first issue of the journal *EcoHealth* in January 2004, the collaboration between these three areas of research activity took a more definite shape. This journal aims to build on the legacy of both *Ecosystem Health* and *Global Change and Human Health* and also intends to cover the area of conservation medicine that was not yet represented by a scholarly journal.

In contrast to the art restoration metaphor, the health metaphor implies a "humans in" approach to ecosystem analysis and assessment. Consequently, the health metaphor is also less negative about technology than the art restoration metaphor: health doesn't depend on some original or authentic state or condition because people actually can feel healthy and function normal with a hearing aid, a bypass, or an artificial kidney. Given these differences, "rehabilitation" seems to provide a far more appropriate term for the health-centered approach than "restoration." Exponents of this approach share its human-inclusive and functionalist perspective with environmental engineers and ecotechnologists.

The health metaphor is a very powerful discursive tool. It is broad enough to encompass a variety of scientific approaches and, like the cybernetic concept of ecosystems, is compatible with mechanic and organic worldviews. It not only facilitates the cooperation between natural, social, and medical scientists but has an important communicative function for the general public as well. It provides a vocabulary of symptoms, syndromes, diagnostic indicators, and so on, with which laypeople are already familiar as potential or actual patients and consumers of health care services.[20]

Over the last years, the health metaphor has gained ground in ecological restoration. As Harris and Hobbes have noticed, the two emergent fields of ecosystem health and ecological restoration have the potential to complement one another comfortably. "If we view the concept of ecosystem health as the diagnostic toolbox and ecological restoration as the treatment toolbox for the management of damaged ecosystems, there is clearly the potential for useful synergy."[21]

Recently, this incorporation of the health metaphor in restoration ecology has provoked criticism from Mark Davis and Lawrence Slobodkin. They reject the idea that communities and ecosystems possess traits such as "health" and "integrity" because this idea is reminiscent of outdated ecological ideas of communities and ecosystems as integrated entities, like Clements' concept of plant communities as super-organisms. They argue that "attributes such as *health* and *integrity* can (only) be meaningfully applied to entities that have been directly shaped by evolution, such as individual organisms. . . . However, communities and ecosystems are not shaped as entities by evolution."[22]

But this line of criticism will not wash. It does not appreciate that the use of metaphors is inevitable and, what is more, that metaphors fulfill indispensable cognitive functions. It is precisely the very inability to provide a "literal" transcription of a metaphor in a scientific theory that constitutes its heuristic power.

The other side of the coin is of course that metaphors can only claim a limited validity. The potential advantages of the health metaphor should not seduce us to embrace it as the one and only truth. Like all metaphors the health metaphor too falls short in some respects. Ecosystems will not, for example, visit a doctor with their complaints. They cannot announce that they are sick and then tell when they are feeling better. Moreover, in the case of ecosystems there is more possibility of a conflict between the health of the whole and the health of the components than in the case of human organisms.[23]

GEOGRAPHY: ISLANDS AND NETWORKS

In his 2004 book *Conserving Words*, Daniel J. Philippon explores how the American environmental movement has been shaped by the seminal works of five famous nature writers: Theodore Roosevelt, Mabel Osgood Wright, John Muir, Aldo Leopold, and Edward Abbey. Each of these writers understood "nature" through a particular metaphor that enabled certain narratives that explained how human beings should interact with nature: frontier (Roosevelt), garden (Wright), park (Muir), wilderness (Leopold), and utopia (Abbey). Philippon calls these metaphors geographical metaphors because each of them refers to nature in terms of a particular place. He concludes his book with a discussion of what he considers to be the latest and most promising geographic metaphor, the metaphor of nature as an island.

In fact, the island metaphor goes back to the New Ecology and to the work of Evelyn Hutchinson in particular. In a pioneering paper published in 1946, "Circular Causal Systems in Ecology," Hutchinson distinguished between two closely related approaches, the "biogeochemical" and the "biodemographic" approach.

Seen from a *biogeochemical* perspective, the entire biosphere appears as a giant cyclical system of energy, matter, and information which is able to maintain a dynamic equilibrium thanks to a series of feedback mechanisms. This perspective was elaborated in particular by Hutchinson's student Howard Odum and his brother Eugene Odum (see the first section on Engineering and Cybernetics).

The *biodemographic* approach deals with groups or communities of organisms, the so-called "populations." In conformity with the cybernetic principle shared by the two approaches, these populations are also perceived as systems attempting to maintain their stability under ever-changing conditions by means of feedback mechanisms. This approach was further elaborated by Robert MacArthur, another of Hutchinson's students. In the 1960s, in collaboration with Edward Wilson, he developed a theory on the biogeography of islands, the "island theory." Using the size of the island and the distance to the mainland as its main parameters, the theory predicts the number of species on a given island. MacArthur and Wilson, too, assumed a dynamic equilibrium: although the taxonomic composition on the island is subject to continuous change, the number of species (which is determined by the rates of extinction and colonization) remains constant. Their 1967 book *The Theory of Island Biogeography* is one of the most frequently cited books in ecology and popular biology.

Although this theory and Odum's systems theory come from the same theoretical background they differ widely with respect to their rhetorical potential. Whereas the metaphor of nature as a clockwork reinforces confidence in our ability to repair damaged ecosystems like we use to repair "the radio or the family car," as Hutchinson once put it,[24] the metaphor of nature as an island, reminds us that there are "limits to growth" and that nature cannot be endlessly exploited. It calls attention to the ongoing fragmentation and ecological isolation by the encroachment of civilization, and all the risks of extinction that come with it.

As Philippon has noted, the island metaphor is intimately related to another metaphor: the network metaphor, that is to the notion of connectivity through, e.g., corridors, ecoducts, stepping stones, and coastal linkages. "The idea of nature as island might at first appear to neglect the idea of connectedness that most clearly defines the modern science of ecology. But this is not actually the case, particularly for island biogeography. Rather, islands provide a means to discuss webs and networks and systems of influence while still preserving a sense that actual *things* are being connected."[25]

While Odum's systems theory was very popular in the sixties and seventies, the island theory had a lightning career from the early eighties onward, especially within European nature policy. It was used to underpin the network-notion, in the (metaphorical) sense that nature areas should be perceived as "islands in a sea of cultivated land." The theory serves as a basis for the attempt to maximize the size of contiguous nature areas and the number of links between them.

The concept of a comprehensive network of nature areas was introduced in Europe with the 1992 Directive on the conservation of natural habitats and of wild fauna and flora (the Habitats Directive). Together with the 1979 Birds Directive, which aims at the protection of endangered bird species by the designation of protected areas, the Habitats Directive constitute the instruments for creating a European ecological network, the so-called *Natura 2000*, that is considered to be the cornerstone for the protection of biodiversity in Europe.

We may credit the so-called greenway movement in the United States with having anticipated the idea of thinking in terms of green networks. The movement started off in the mid 1860s when Frederick Law Olmsted and his colleagues began designing so-called parksystems in and around cities, systems that were composed of urban parks connected by parkways. After fading from public consciousness during the second third of the previous century, the greenway concept enjoyed a popular revival in the early 1970s. In the context of the new environmental movement, the greenway was reincarnated as an innovative conservation strategy, with clear recognition of the ecological importance of the green connections and environmental corridors.[26]

Currently, the network concept is rapidly gaining still more importance due to the growing need for migration of many species struggling to escape extinction as a result of climate change, and of global change in general.

Although the network metaphor is no more true or accurate than any of the other metaphors, it may well turn out to be more useful as a communicative device in scientific discourse, public debate, and political decision making about restoring nature. The network metaphor does not only connect areas, populations, and habitats, but it also encourages the creation and establishment of linkages between people, disciplines, and practices. It enables communication and cooperation between environmentalists and ecologists over the whole restoration spectrum, from the islands and fragments of biodiversity, and the corridors between them to the buffer zones and the wider surroundings.

TOWARD A MULTIPLE VISION

As we have seen in the previous sections, every metaphor is restricted in range and in relevance. Metaphors are like searchlights that highlight certain aspects and features, while blocking out others. According to Lakoff and Johnson, each metaphor is true for certain purposes, in certain respects, in certain contexts. As Sara Ebenreck has written, "Rather than proceed as if any one metaphor is the finally correct metaphor, ethicists conscious of the constructive imagination at work in these basic metaphors might be more aware of the limits of any metaphorical construction and more open to the experiences and values embodied in alternate metaphoric constructions of the Earth." Moreover, the search for the one best metaphor is not without pitfalls and can lead to what Mark Meisner has called "a sort of perceptual hegemony." This is the case if a metaphor ceases to be perceived as metaphor and is taken literally, so that we are no longer able to recognize that it represents but a singular perspective.[27]

In order to prevent such one-sidedness, we should adopt what Donald Schön and Martin Rein used to call a *double vision*: "the ability to act from a frame while cultivating awareness of alternative frames." We should learn to "squint" so to speak in order to see things from different angles simultaneously, or we should develop what philosopher of technology Don Idhe has called a "compound eye."[28]

What would be the outcome of employing such a double—or multiple— vision on ecological restoration? Given the similarities and differences between metaphors from cybernetics, aesthetics, and medicine, how should we judge the value and meaning of these metaphors? And how do they relate to one another?

Environmental philosopher Baird Callicott and his colleagues Larry Crowder and Karen Mumford have provided us with an interesting possibility to arrange the most important metaphors used in restoring nature. They distinguish two contemporary schools of conservation philosophy: compositionalism and functionalism. According to Callicott and co-workers the compositionalist emphasis is on ecological restoration, the process of returning a biotic community to its original condition of biological diversity and integrity. On the other hand the functionalist emphasis is more on ecological rehabilitation, the process of returning an ecosystem to a state of health. Callicott and co-workers consider compositionalism and functionalism as two ends of a continuum: the compositionalist emphasis on the ecological restoration of biological integrity and diversity is appropriate for the management of the less severely degraded areas such as wilderness areas, national parks, and state parks. The functionalist emphasis on the ecological rehabilitation of ecosystem health is more suited for the much greater part of the world that is inhabited and economically exploited by humans.

Using the island metaphor, Callicott *cum suis* argue that there is a complementary and dialectical relationship between these approaches:

> The preservation of islands of biological diversity and integrity and ecological restoration necessarily occurs at present in a humanly inhabited and economically exploited matrix. Hence the success of nature preservation and restoration necessarily depends on ecologically rehabilitating and maintaining the health of these matrices. On the other hand the maintenance of ecosystem health in humanly inhabited and economically exploited areas depends upon the existence of reservoirs of biological diversity and integrity.[29]

With the notion of a spectrum of conservation philosophies and nature management styles between the poles of less and more severely degraded areas, this quote suggests an answer to our question with respect to a multiple vision on ecological restoration. This notion should, however, be expanded to include environmental engineering and ecological engineering (or ecotechnology) as well. After all, metaphors from the domain of engineering and cybernetics share a marked functionalism with metaphors from the domain of medicine and health care, in apparent contrast to metaphors from the field of art and aesthetics which have an obvious compositionalist emphasis (Figure 22.1).

DISCUSSION

The notion of a spectrum or scale ranging from less to most severely degraded areas can be useful in addressing potential problems of communication, coordination, and cooperation between various scientific (sub) disciplines, management practices, and the general public. On the other hand, this notion is also problematic because in many cases there is simply no agreement on what should count as degraded or not. Take the so-called Chicago Restoration Controversy that broke out as soon as plans were made public to clear forestland to create prairies and open woodlands. The restoration stewards, who considered the actual landscape as "degraded" and "overgrown," invited their critics to visit some of their preserves in the hope to win them over. But this was unsuccessful. Whereas the restoration stewards valued the recently burned woodlands as beautiful sites that were

	Sources of Metaphor			
Most Degraded Areas	Engineering & Cybernetics	Medicine & Health Care	Art & Aesthetics	Less Degraded Areas
	Geography: Islands & Networks			

Figure 22.1 Sources of Metaphors on a Scale from Most to Less Degraded Areas.

"finally opened up so that the native understory could return," the critics saw them as "virtually wastelands where the soil is sterile and nothing but charred stumps remain." For one group it was an exciting and promising site; for the other it was a sad and scary portent of the future of the land.[30]

It is clear from this and many other controversies that what people consider as degraded areas depends on their perceptions of nature, and in particular on the metaphors they use to express their attitude toward nature. As we saw in the previous section, Callicott and co-workers also specify their notion of degradation with the help of a metaphor, namely the metaphor of nature areas as islands that are threatened to be engulfed by the sea of cultivated land.

At this point we risk running in a circle: the possibility of classifying metaphors according to the kind of site (more to less degraded) seems to depend on the possibility of yet another metaphor. This suggests that it is impossible to break out of the cycle of metaphors that allows us to find criteria for evaluating them. But (and this the first step to resolving the problem) this circle should not be seen as a vicious circle that one must break out of, but as an hermeneutic circle of the sum and its parts, of text and context, that we must try to enter in the right way.

The force of metaphors depends on the whole system, on the "ecology" of metaphors, that together constitute the cultural context and cultural climate of a particular era. Whether metaphors can acquire significant cognitive, discursive, and normative power is determined by the degree to which they resonate with the period's dominant social values and visions. As we have seen, the island metaphor and adjacent network metaphor seem to have the best credentials for today's professionals and larger public. Given this "working consensus," I believe that my proposal to arrange the different metaphors based on Callicott's notion of a spectrum from less to most degraded areas seems justified.

Of course, this proposal cannot claim absolute validity, but it is context specific and relative with respect to time, place, and circumstance. We should be cognizant of our hermeneutic situation, but we should also proceed pragmatically while guarding against reflexivity degenerating into hyper-reflexivity. As Ludwig Wittgenstein so aptly sums up his essay collection, *On Certainty*: "We just *can't* investigate everything, and for that reason we are forced to rest content with assumption. If I want the door to turn, the hinges must stay put."[31]

CONCLUSION

The most common metaphors in framing ecological restoration have been explored here in some detail. Acknowledging the fact that every metaphor inescapably falls short in some respect, I argue that the search for the one best metaphor is a dead end and that we should instead develop a multiple vision. This chapter aims to create some order out of the chaos of different and often diverging metaphors. My proposal to arrange these metaphors according to a scale from less to most degraded areas (however incomplete or imperfect) could improve communication and cooperation between people from different disciplines, between practitioners and theoreticians, and between experts and laypeople.

NOTES

1. S. C. McCutcheon and W. J. Mitsch, "Ecological and Environmental Engineering: Potential For Progress," *Ecological Engineering* 3 (1994): 107–09.
2. S. C. McCutcheon and T. M. Walski, "Ecological Engineers: Friend Or Foe?," *Ecological Engineering* 3 (1994): 109–12; W. J. Mitsch, "Ecological Engineering: A Friend," *Ecological Engineering* 3 (1994): 112–15.
3. P. J. Taylor, "Technocratic Optimism," *Journal of the History of Biology* 21:2 (1988): 213–44.
4. S. Pepper, *World Hypotheses: A Study in Evidence* (Berkeley: University of California Press, 1942).
5. C. L. Kwa, "Representation of Nature Mediating between Ecology and Science Policy: The Case of the International Biological Programme," *Social Studies of Science* 17:3 (1987): 413–42.
6. H. Bergson, *L'Evolution Créatrice* (Paris: Alcan, 1907); A. N. Whitehead, *Science and the Modern World* (New York: Macmillan, 1925); C. Merchant, *The Death of Nature: Women, Ecology and the Scientific Revolution* (New York: Harper and Row, 1980); M. Berman, "The Cybernetic Dream of the Twenty-First Century," *Journal of Humanistic Psychology* 26:2 (1986): 24–51.

7. M. Meisner, "Metaphors of nature: Old vinegar in new bottles?," *Trumpeter* 9:4 (1992), 8.
8. J. Keulartz, *The Struggle for Nature. A Critique of Radical Ecology* (London and New York: Routledge, 1998), 149.
9. D. Worster, *Nature's Economy: A History of Ecological Ideas* (Cambridge: Cambridge University Press, 1977), 314.
10. SER Science & Policy Working Group, *The SER International Primer on Ecological Restoration* (2004), available at: http:/www.ser.org, 12.
11. P. H. Gobster and R. B. Hull, eds., *Restoring Nature: Perspectives from the Social Sciences and Humanities* (Wash., D.C.: Island Press, 2000); W. Throop, ed., *Environmental Restoration: Ethics, Theory, and Practice* (Amherst, NY: Humanity Books, 2000); R. Elliot, "Faking Nature," *Inquiry* 25 (1982): 81–93.
12. R. Elliot, "Faking nature," in *Environmental Ethics. An Anthology*, ed. by A. Light and H. Rolston III (Cambridge: Blackwell Publishers, 2003), 396.
13. M. Sagoff, "On Restoring and Reproducing Art," *The Journal of Philosophy* 75:9 (1978), 457.
14. A. Light, "*Faking Nature* Revisited," in *The Beauty Around Us: Environmental Aesthetics in the Scenic Landscape and Beyond*, ed. By D. Michelfelder and B. Wilcox (Albany, NY: SUNY Press, 2003).
15. H. Rolston, "Restoration," in Throop, *Environmental Restoration*, 127–32; see, for example, W. R. Jordan, III, "Restoration and Management as Theater," *Restoration & Management Notes* 5: 2 (1987); W. R. Jordan, "Sunflower Forest," in Throop, *Environmental Restoration*, 215.
16. A. Leopold, *A Sand County Almanac* (New York: Oxford University Press, 1949), 240.
17. D. Rapport et al., "Ecosystem Health: The concept, The ISEH, and the important tasks ahead," *Ecosystem Health* 5:2 (1999), 84.
18. L. Kristjanson and R. Hobbs, "Degrading landscapes: Lessons from palliative care," *Ecosystem Health* 7:4 (2001): 203–13.
19. P. Martens et al., "Globalisation, Environmental Change and Health," *Global Change & Human Health* 1:1 (2000): 4–8.
20. D. Rapport et al., eds., *Ecosystem health* (Malden, MA: Blackwell Science, 1998).
21. J. J. Harris and R. J. Hobbs, "Clinical Practice for Ecosystem Health: The Role of Ecological Restoration," *Ecosystem Health* 7:4 (2001), 200.
22. M. A. Davis and L. B. Slobodkin, "The Science and Values of Restoration Ecology," *Restoration Ecology* 12:1 (2004), 1.
23. M. Hammond and A. Holland, "Ecosystem health: Some prognostications," *Environmental Values* 4 (1995), 285, 286.
24. Kwa, "Representation of Nature . . . ," 427.
25. D. J. Philippon, *Conserving words. How American nature writers shaped the environmental movement* (Athens: The University of Georgia Press, 2004), 270.
26. J. G. Fabos and R. L. Ryan, "An Introduction to Greenway Planning around the World," *Landscape and Urban Planning* 76 (2006): 1–6.
27. G. Lakoff and M. Johnson, *Metaphors we live by* (Chicago: University of Chicago Press, 1980), 14; S. Ebenreck, "Opening Pandora's box: Imagination's role in environmental ethics," *Environmental Ethics* 18 (1996), 14; Meisner, "Metaphors of nature."
28. D. Schön and M. Rein, *Frame reflection* (New York: Basic Books, 1994), 207; D. Idhe, *Philosophy of technology: An introduction* (New York: Paragon, 1993).

29. J. B. Callicott et al., "Current normative concepts in conservation," *Conservation Biology* 13:1 (1999), 32.
30. R. M. Helford, "Constructing nature as constructing science: expertise, activist science, and public conflict in the Chicago wilderness," in Gobster and Hull, *Restoring Nature*, 130.
31. L. Wittgenstein, *Über Gewissheit* [On Certainty], ed. by G. E. M. Anscombe and G. H. von Wright (Oxford, UK: Basil Blackwell, 1969–1975), proposition 343.

23 Rewilding the Restorer

David Kidner

CONVENTIONAL UNDERSTANDINGS OF RESTORATION

The Science & Policy Working Group of the Society for Ecological Restoration International defines restoration as "the process of assisting the recovery of an ecosystem that has been degraded, damaged, or destroyed."[1] This definition seems to carry certain implicit assumptions which merit examination. We know, for example, that industrial humanity has the power to "degrade, damage, or destroy" ecosystems; but does this imply a complementary power to "assist their recovery"? Can the same scientific knowledge that is used to convert ecosystems into "raw materials" for industrial processes also be used to reestablish wild nature; or does the latter task *also* require a different type of knowledge? And what do we understand by "recovery"? Furthermore, what qualities are required of the restorer, coming as we do from an industrialist context that has often proved destructive to wild nature; and what are our motives in attempting to restore ecosystems? In this chapter, I will explore some of these seldom addressed questions.

THE UNWILDING OF EXPERIENCE

The story of our psychological distancing from the world and the subjugation of wildness by industrialist ideologies is well known, and I will not repeat it here. Many writers have focused on the Enlightenment and the "scientific revolution" as particularly significant eras in the splitting of an autonomous, thinking individual from a world conceived of as understandable and controllable through thought, although one can trace the roots of these changes at least back to Plato.[2] This process of separation left an imprint which profoundly affects the ways we think and behave today, not least with respect to our attitudes to the natural world. I want to focus here on one element of this historical imprint—the repression, and possible restoration, of wildness both in ourselves and in the outside world.

Although in conventional terms the disappearance of wildness from the "external" world is viewed as having little relation to our own self-definition, the work of various social theorists suggests the existence of this relation, albeit often unwittingly. Freud's work, for example, incorporates an ideology of modernization, drawing explicit parallels between the domestication of the physical world and the development of the "civilized," thinking, self. In recognizing the conflict between a (largely unconscious) embodied sense of self—the "Id"—and a consciously foregrounded conceptual sense of self (the "Ego"), he sees the relation between the two as "that of a rider to his horse. The horse supplies the locomotive energy, while the rider has the privilege of deciding on the goal and of guiding the powerful animal's movement."[3] Wildness, both in the self and in the outside world, is viewed as structureless energy, ready to be harnessed and used by means of human intelligence. Pursuing an unwilding agenda both in the external world and within the personality, Freud declares that "[w]here Id was, there Ego shall be," comparing it with "the draining of the Zuider Zee." Elsewhere, he proposes a "dictatorship of the intellect," suggesting that the solution to our "discontents" lies in "going over to the attack against nature and subjecting her to human will."[4] Furthermore,

> [in] countries which have attained a high level of civilisation . . . we find that everything which can assist in the exploitation of the earth by man and in his protection against the forces of nature . . . is attended to and effectively carried out. . . . Wild and dangerous animals have been exterminated, and the breeding of domestic animals flourishes.[5]

Freud here exemplifies the spirit of his, and our, age: one in which the dominance of technological consciousness and the domestication of the landscape are two facets of a common project involving the rationalization of nature.[6] Just as domesticated landscapes tend to be monocultures of few geometrically arranged species, so rational thought is a simplification of our wild capacities, achieved by foregrounding certain selected factors and relegating others to insignificance. A related impetus to dominate and restructure pervades such fashionable contemporary approaches as cultural constructionism which view landscapes as humanly created, implying that non-human influences are insignificant. As one participant at a recent conference complained to me, "Why are people talking about the natural world? It's all cultural these days!" Another writer argues that because shards of ancient pots have been found deep in the Amazon jungle, this demonstrates that even such remote areas are "not just construed but actually created by humans. The Amazon wilderness, in the sense of a place undetermined by humankind, does not exist."[7] While not without a grain of truth, such claims grossly overstate human powers while denying the vital contributions of other species. These trends of contemporary thought express a common hostility to the wild other, redefining wildness in ways that facilitate its assimilation by an industrialist agenda. Rewilding, as an attempt to reverse some of the destructive effects of this rationalising project, will therefore need not only to recognize the value of wild ecosystems and their component species, but also to lay bare the shared historical roots of contemporary individuality and the rationalized landscape. As John Rodman puts it, our domestication of nature and

> our division of ourselves into a "human" part that rules or ought to rule and a "bestial" part that is ruled or ought to be ruled are by now so hopelessly intertwined that it seems doubtful that we could significantly change the one without changing the other. I say this not to postpone efforts to change "the system" until we have revolutionised consciousness, for I cannot imagine the latter occurring in isolation, so much do the "inner" and "outer" interpenetrate one another, but because I agree with Marcuse that "revolutions" fail when led by people who perpetuate in their character the authority structure of the old regime.[8]

Like Rodman, I do not see the "liberation of nature" as "reducible to human liberation, but only as inseparable from it."[9] Consequently, if our efforts at restoration are tacitly based around conceptual templates that are experienced as prior to and unchangeable by our experience of particular ecosystems, it is likely that we will reproduce in some form a humanly dominated landscape rather than a genuinely wild one. It is perhaps significant that among peoples who actually inhabit relatively wild environments, thought is typically not prior to experience of the landscape, but is rather engaged in a sort of ongoing dialectic with it—a point I will elaborate on later.

As conventionally viewed, human and ecological health are dealt with by different disciplines, understood through different theories, and ascribed to separate causes, turning the relation between these two forms of well-being into a sort of zero-sum game in which our needs must be "balanced" *against* those of the environment. But the separation of human and ecological health may owe more to intellectual history than to physical realities; and it is interesting that the available evidence to date suggests that the capacity to confer health operates most strongly in the other direction: that is, while our record in successfully restoring ecosystems is somewhat patchy,[10] physically experiencing nature or simply viewing pictures of it has been found to directly benefit both psychological and physical health.[11] These findings suggest that our interactions with nature do not simply reflect the use of human intelligence either to exploit or restore the natural world, but may more realistically involve a mutual dialectic of influence and health. What we refer to as *rewilding* may be the application to one specific external arena of an epistemological upheaval which at a less conscious level also engulfs ourselves.

From this perspective, the rejuvenation of near-extinct ecosystems is not *only* an ecological event, but one which resonates with many areas of knowledge and experience. The way we are drawn to the past may not be so much a "romantic fantasy" of a "mythical Eden," but rather a deeply sensed (if not conceptually lucid) recognition that the past is intrinsically embodied in our genetic makeup and that of the flora and fauna of an ecosystem, reflecting the long process of co-adaptation to a particular sort of world. From this point of view, wildness is an ever-*present* human and ecological potential that strives to be expressed, not an ungrounded fantasy of another era; and this intuition expands our understanding of rewilding from the restoration of a past ecosystem to the wider facilitation of a (human and non-human) potential which already exists, stunted and repressed by current ideological actualities.[12]

PITFALLS OF UNWILD THOUGHT

On an evolutionary time-scale, conceptual thought has evolved very recently, often in relation to a "civilizing" agenda which involves controlling and taming various aspects of the physical world. Our ability to sense and respond intelligently to the world, however, has evolved over much longer time-spans. The notion of relying solely on *conceptual* thought to "re-wild" the world therefore has a certain paradoxical undertone. Is it not plausible, then, that any attempt at rewilding will also require some sort of input from those archaic, arational types of sensing which evolved to be consistent with a pre-industrial landscape? If this is the case, then rewilding would demand another, parallel sort of restoration: the recovery of those often repressed forms of awareness which are inherent in our embodied character as evolved creatures, and which are more suited than conceptual thought to dealing with complex systems.[13]

Science is such a powerful and effective means of understanding that we may forget that it omits significant aspects of both the world and ourselves—most importantly for present purposes, those we might classify as "wild." It is the very success of science that causes us to see it not merely as a model *of* nature, but as a model *for* nature, implying the exclusion of whatever is not scientifically recognizable. In Jean-Pierre Dupuy's terms, "nature is taken to imitate the very model by which man tries to imitate it."[14] This (usually unrecognized) epistemological reversal is particularly problematic for rewilding, which attempts to reassert the priority and the very existence of precisely those wild aspects of the world which conceptual thought so often overlooks. Some of the problems which arise from this forgetting of the historical and ontological contexts of conceptual thought include the following:

1. The concept of restoration, in common with other "re-words," as David Lowenthal points out in a previous chapter, carries an implication

of revivifying some previous state. *Conceptually*, this is straightforward: as cognitively capable adults, it is simple for us to press the "rewind" button and mentally recreate past situations. In the natural world, however, the past is often *incorporated* rather than recreated; and the record of attempts to reinstate some past ecological situation is not impressive.[15]

However, this does not imply that restorers can ignore the past and focus on our present conceptions of the sort of ecosystem we would like to create. For example, if "ghost" species which are essential to the reconstitution of the ecosystem are extinct, reassembly will be impossible even if all the constituent species of the target ecosystem are present.[16] More generally, the trajectory of ecosystems has been shown to be heavily dependent on the *history* of the system, and particularly on the *order* in which species are reintroduced. As Daniel Botkin puts it: "The biosphere has had a history and what it will be tomorrow depends not only on what it is today, but also on what it was yesterday. Like an organism, the biosphere proceeds through its existence in a one-way direction, passing from stage to stage, each of which cannot be revisited."[17] The structure of ecosystems, in other words, includes a *temporal* dimension. This implies a shift of emphasis for restoration—away from the (conceptually manageable but ecologically unrealistic) notion of recreating a past state towards the (conceptually more elusive but ecologically realistic) restitution of a healthy trajectory that incorporates and builds on elements from the past. This suggests a more delimited role for the human restorer and a correspondingly greater reliance on the inherent intelligence stored within reintroduced species' genetic makeup, together with more humble expectations of our conceptual models' predictive abilities. These lowered predictive expectations, however, do not simply reflect regrettable inadequacies of our scientific models but are, rather, welcome signs that the wild ecosystems we are restoring are indeed wild, and so are able to outstrip our cognitive capabilities.

2. The tacit assumption that the natural world is made up of discrete "things" reflects our confusion between concepts and reality. The cognitive world is indeed made up of separate entities; but the natural world may be better modelled as a *nexus of relations*—between plants and pollinators, predators and prey species, mutualists, parasites, and host species, all of which are involved in often very complex and conceptually opaque changes over time. Because of these conceptual difficulties, there is often a cognitive slippage towards the reification of more readily identifiable ecosystem "components," and the abandonment of less easily identifiable aspects.

For example, Jack Turner notes that there is a shift "from wildness to wilderness, then biological communities, and recently, to biodiversity. Wildness as a quality and its relation to other qualities is now rarely discussed. This shift was broadly materialist—a move from quality to quantity, to acreage, species, and physical relations."[18] This slippage away from the conceptually elusive *quality* of wildness to the conceptual manageability of measurable quantities and clear delineation of entities is itself de-wilding, since it moves us from the resonant *experience* of nature towards a simpler *conceptual* taxonomy of characteristics that are abstracted from the world and placed in a cognitive context. Tim Ingold points to an example of this shift: "every creature is specified in its essential nature through the bestowal of attributes passed down along lines of descent, independently and in advance of its placement in the world."[19] Consequently, ecological relation becomes secondary to individuality, so that creatures are *firstly* defined according to their "intrinsic" attributes, and then, individuality and separateness having been safely established, can be allowed a degree of (scientifically understood) relation.

Similarly, Gary Paul Nabhan points out that focusing on biodiversity can distract us from the need to attend to "other levels of biological organisation [such as] genetic variations within populations, variability between populations, habitat heterogeneity, ecosystem diversity that are more difficult to monitor or measure." A related problem is the assumption that greater biodiversity should *necessarily* be equated with a "healthier" ecosystem. A landscape is not necessarily "healthier" if our interventions increase the number of species that are present. The grammar of nature contains pauses and absences as well as presences; and there is a danger of creating wildlife parks rather than well-functioning ecosystems.[20]

3. There is a danger, too, that conceptual clarity is imposed on landscapes in ways that are sometimes damaging. For example, Frans Vera has shown in an earlier chapter that our attempts to restore, say, a wild forest may be tainted by unrealistically clear boundaries between different types of ecosystem. Mediaeval forests may have been much more open than the term "forest" currently implies.[21] And the clear conceptual distinctions between, say, "forest" and "grassland" may, if unthinkingly transposed into the practice of restoration, result in an equally clear, but unnatural, discontinuity between these two types of landscape. Furthermore, such distinctions often imply a static view of ecosystems that does unwitting violence to their dynamic character. For example, "woodland" may naturally involve an alternation between woodland and savannah.[22] In "restoring" woodland, then, there is a danger that we may unintentionally be eliminating the temporal dynamics of ecosystems.

Similarly, Ellen Wohl points out that river restoration projects sometimes problematically focus on either ecosystemic function or appearance. In the latter case, there is a danger of creating a river that

> preserves a simplified albeit attractive form but has lost function because the hydrologic and geomorphic processes no longer create and maintain the habitat and natural disturbance regime necessary to ecosystem integrity.[23]

At the other extreme, as Eric Higgs points out in this volume, restoration may focus on such indices as "efficiency," as is the case if tree planting is seen as a means of carbon sequestration. Such conceptual foci are over-simplifications of necessary complexities, and are potentially damaging to wildness.

My own views concerning conceptual clarity were reinforced by an experience I had a dozen or so years ago. My office was in Clifton Hall, a large Georgian pile on the outskirts of Nottingham, adjacent to several acres of heavily overgrown landscaped gardens. During my lunch break, I used to explore the just-discernable paths that penetrated the dense chaos of tangled exotic shrubs and trees, an oasis of wildness in the midst of a somewhat drab, domesticated landscape. One day I was shocked to discover that a 50-yard strip had been bulldozed along the entire length of the gardens. I rang up the local council and was eventually transferred to the person responsible for this barbarous act. He explained enthusiastically that he was organizing the "restoration" of the gardens, removing all the overgrown exotics and replacing them with deciduous native species. Such acts prioritize an *idea* of wildness over the physical embodiment of wildness, as well as implicitly disregarding the capacity of a domesticated landscape to rewild itself over time.

4. Much writing on restoration seems to be based on the assumption that if only we knew enough about ecosystems, we could make interventions that are precise enough to ensure certain outcomes. This, however, may reflect a misreading of the world. Uncertainty and probability do not *only* reflect our lack of knowledge; they may also "reflect an indeterminacy in the process itself" which may be an essential aspect of wildness.[24] The notion that the world is in principle entirely understandable is a prejudice stemming from the replacement, especially since the seventeenth century, of the view that nature is a mystery by the portrayal of nature as a machine (to add to Keulartz's metaphorical discussion in the previous chapter). As is the case with most swings of the epistemological pendulum, the former view may not be entirely devoid of merit. As Robert Ulanowicz summarizes Karl Popper's viewpoint,

> the deterministic realm where forces and laws prevail is but a small, almost vanishing subset of all real phenomena, [which] are suffused with indeterminacies that confound efforts at deterministic prediction. Popper's is not the schizoid world of strict forces and stochastic probabilities, but rather a more encompassing one of conditioned probabilities, or deep-seated *propensities* that are always influenced by their context or environment.[25]

Wildness is indeed a tangled jungle of such "propensities," inherently resistant to conceptual penetration. This is not, by the way, an argument for some sort of postmodern relativism which suggests that "any nature is as good as any other," implying that our aims should be determined by human preferences. It is, rather, an argument for human humility in the

Figure 23.1   Here are samples of an unrestored garden plot (above) and a nearby plot that was "restored" approximately 12 years previoulsy (below), both near Nottingham. (Photographs by David Kidner)

face of a reality that is, in the words of an oft-quoted aphorism, not only more complex than we know, but more complex than we *can* know. This leaves scientists with a choice: either to admit that our cognitive models are no match for the complexity of nature, or else to claim that nature itself is essentially random and disordered. Predictably, since we generally prefer the pretence of omniscience to the admission of fallibility, the latter is the more popular option.

The inadequacy of this preference, however, becomes readily apparent in restoration practice; and a more realistic choice is a *dialectical* approach to restoration which draws us into the ecosystem rather than allowing us to stand outside it and plan our interventions from afar. In other words, we may tentatively try a particular intervention, then see how the system reacts before deciding upon our next move. This, as it happens, is rather consistent with the attitude of many indigenous groups. For example, Jonathan Long and his colleagues, discussing the approach to restoration taken in the White Mountain Apache reservation, observe that "because nature directs the recovery, practitioners may not plan a full course of treatment until observing how a site responds." In the words of one tribal member, "You go to a place and do some work for it. You let it rest, and then you come back to it to see what it has done. Then it thanks you."[26] Such a dialectical approach opens itself to the intelligence—or, to use Ulanowicz's term— *propensities* of the system, so that aligning our own intelligence with that of the system results in a sort of *cooperative* effort. Seen from this perspective, the *conceptual* distinction between preservation and restoration tends to diminish or disappear, to be replaced by a common emphasis on *participation*.[27]

5. Robert Hilderbrand and his colleagues have pointed out that attempts to "fast forward" the recovery of ecosystems are often unsuccessful.[28] The capacity for "fast forwarding," like the possibility of "rewinding," is a *conceptual* property of operational thought, not a characteristic of the natural world; and the attempt to speed up natural processes may reflect a dissociation from nature and the influence of the industrialist ideology of rapid change and "development." As restorationists realize this, so we adapt our thinking towards a more continuous, gradual approach. As Steve Windhager puts it, "[o]riginally we thought we could just repair the past damage, get the ball rolling again, and then sever our ties and let the restored system go on without us. What we have found, however, is that our ongoing involvement is necessary to ensure the continued health of the ecosystem. Originally this was taken as a failure of restoration, but I have come to look at it as a success."[29] What seems to be happening here is that our epistemology—and hence the boundaries of the rewilding process—are being shaped to *include us*, within a wider process that is mostly beyond conscious intention.

6. Frequently, environmental fashion involves an alternation between conceptually straightforward beliefs that are equally misleading. For example,

the notion that wilderness is necessarily unpeopled has been largely replaced by the equally problematic "realization" that wilderness is "culturally constructed." Steve Packard astutely points out that what the first viewpoint preserves is "an idea of nature;"[30] but the same point is at least as strongly applicable to the second, constructionist viewpoint. Another example is that of the "climax state" as a desirable endpoint for restoration. The reaction to this view—that nature involves constant change, sometimes being used to support the implication that *any* change is acceptable—is just as improbable. Similarly, the movement away from "noble savagery" towards the just as unrealistic view that all peoples are equally destructive of their natural environments—so that, as Alf Hornborg sardonically puts it, "there emerges the new but implicit message that we have always been capitalists,"—is more symptomatic of the lack of connection between academic convention and cultural realities than of any profound new insights.[31] Sometimes such swings reflect the larger currents of political ideology. For example, Temperton and Hobbs argue that "scientists from socialist countries have historically placed more emphasis on mutualism and symbiosis than their counterparts in capitalist countries;" and Colinvaux dismisses the notion of ecosystemic behavior as better explained by "private enterprise."[32] Such examples reflect conceptual thought's tendency to slip into lucid, well- defined, and entirely misleading polarizations of complex situations; and any genuine rewilding will anchor itself through sensed connections to specific landscapes rather than allowing itself to drift rudderlessly in the treacherous eddies of academic fashion.

By now, I suspect, many practicing restorationists will be muttering "Of course! That's what I do anyway. . . ." But that's the point. While a Martian reader of the restoration literature might understand the field as being dominated by a purely scientific viewpoint, in practice restoration is leavened with a fair dose of intuition, passion, and embodied insight. Such qualities, however, often lead a shadowy, unacknowledged existence within a world in which the overt face of restoration is necessarily one of scientific rationality. While restorationist *discourse* is governed by scientific rationality, our *behavior* as restorers can be viewed as symptomatic of motives which are less consciously acknowledged.

These limitations of conceptual thought, which illustrate that cognition involves a reduction and often a distortion of ecosystemic realities, should make us cautious about leaning too heavily on cognition. As I will show in the next section, anthropological work leads us towards the same conclusion.

CROSS–CULTURAL INSIGHTS

Like non-industrial peoples, we have evolved to interact with the natural world through *all* our faculties; but the industrial era has seen a rapid constriction of thought towards a form of "rationality" which derives its enormous power from its ability to focus on a few variables and to exclude "irrelevant" ones. This very focus, however, may also be a limitation in our attempts to restore natural systems, which cannot realistically be reduced to a few interacting variables. Although it is important not to polarize the differences between non-industrialized peoples and ourselves—especially given the global influence of industrialism—it may be that the embodied, world-related character of what Medin and Atran refer to as "folkbiological" knowledge holds useful lessons for the rewilder.[33] As Hugh Brody says of the Innu of Labrador, "reasoning is subliminal, and therefore has the potential to be more sophisticated, more a matter of assigning weight to factors, than can be the case with linear logic."[34] In a similar vein, Raymond Rogers' remarks about wild animals conveys a suspicion about constructs and principles that can equally be applied to tribal peoples and rewilders:

> It is the wholeness of the wild animal that distinguishes it from the experientially deprived domesticate. It is the wholeness of the wild animal that makes ethical constructs unnecessary—indeed, probably unthinkable. Why create an abstract set of rules and guidelines when you are already doing all the right social things, and always have? Rules and guidelines are for domesticates.[35]

We need to recognize that attempting to restore a wild ecosystem differs fundamentally from manufacturing an artifact. If we are, say, designing a bridge, then focusing on a few selected variables such as tensile strength and gross weight may well be appropriate. The rewilder, however—like the inhabitant of the non-industrial world—is not designing something that can be simply defined, but is trying to rejuvenate a natural system that embodies a multitude of interacting processes and variables. As Ernest Gellner notes, our behavior in industrial societies is often

> governed by a single aim or criterion . . . A man making a purchase is simply interested in buying the best commodity, at the least price. Not so in a multi-stranded social context: a man buying something from a village neighbour in a tribal community is dealing not only with a seller, but also with a kinsman, collaborator, ally or rival, potential supplier of a bride for his son, fellow juryman, ritual participant, fellow defender of the village, fellow council member. . . . All these multiple relations will enter into the economic operation, and restrain either party from looking only to the gain and loss involved in that operation, taken in isolation. In such a many-stranded context, there can be no question of "rational" economic conduct, governed by the single-minded pursuit of maximum gain. . . . When there is a multiplicity of incommensurate values, some imponderable, a man can only *feel*, and allow his feelings to be guided by the overall expectations or preconceptions of his culture. He cannot calculate.[36]

Industrialism and its artifacts are built on the rejection of embodied experience; and Descartes and his successors saw this as essential if we are to become "masters and possessors of nature." In contrast, rewilding is not simply an attempt to reconstitute a different sort of ("wild") artifact, but involves a re-embracing of precisely those (human and more-than-human) qualities which have been extinguished during the *un*-wilding of the world and the isolation of the "individual" from the "environment." The wholeness which Rogers emphasizes, then, is more than an *individual* wholeness: it is one which extends into the world, implicitly questioning our isolation *from* the world and transforming the character of selfhood.

This implies not only an alternative *knowledge*, but also a different way of *being*. Rather than withdrawing into a realm of manufactured thought, language, and sociality, the hunter–gatherer engages with the world with all the faculties that have evolved for exactly this task, including those we push aside in our drive toward single-minded rationality. Among such peoples, Tim Ingold suggests, knowledge of the world is gained

> by moving about in it, exploring it, attending to it, ever alert to the signs by which it is revealed. Learning to see, then, is a matter not of acquiring schemata for mentally constructing the environment but of acquiring the skills for direct perceptual engagement with its constituents, human and non-human, animate and inanimate.[37]

In other words, prolonged intimate contact with a landscape—unsurprisingly—seems to harmonize thought and embodied experience with the world. In Ingold's terms,

> In the hunter–gatherer economy of knowledge . . . it is as entire persons, not as disembodied minds, that human beings engage with one another and, moreover, with non-human beings as well. They do so as beings in a world, not as minds which, excluded from a given reality, find themselves . . . having to make sense of it.[38]

Ingold gives many examples of peoples who live within what he refers to as a "sentient ecology," involving "a knowledge not of a formal, authorised kind . . . [but one] based in feeling, consisting in the skills, sensitivities, and orientations that have developed through long experience of conducting one's life in a particular environment. . . . Another word for this kind of sensitivity and responsiveness is *intuition*."[39] Put differently, restoration has as much to do with the heart as with the mind, as much with art as technology.

It is sobering to realize that those non-industrial peoples who have a sustainable relation to the natural world often assign less priority to conceptual understanding than we do, suggesting that scientific rationality may not always be an advantage in this respect. In a study of three Guatemalan forest dwelling groups, for example, Scott Atran found that although the Itzaj of the Petén Maya lowlands are better conservationists, and recognize more mutualist relations, than other groups he studied, they have little in the way of social organization that would encourage conservation, and seem rather individualistic. But a more intimate familiarity with the Itzaj reveals that they embody an "emergent knowledge structure." As Atran explains, an

> emergent knowledge structure is not a set body of knowledge or tradition that is taught or learned as shared content. . . . The general idea is that one's cultural upbringing primes one to pay attention to certain observable relationships . . . [For Itzaj,] there is no "principle of reciprocity" applied to forest entities, no "rules for appropriate conduct" in the forest, and no "controlled experimental determinations" of the fitness of ecological relationships. Yet reciprocity is all pervasive and fitness enduring.[40]

It is worth emphasizing that the focus here is on "certain observable relationships": in other words the relationships *are in the world*, not (primarily) in the mind and *applied to* the world. This atunement to the landscape, as Ingold points out, is often inverted by theorists, so that "[a]stonishingly . . . meanings that people claim to discover *in* the landscape are attributed to the minds of people themselves and are said to be mapped *onto* the landscape."[41] The lesson for rewilding is clear: if rewilding is to have a chance of success, we need to attend to natural realities and to beware of imposing conceptual prejudices.

THE HEART AND MIND OF RESTORATION

Despite the partial colonization of our minds by an industrialist ideology which assumes our emotional self-exclusion from the natural order, we still retain an embodied capacity for relation, albeit often mutely. A good illustration of this is provided by Robert Ryan's work, which demonstrates the gulf between a felt and a conceptual understanding of the natural world. Ryan studied the attitudes of various groups towards an urban natural area, finding that those people who actively *used* the area for recreational purposes developed a "place-specific attachment," favoring minimal management and "letting nature take its course." "To many of them," Ryan states, "any tree or shrub, regardless of species, has a place in these natural areas and should be allowed to grow undisturbed."[42] On the other hand,

> staff and volunteer restorationists expressed a more conceptual attachment; they were attached to a particular type of natural landscape such as a prairie rather than to a specific place. For these study participants,

seeking another site would be an acceptable option if the one they were working on might change in a negative manner. This lack of place dependency suggests a substitutability of natural areas that is not shared by the other users.[43]

Those expressing a conceptual attachment also held positive attitudes towards more intensive management practices such as cutting exotic trees and using herbicides. "Conceptual attachment" therefore seems to involve an attachment to *ideas* rather than physical realities, and so seems to exist within a Platonic realm of 'forms' rather than reaching out into the landscape itself. A consequence is the enthusiasm for a more active approach to shaping the landscape towards a conceptually held ideal, "fuelled more by such concepts as ecosystem integrity, wildlife habitat, and species diversity than by the actual place itself."[44]

On the other hand, the "place specific" attachment—as the term suggests—is a more embodied relation to a particular place, and has much in common with a relationship to another *person*. Indeed, the psychological aspects of the attachments to place and to a person are very similar,[45] suggesting that at an embodied level the ontological discontinuity between the human "subject" and the non-human "object" has dissolved. Just as most of us would not be happy to swap our particular spouse or partner with another who had "similar characteristics," so the users of natural areas in Ryan's study were reluctant to transfer their attachment to another area. This fidelity to place suggests that the direction of influence between place and person is reciprocal rather than one-way—in other words, that the person's experience is changed by the attachment as well as the place being affected by the person's actions. Furthermore, since the attachment is to a particular place, rather than to an idea of what the place *should* be, it is likely that the person becomes a conservative force *within* the ecosystem rather than an external influence *outside* it. These differences between the "conceptual" and the "place specific" attachments—and here I am extrapolating from Ryan's study—suggest the presence of two different epistemologies, which I will refer to as the "conceptual" and "embodied" epistemologies, the main characteristics of which I will summarize in Table 23.1.

I emphasize that both these epistemologies exist within all of us to varying extents, implying that some of the uncertainties about restoration are expressions of a pre-existing, culturally chronic conflict within the contemporary person. Many psychological and psychotherapeutic approaches have recognized this conflict; and these include Freud's distinction, mentioned earlier, between Id and Ego; Carl Rogers's account of the "incongruence" between "Experience" and "Self-concept"; and Carl Jung's discussion of the collective and egoic aspects of self.[46] These aspects of our being express the conflict between the two great systems that currently govern our lives: nature and industrialism. Unfortunately, contemporary theories such as

Table 23.1 Conceptual and Embodied Eepistemologies

Conceptual epistemology	Embodied epistemology
Fundamental discontinuity between restorer and restored	Restorer and restored become integrated
Tendency towards actively changing the landscape	Tendency towards leaving things alone
Attachment to ideas of ideal landscape	Attachment to one particular place
Experience of place as object	Experience of place as subject
Distinction between existing and "ideal" states	Feeling for history of ecosystem
Dualistic approach	Monistic approach
I – It relation	I – Thou relation
Understanding	Empathy and resonance
Control and mastery	Openness to the landscape's own direction
Only we are intelligent	Landscape as intelligent

social constructionism often deal with this conflict by ignoring our embodiedness, or redefining it as a by-product of discourse, paralleling the redefinition of wilderness as a "product of civilization."[47] In other words, just as industrialism eliminates wildness by redefining natural entities as "raw materials" for industrial processes, so the social sciences often achieve the same ends by redefining our own natural propensities as the products of conceptual or linguistic processes.[48] Any approach that relies on this sort of ideological railroading will not only be incapable of contributing to the rewilding project, but will be covertly hostile to it.

Because of the conscious dominance of the conceptual orientation, our beliefs—and indirectly our actions—incorporate many of the characteristics of cognition, which may or may not accurately represent those of the natural world. Nevertheless, at an embodied and only partly conscious level, we retain a strong "gut level" feeling that often diverges from our rational understanding of what needs to happen. An example is Eugene Hargrove's discussion of Ian Douglas-Hamilton's research in a Tanzanian national park. In this case, elephants were demolishing most of the trees, and one obvious course of action was to cull the elephants; but none of the rangers wanted to shoot the elephants, feeling that they had great intrinsic value. Nevertheless "they did not believe that their feelings could be part of a professional justification for not shooting the elephants. Given that such justifications were closed off for him, Douglas-Hamilton concluded that he was supposed to find some facts that would independently justify this position so that aesthetic considerations would not have to be mentioned."[49] What this suggests is that in a hostile ideological climate, the embodied motivations that drive restoration tend to shelter behind a more prominent conceptual orientation that is not entirely congruent with it.

Psychological studies of the motivations underlying restoration also suggest that behind the dualistic view of the effect of an (active) restorer on a (largely passive) landscape, a more complex, and more reciprocal, drama is being played out. In one study of satisfaction among volunteer restorationists, for example, the two most well-defined factors were: "Meaningful Action"— indicated by such items as "feeling I am doing the right things," "a sense of accomplishment," "feeling I can play a role in nature," and "Personal Growth"—indicated by items such as "being a part of something profound," "changing my life," "restoring or contributing to my spirituality."[50] These findings suggests that the motivations underlying restoration have at least as much to do with factors related to the self as with those that refer to the landscape. As one volunteer remarked—perhaps with unconscious insight— "there is a sense of communion; it is fulfilling and self-transcending."[51] In other words, while the restoration of the landscape is important, so also is the restoration of personal meaning, wholeness, and a sense of being part of nature—precisely what has been lost through our separation from the world. I am not, however, suggesting that Miles's results indicate that people volunteer for restoration programs in order to fulfil egoic needs as well as to benefit the environment. What is implied is, I think, rather more profound than this, amounting to a rejection of the dualistic perspective which separates "subject" from "object" and individual needs from those of the ecosystem, and an acceptance of a more monistic epistemology which begins to heal a historically sedimented sense of loss. The lack of wholeness we are trying to repair, in other words, is not just "out there" in the ecosystem, but also within us; and the recognition of this mutual loss is the first step towards the recovery of a shared underlying unity.

CONCLUSION: REWILDING THE RESTORER

Rewilding, drawing on our own embodied evolutionary history within a wild world, expresses the deeply felt recognition of the difference between a healthy world and a damaged one, guiding us towards fostering a world we feel at home in. To take this path is in effect to move away from the industrial system, identifying more closely with the natural order and becoming part of it. We cannot adequately do this without recontextualizing those dominant ways of thinking that have developed in conjunction with industrialism, seeing them again as *components* of a holistic sensing capacity which also includes emotional and intuitive aspects as well as the embodied experience of wild nature. As Donlan and his colleagues ask:

Are you content with the negative slope of our current conservation philosophy? Are you willing to risk the extinction of the remaining megafauna should economic, political, and climate change prove catastrophic for Bolson tortoises, cheetah, camelids, lions, elephants and other species within their current ranges? Are you content that your descendents might well live in a world devoid of these and other large species? Are you willing to settle for an American wilderness that is severely depauperate relative to just 100 centuries ago?[52]

These crucial questions are not simply about *rational* choices concerning an *external* nature, but also refer to deep-seated *emotional* preferences and archaic needs that define us as humans rather than cyborgs, earth-dwellers rather than aliens, and whole creatures rather than disembodied minds. They are, in other words, questions which implicitly transcend the isolation of our own nature from that of other life forms; and in this transcendence, we rediscover our identity as natural beings within a natural world.

NOTES

1. Society for Ecological Restoration International online: http://www.ser.org/content/ecological_restoration_primer.asp. Visited on March 2, 2009.
2. See, for example, Carolyn Merchant, *The Death of Nature: Women, Ecology, and the Scientific Revolution* (Harper and Row, 1980); Val Plumwood, *Feminism and the Mastery of Nature* (London: Routledge, 1993).
3. Sigmund Freud, "The dissection of the psychical personality," in James Strachey, ed., *The Standard Edition of the Complete Psychological Works of Sigmund Freud*, Vol. 22 (London: Hogarth Press, 1964), 77.
4. Freud, "The dissection of the psychical personality," 80; Sigmund Freud, *Civilisation and its Discontents*, in Strachey, ed., Standard Edition Vol. 21, 77.
5. Freud, *Civilisation and its Discontents*, 92.
6. Note this conclusion does not rest on the accuracy or otherwise of Freud's view of the psyche. I am using Freud's work here merely as an illustration of the pervasiveness of an ideology which structures both self and world.
7. Charles C. Mann, "Three trees," *Harvard Design Magazine*, Winter/Spring 2000, 31.
8. John Rodman, "The liberation of nature?," *Inquiry* 20 (1977), 104.
9. Rodman, "The liberation of nature?," 104.
10. Julie L. Lockwood and Stuart L. Pimm, "When does restoration succeed?," in E. Weiher and P. A. Keddy, eds., *Ecological Assembly Rules: Perspectives, Advances and Retreats* (Cambridge: Cambridge University Press, 1999): 363–92.
11. See, for example, Peter H. Kahn, *The Human Relationship with Nature: Development and Culture* (Cambridge, Mass.: MIT Press, 1999); Howard Frumkin, "Beyond toxicity: Human health and the natural environment," *American Journal of Preventive Medicine* 20:3 (2001): 234–40; Rachel Kaplan and Stephen Kaplan, *The Experience of Nature: A Psychological Perspective* (Cambridge: Cambridge University Press, 1989).

12. See my "Industrialism and the fragmentation of temporal structure," *Environmental Ethics* 26:3 (Summer 2004): 135–53.
13. For example, Ap Dijksterhuis, among others, has demonstrated that "decisions about complex matters can be better approached with unconscious thought." See Ap Dijksterhuis et al., "On making the right choice: The deliberation-without-attention effect," *Science* 311: 5763 (2006): 1005–1007.
14. Jean-Pierre Dupuy, *The Mechanisation of the Mind: On the Origins of Cognitive Science* (Princeton: Princeton University Press, 2000), 29–30.
15. Lockwood and Pimm, "When does restoration succeed?"
16. Stuart Pimm, *The Balance of Nature?* (Chicago: Chicago University Press, 1991).
17. Daniel B. Botkin, *Discordant Harmonies: A New Ecology for the Twenty-First Century* (New York: Oxford University Press, 1990), 159.
18. Jack Turner, "The quality of wildness: Preservation, Control, and Freedom," in David C. Burks, ed., *The Place of the Wild* (Washington DC: Island Press, 1994), 175–76.
19. Tim Ingold, *The Perception of the Environment: Essays in Livelihood, Dwelling, and Skill* (London: Routledge, 2000), 217.
20. Gary Paul Nabhan, "Cultural perceptions of ecological interactions," in Luisa Maffi, ed., *On Biocultural Diversity: Linking Language, Knowledge, and the Environment* (Wash., DC: Smithsonian Institution Press, 2001), 146; see the summary in Lila Guterman, "Have Ecologists Oversold Biodiversity," *Chronicle of Higher Education*, Oct. 13, 2000, available at http://chronicle.com/free/v47/i07/07a02401.htm.
21. Frans Vera, *Grazing Ecology and Forest History* (Wallingford: CABI Publishing, 2000).
22. Lindsey Gillson, "Testing non-equilibrium theories in savannas: 1400 years of vegetation change in Tsavo National Park, Kenya," *Ecological Complexity* 1 (2004): 281–98.
23. Ellen Wohl, "Compromised Rivers: Understanding historical human impacts on rivers in the context of restoration," *Ecology and Society* 10:2 (2005), available at http://www.ecologyandsociety.org/vol10/iss2/art2/.
24. Robert Ulanowicz, *Ecology, the Ascendant Perspective* (New York: Columbia University Press, 1997, 64.
25. Ulanowicz, *Ecology, the Ascendant Perspective*, 64.
26. Jonathan Long, Aregai Tecle, and Benrita Burnette, "Cultural foundations for ecological restoration on the White Mountain Apache reservation," *Conservation Ecology* 8:1 (2003) available at http://www.consecol.org/vol8/iss1/art4/.
27. Cf. Herbert W. Schroeder, "The restoration experience: volunteers' motives, values, and concepts of nature," in Paul H. Gobster and R. Bruce Hull, eds., *Restoring Nature: Perspectives from the Social Sciences and Humanities* (Wash., DC: Island Press, 2000), 261.
28. Robert H. Hilderbrand, Adam C. Watts, and April M. Randle, "The Myths of restoration ecology," *Ecology and Society* 10:1 (2005), available at http://www.ecologyandsociety.org/vol10/iss1/art19/.
29. Steve Windhager, "Restoration: Rehabilitating nature and ourselves," *Texas Restoration Notes* 5:2 (Winter 2000–01), 3, 5, available at http://www.wildflower.org/?nd=articles_res&view=full&key=2.
30. Steve Packard, "Restoring oak ecosystems," in William Throop, ed., *Environmental Restoration: Ethics, Theory, and Practice* (Amherst: Humanity Books, 2000), 152.
31. Alf Hornborg, "Ecological embeddedness and personhood: Have we always been capitalists?," in Ellen Messer and Michael Lambek, eds., *Ecology and*

the Sacred: Engaging the Anthropology of Roy A. Rappaport (Ann Arbor: University of Michigan Press, 2001), 90.

32. Vicky M. Temperton and Richard J. Hobbs, "The search for ecological assembly rules and its relevance to restoration ecology," in Vicky M. Temperton, Richard J. Hobbs, Tim Nuttle, and Stefan Halle, eds., *Assembly Rules and Restoration Ecology* (Washington DC: Island Press, 2004); Colinvaux quoted by Donald Worster, "The Ecology of Order and Chaos," *Environmental History Review* 14 (1990), 12.

33. Douglas L. Medin and Scott Atran, eds., *Folkbiology* (Cambridge: MIT Press, 1999).

34. Hugh Brody, *The Other Side of Eden: Hunter–Gatherers, Farmers, and the Shaping of the World* (London: Faber and Faber, 2001), 269.

35. Raymond A. Rogers, *Nature and the Crisis of Modernity* (Montréal: Black Rose Books, 1994), 92.

36. Ernest Gellner, *Plough, Sword, and Book: The Structure of Human History* (Chicago: University of Chicago Press, 1992), 44.

37. Ingold, *The Perception of the Environment,* 55.

38. Ingold, *The Perception of the Environment,* 47.

39. Ingold, *The Perception of the Environment,* 25.

40. Scott Atran, "The vanishing landscape of the Petén Maya Lowlands." In Maffi (ed.), *On Biocultural Diversity,* 166–67.

41. Ingold, *The Perception of the Environment,* 54.

42. Robert L. Ryan, "A people-centered approach to designing and managing restoration projects: insights from understanding attachment to urban natural areasi" in Gobster and Hull, *Restoring Nature,* 215.

43. Ryan, "A people-centered approach to designing and managing restoration projects," 213.

44. Ryan, "A people-centered approach to designing and managing restoration projects," 215.

45. See several of the chapters in Susan Clayton and Susan Opotow, eds., *Identity and the Natural Environment: The Psychological Significance of Nature* (Cambridge: MIT Press, 2003).

46. Carl R. Rogers, *Client-Centered Therapy* (Boston: Houghton Mifflin, 1951), Chapter 11.

47. William Cronon, "Introduction," in William Cronon, ed., *Uncommon Ground: Rethinking the Human Place in Nature* (New York: Norton, 1996), 69.

48. It could be pointed out that some theories such as evolutionary psychology *do* focus on the biological aspects of selfhood. My position is that *any* form of reductionism, whether social or biological, does violence to the wholeness of human being, and that a fully interactionist account is needed.

49. Eugene Hargrove, "Taking environmental ethics seriously," in Dorinda G. Dallmeyer and Albert F. Ike, eds., *Environmental Ethics and the Global Marketplace* (Athens, GA: University of Georgia Press, 1998), 18.

50. Irene Miles, "Psychological benefits of volunteering for restoration projects." *Ecological Restoration* 18:4 (Winter 2000): 218–27.

51. Miles, "Psychological benefits of volunteering for restoration projects," 223.

52. J. Josh Donlan et al., "Pleistocene Rewilding: An Optimistic Agenda for Twenty-First Century Conservation," *The American Naturalist* 168:5 (Nov. 2006), 674.

Part VI

Implementation
Rewilding, Regardening, and Renaturing

24 Implementing River Restoration Projects

Daniel McCool

As the preceding chapters make clear, much of the literature on restoration focuses on the resource itself—the land, rivers, wetlands, plains, ecological communities, etc. But some of the most daunting challenges to restoration are political rather than natural. Much of the political debate over restoration focuses on whether a restoration project should be an effort to re-create a wild, natural ecosystem, or engineer a park-like quasi-natural area, or some combination of both. The decision as to how "natural" a restoration project will be is determined by both ecological and political factors. When it comes to implementing restoration, the realm of the possible must deal with everything from invasive species, to extinction, to opposition from organized interests. Given the limitations, it is somewhat amazing that any restoration projects are successful, yet many do succeed in the face of limited resources and political adversity.

This chapter will examine recent river restoration projects to illustrate three different approaches to "naturalness" that are discussed throughout this book. For this chapter, "rewilding" of rivers denotes projects that focus almost exclusively on restoring a section of river to its natural, pre-Columbian state. "Regardening" refers to projects that create a park-like setting with obvious human-designed features. The term "renaturing" pays attention to a combination of natural elements and human-made features in a restoration project. A river restoration project will be used to illustrate each of the three types of restoration.

Before we begin a discussion of the specific projects with case studies in the United States, it is essential to briefly review the contemporary politics of rivers and the changing currents of U.S. water policy. The first section of the chapter will examine the broader political context of river restoration. This will be followed by a discussion of the three types of restoration, using specific projects illustratively. A concluding section will speculate about the future of America's rivers, and the possibilities for restoration.

THE CHANGING POLITICAL CONTEXT

Much has been written about the waste and inefficiency of traditional water projects.[1] Some authors have focused primarily on the negative environmental impacts of supply-centered water development.[2] Others have specifically criticized the economic inefficiency of heavily subsidized water projects.[3] In addition, there have been innumerable critiques regarding specific projects or river basins. Even the Corps of Engineers acknowledged the need for change.[4] It would not be an exaggeration to say that a whole literature has developed devoted exclusively to criticizing traditional structural water policy, and demanding fundamental changes.

These changing priorities have renewed the discussion of what constitutes restoration. The politics of restoration inevitably lead to compromise, which often determines how natural a restoration project strives to be. Thus, the practical reality of river restoration is a negotiated definition of restoration that determines the extent to which a project will combine naturalness with human-made elements. This balancing of natural versus human-made often changes as restoration proposals work their way through the political process.

In researching dozens of restoration projects throughout the United States, I have found that most of them are typically initiated by an individual or small group I refer to as an "instigator." These are people with a vision of a river's future that is not widely shared at the time of origination, but otherwise they are fairly typical people and usually not considered to be politically powerful—housewives, students, middle-class working people, usually with no political experience. They often initiate the concept of a project with an emphasis on rewilding, but often must accept a greater degree of human intrusion as the price of political success. Thus, restoration projects vary in their "naturalness" over both time and space. This will be evident in the three projects that are discussed in the following section.

REWILDING

No American river restoration project is purely a case of returning a river to a pre-Columbian state. As the previous chapters have indicated, there is widespread disagreement, not only as to how to accomplish such a feat, but how to even define what is truly natural. However, some river restoration projects have a clearly established goal of returning a stretch of river to a relatively pristine condition. The techniques that are utilized are those that mimic nature, or speed up natural processes, rather than replacing nature with an engineered design.

One of the most ambitious restoration projects that strives to achieve naturalness is the removal of Condit Dam on the White Salmon River in Washington state. The White Salmon rises in the Mount Adams Wilderness Area of the Cascade Range and courses due south for forty-three miles to the Columbia River. Historically the White Salmon had bountiful anadromous fish runs. But in 1913 a power company built Condit Dam three miles upstream from the Columbia River. A wooden fish ladder was constructed, but it was hopelessly inadequate and soon collapsed. As a result, the anadromous fish runs in the river above the dam disappeared.

PacifiCorp, the current owner of the dam, applied for re-licensing by the Federal Energy Regulatory Commission (FERC) in 1991. In 1996 the agency responded by insisting that the company resolve the fish passage issues in order to get the dam re-licensed. Condit Dam is a concrete wall dam, 125 feet high, situated in a narrow mountain valley. Building fish passage around the dam would be difficult and expensive. A PacifiCorp analysis concluded that fish passage would cost over $30 million, but that dam removal would only cost $17 million. Company officials then entered into negotiations with FERC, environmental and fishing groups, and other stakeholders to work out a way to remove the dam.

The proponents of dam removal included a vast array of interest groups. One of the leading instigators was Phyllis Clausen, the proverbial grandmother in tennis shoes who helped organize Friends of the White Salmon River. The upper stretch of the river had already been designated as a Wild and Scenic River. In recent years the White Salmon had become very popular for white-water recreation, supporting several commercial rafting companies. The lower river had great potential for additional whitewater recreation, but was blocked by the dam. In addition, several home owners in the area become vocal advocates for removing the dam and restoring the fish runs. Most proponents of the dam focused on restoring the river channel to a natural state and bringing back the native fish, and many began advocating for Wild and Scenic River status for the lower river after the dam is removed and the river corridor restored.

After considerable negotiation, PacifiCorp signed a settlement agreement in 1999, agreeing to remove the dam. That agreement was modified in 2005 to move back the date of initiation of removal by two years to give the company more time to generate funding for removal and find alternative power sources. The plan is to blast a hole in the dam fifteen feet in diameter and drain the reservoir rapidly—in less than six hours. The dam will then be cut into chunks and trucked out. The company will also remove the power house and two penstocks—one of which is made of wooden staves. The entire site will then be revegetated with native plants.[5]

There are three factors that made it politically possible to strive for the very ambitious goal of rewilding for the lower White Salmon River. First, there was a pristine context; the upper White Salmon is a Wild and Scenic River, and the lower stretch, while not wilderness, is still largely rural. Thus, rewilding was regarded as a way to expand upon, and take advantage of, the existing natural features in the river basin. Second, there was a convincing economic rationale for rewilding; many locals could see that the river would generate more economic activity if it was in a natural state. Money talks, even when it comes to rewilding. And third, there was a very well organized pro-restoration movement with widespread support among many diverse stakeholders and an effective instigator. And, there was a powerful federal agency, FERC, that was pushing to restore fish runs in the most effective way possible, which in this case meant dam removal. To be sure, there was organized opposition by small groups of locals, but they were a distinct minority. The restoration of the White Salmon will inevitably become a sort of template of how an ambitious rewilding project can be accomplished.

REGARDENING

Not everyone wants a wilderness river in their backyard. Rivers serve many purposes, and can be manipulated and modified to meet the various needs of humankind. Many of these needs require some form of development, usually in the form of structural modifications to rivers. Yet even a much modified river can offer natural benefits, which can be especially valued in a dense urban setting. In some crowded urban and suburban settings, a river and its ribbon of riparian land offer the only open space, the only greenery, and the only place to escape the noise and tension of an urban environment. Thus, urban river restoration projects often offer the best examples of regardening—restorations that look more like parks than wilderness, but offer natural amenities.

Urban river restorations have become quite popular in recent years, and they vary widely in both quality and design. Some, such as Cincinnati's new waterfront, are largely structural—for example its "Serpentine Wall." Others combine river parks with historic buildings and structures; the James River in Richmond, Virginia, and the Potomac Gorge are two examples. And others focus on river recreation, such as Reno, Nevada's kayak park, and Tempe, Arizona's "instant lake" on the Salt River created by inflatable dams. And other cities use a river as a centerpiece for a park-like setting for retail business, such as San Antonio, Texas' Riverwalk. In this section we will examine an ambitious river restoration effort in Boston that is an excellent example of regardening.

Since its inception, Boston has been defined by the Charles River and its outlet to the sea. The industrial revolution in the United States began in this watershed, and as a result the Charles River has seen over 300 years of use and abuse. Much of the early history of this nation was made within sight of the Charles River. The prestigious area of the city known as Back Bay was created as a landfill in the river. By the late 1800s, the river was being crowded by development, and water quality had plummeted to that of a cesspool. A visionary by the name of Charles Eliot, the son of Harvard's president and a protégé of Frederick Law Olmsted, could see that the river corridor, if properly planned, could become a centerpiece for the Boston area—a place where all classes of Bostonians could mingle and escape from the pressures of the city. He convinced city fathers to purchase riparian lands and create a linear river park nearly all the way through Boston. The result was a park system that was a great public amenity, and thousands of Bostonians thronged the river's beaches on weekends. Photographs from that era show huge crowds of people swimming in the river and enjoying the beaches near Cambridge.

Unfortunately, Eliot's vision was abandoned as the city grew. Development encroached upon the parklands, and water quality continued to deteriorate. The river became a health hazard, and in 1955 the state made it illegal to swim in the Charles River. The beach parks were abandoned, and much parkland was destroyed by a freeway built in 1951. Like so many eastern cities, Boston had turned its back on its river.

The idea of reviving Eliot's river parks began soon after Governor Michael Dukakis made a national issue out of cleaning up Boston Harbor in the 1980s. One of the instigators in this movement was Renata von Scharner, a teacher at Radcliffe with an expertise in planning and architecture. She started the Charles River Conservancy, and began lobbying for the "stewardship and renewal of the Charles River Parklands from the Boston Harbor to the Watertown Dam."[6] The city of Boston gradually began renewing its relationship with the river. Improvements in water quality, the creation of new parks along the river, and new upscale development in the area all gave a boost to the restoration effort.

To be sure, this is not rewilding; 350 years of Euro–American habitation have made that impossible. For example, although the quality of the water in the river is now considered good enough for swimming (at least by some hardy souls), the river bottom is still a toxic stew that is best left untouched. And busy highways now crowd the river corridor. Thus, restoration of the Charles River is a process of utilizing bits of open space and greenery, buying run-down property and returning it to the park system, and encouraging river-sensitive development along the river to bring people to the water's edge and get the river involved in their lives. Not everyone is pleased with the possibility of additional development along the river, but as Ms. Von Scharner explained in an interview, such development can actually aid restoration efforts: "There are people who talk about returning the river to a natural state, and making it a wildlife area. But this just isn't feasible. The best way to protect the river is to encourage development that is compatible with the river. If people use the river, they will learn to love the river."[7]

The Charles River is an example of making the impossible happen. At one time the Charles River was so polluted it was a health hazard to even be close to it. Initially, Boston thrived because of the Charles River; later, it thrived in spite of it. But a new-found awareness of rivers, and a realization of their impact on quality of life, led visionaries to dream of a river that was an asset rather than a liability to their community. This vision is not some hopeless yearning for returning to a past era—an era that is gone, never to return. Rather, it is a way of incorporating natural elements into a heavily engineered environment and looking for ways to symbiotically combine development and altered landscape with a living river.

RENATURING

In a nation of 300 million people, broadly distributed, the United States is a difficult place to find watersheds that are sufficiently free of human intrusion to make a rewilding project feasible. In most cases, river restoration projects are attempts to restore partial natural function in the midst of human impact. In this section we will examine a large-scale restoration project in Florida that is attempting to bring a river back to life within the context of a structured human environment. The Kissimmee River restoration project illustrates the political compromises that must be made if any sort of restoration is to succeed.

The Kissimmee River drains a lake with the same name and flows south through the heart of Florida to Lake Okeechobee. At one time the river meandered widely through a dense jungle of semi-tropical forest and wetlands. It was a haven for wildlife, and watered an ecological niche unlike any other place outside of Florida. It was also part of the "river of grass" that delivered water and wildlife to the Everglades. And, like all natural rivers, it often spilled out of its meanders and covered its floodplain with thick, brown nutrient-rich water. To the wildlife in the area, it was a natural refuge.

But to the Corps of Engineers, it was an abomination. In the mid-1950s the Corps developed plans to turn 100 miles of meandering river into a 56-mile canal 30 feet deep and 300 feet wide. The spoil—the vegetation and soil dug out of the channel—would simply be piled in heaps along the new canal. To generate public support, the Corps produced a short film titled "Waters of Destiny," which explained that the "maddened forces of nature" and the "crazed antics of the elements" would be conquered by the proposed project. With strong support from real estate interests, the Florida cattle industry, and area farmers, the Corps succeeded in getting authorization for the project, which was completed by the early 1960s. The new project did reduce flooding in the river's floodplain, but it did so by turning the plain into a virtual dead zone. Now sparsely vegetated, it was primarily used for grazing cattle.

The Kissimmee Channel never lived up to its promised economic activities, and as time passed it became more evident that its impact on fish and wildlife—one of the mainstays of Florida's tourist-based economy—was devastating; an estimated 90 percent of migratory waterfowl simply disappeared. A small group of local citizens began to agitate for a restoration of the Kissimmee floodplain. These instigators faced the usual litany of catcalls, but they had three things going for them. First, a restored river aligned nicely with the state's dominant economic activity, tourism. Second, the Corps of Engineers, by the 1970s, was running out of big projects and facing a barrage of political opposition; the agency began looking for new projects that would not antagonize the environmental community. The Corps realized that a massive restoration effort, with a budget exceeding half a billion dollars, would fatten the agency's budget significantly. And third, the impending demise of the Everglades—a national park—gave increased impetus to all projects that could help save the massive wetlands in the park.

At first glance one might think that the Kissimmee restoration would clearly be a candidate for rewilding. But encroaching development made that impossible; the Kissimmee had to be saved by working within the parameters of existing development, and that meant that a lot of project funding would be allocated to pouring concrete. This helps explain the massive price tag for the restoration: a whopping $365 million (and probably more if the usual cost-overruns occur). So, the Kissimmee River "restoration" includes building new bridges, box culverts, floodgates, and even more channels.

For example, the area around Istokpoga Lake was developed for farming, and is now a principle growing area for flowers. To prevent flooding, the Corps channelized Istokpoga Creek about the same time that they developed the Kissimmee. But a restored Kissimmee River cannot absorb as much floodwater, so the Istokpoga channels have to be enlarged to increase their storage capacity, and gated to prevent backflow. Thus the Corps is pouring concrete in the name of restoration.[8] By building these structures, the Corps was able to return part of the river to its natural condition. It could not happen any other way.

Why is the Kissimmee River restoration project an example of renaturing, rather than a more complete return to nature characterized as rewilding? There are three reasons. First, the river is in the eastern United States, which is densely populated and has been subjected to heavy development for many years. There is very little landscape in the East that could be classified as wild. Thus, the ecological context was one that mixed together numerous natural and human-made elements. Second, although restoration proponents put together impressive political coalitions, their opponents were well organized too, and were not about to relinquish the advantages they receive from river development. No agreement could be negotiated without their approval. And finally, it was physically possible to meet the basic goals of the restoration project without the full restoration of the river basin. There was sufficient water, land, and riparian area to bring the river back to life without restoring every part of the river basin; this sufficiency in natural resources made it easier to reach a compromise.

CONCLUSION

The recent shift toward river restoration has renewed and enlivened the debate over what constitutes restoration. Is it an effort to re-create a river's natural setting? Or, given economic and ecological realities, is it an effort at what Marcus Hall calls "reparative gardening,"[9] where a river is re-designed and developed to improve its esthetic appeal and its ability to serve ecological and human functions?

Given the dissatisfaction with past river policy, and the demand for living, healthy rivers, there is a great potential for a fundamental departure from traditional approaches to river management. It is my contention that the United States is on the cusp of a new era in water policy that will focus on river restoration. Instigators will continue to lead us into this new era. Public conceptualization of rivers and their relationship to humans is undergoing a fundamental change. This changing perception will result in a new approach to water resources that focuses on the public benefits of healthy, relatively intact river ecosystems. In other words, water policy will be replaced by river policy. Changing public values and political priorities have created a basis for new conceptualizations of what we want from our rivers, and how we can restore them.

Environmentalists are fond of saying that extinction is forever. That is true, but dams are not. Other than the loss of native species, almost any river can be restored, given the right set of incentives (see: *American Rivers* 2001).[10] In most cases, restoration requires a package of trade-offs, making even the most ambitious restoration efforts only a partial attempt to rewild. The only force on Earth that can totally restore a river is nature. But, with the right combination of "crazy" instigators and economic incentives, many rivers are becoming more natural. Removal of dams is only one tool to achieve that goal; many river restoration projects take place within the context of dams either upstream or down, or both. Still, changing public sentiment and an evolving economy are pushing the United States into a new era of water policy that will consider rivers, not as something to be disaggregated into extractable resources, but as a natural ecosystem offering a wealth of goods and services to humankind.

NOTES

1. Barbara Andrews and Marie Sansone, *Who Runs the Rivers? Dams and Decisions in the New West* (Stanford Environmental Law Society, 1983); Mark Reisner, *Cadillac Desert* (New York: Viking, 1986); Robert Gottlieb, *A Life of Its Own: The Politics and Power of Water* (New York: Harcourt Brace, 1988); Sarah Bates et. al., *Searching Out the Headwaters: Change and Rediscovery in Western Water Policy* (Washington, D. C.: Island Press, 1993); Daniel McCool, *Command of the Waters: Iron Triangles, Federal Water Development, and Indian Water* (Berkeley: University of California Press, 1987; re-issued by Tucson: University of Arizona Press, 1994); Daniel McCool, *Native Waters: Contemporary Indian Water Settlements and the Second Treaty Era* (Tucson: University of Arizona Press, 2002).
2. Philip Fradkin, *A River No More* (New York: Alfred Knopf, 1981); Tim Palmer, *Endangered Rivers and the Conservation Movement* (Berkeley: University of California Press, 1986); Patrick McCully, *Silenced Rivers: The Ecology and Politics of Large Dams* (London: Zed Books, 1996); M. J. Collier, R. H Webb., and J. C. Schmidt, "Dams and Rivers: Primer on

the Downstream Effects of Dams," *U.S. Geological Survey Circular* 1126 (1996); Elizabeth Grossman, *Watershed: The Undamming of America* (New York: Counterpoint Press, 2002); Amy Vickers, *Handbook of Water Use Conservation* (Amherst, MA: Waterplow Press, 2002).

3. John Ferejohn, *Pork Barrel Politics: Rivers and Harbors Legislation, 1947–1968* (Stanford: Stanford University Press, 1974); Terry Anderson, *Water Crisis: Ending the Policy Drought* (Baltimore: Johns Hopkins Press, 1983); Richard Wahl, *Markets for Federal Water: Subsidies, Property Rights, and the Bureau of Reclamation* (Washington, D. C.: Resources for the Future/ Johns Hopkins University Press, 1989); World Commission on Dams, 2000, "Dams and Development: A New Framework for Decision-making," Overview Report, 2002. www.dams.org.

4. Corps Operating Principles. 2002. U. S. Army Corps of Engineers, "Environmental Operating Principles." Announced March 26, 2002 by Chief Engineer Lt. General Robert Flowers. www.hq.usace.army.mil/cepa/envprinciples.htm.

5. U.S. Department of the Interior, "Babbitt Announces Plan for Condit Dam Today," Press Release, Sept. 22, 1999, www.doi.gov/news/990922.html; Central Cascades Alliance, "Condit Dam Background," www.cascades.org/what/background/htm accessed 2003; American Whitewater, "Condit Dam (White Salmon River WA) Removal Agreement," www.americanwhitewater.org/archive/article/4/; Washington State Department of Ecology, "Condit Dam Removal," State Environmental Protection Agency, Supplemental Environmental Impact Statement, Ecology Publication # 05–06–022, 30 Sept. 2005.

6. "Charles River Parklands," *Newsletter of the Charles River Conservancy* (Cambridge, MA, Fall 2005): 1.

7. Renata von Scharner. 2005. Interview with the author, Boston, MA. April 6.

8. South Florida Water Management District, "River Restoration Takes Flight," *Water Matters* (February/March 2003): 1–3. See also: Special Issue, Kissimmee River Restoration, *Restoration Ecology* 3:5 (Sept. 1995).

9. Marcus Hall, *Earth Repair: A transatlantic history of environmental restoration* (Charlottesville: University of Virginia Press, 2005), 13.

10. See *American Rivers*, 2001, Compilation of removed dams, www.americanrivers.org.

25 Cloning in Restorative Perspective

Eileen Crist

Where does all this work [on cloning] fit into the agenda of ecological restorationists? Obviously, if cloning of rare and endangered animals ever becomes commonplace restorationists will be called upon to provide suitable habitats for their long-term well-being.
—Dave Egan, Society for Ecological Restoration International

Even as conservationists remain wary of the approach that cloning represents, the potential of cloning for conservation is receiving increasing attention in scientific circles and the public.[1] Can cloning be used to rescue species or bring back extinct animals? If the answer is potentially affirmative, doesn't the technology merit the involvement of conservation scientists and restoration ecologists? I argue that despite its limitations cloning represents a tenable conservation tool, especially one that can be prepared for by preserving cell-lines of endangered species for future efforts. If undertaken in conservation contexts, and with the interests of the animals and their habitats in mind, cloning species that people have extinguished or decimated can be a justifiable strategy.

Does cloning extinct and endangered animals have the potential to substantively redress biodiversity losses? The answer is a resounding no: when all facets of present-day biodepletion are tallied—mass extinction, unrecoverable losses of species, subspecies, and genetic variation, destruction of ecosystems, and habitat fragmentation—it becomes clear that the implementation of cloning technology is largely inconsequential. A more reasonable question to pose is whether cloning can make a limited contribution to restoration efforts. I offer cautious support for using cloning in conservation practice—for there are compelling reasons to be suspicious of cutting-edge technologies as proffered solutions to the destruction of biodiversity. Before discussing why I support cloning as a limited but potentially effective tool, I briefly summarize the most powerful criticisms of why it is irrelevant or even detrimental to restoration efforts.

Conservationists tend to be suspicious of this still-experimental technology for at least five reasons. First, the biodiversity crisis is too multi-dimensional for any purported technological solution; even focusing on extinction alone, it is happening at a magnitude and rate that only a profound change in the relationship between humanity and the natural world can turn things around.[2] Second, protecting landscapes and their interconnectivity constitutes the soundest approach to sustaining species, populations, ecosystems, biological processes, behavioral patterns, and genetic variability; cloning can amount to a distraction from the scientific principles of, and appropriate investment in, conservation efforts.[3] Third, high-tech approaches to ecological problems reinforce the conceit that technological solutions or replacements can atone for the damage inflicted on the natural world; to turn to a technological fix—especially one as chock-full of hubris as creating life by manipulating cells across organisms and species—is redolent with folly.[4] Fourth, cloning harbors the peril of fostering false security in the public by encouraging the illusion that science can fix extinction after the fact;[5] indeed, cloning endangered or extinct animals is often reported under grossly misleading but catchy "end of extinction" headlines. Last but not least, cloning for conservation encourages more interference with, and management of, wild nature, promotes the further erasure between the natural (wild) and the artificial (man-made), and may risk unintended consequences.[6] To top off these grievances, cloning endeavors are unreliable—resembling experiments that "evolve haphazardly" in Quammen's apt words;[7] such experiments often involve animal suffering, which argues for postponing applications of the technology.

These are compelling concerns that any argument favoring the use of cloning must grapple with. But well-founded and thoughtful critiques of technological approaches to biodiversity conservation and restoration efforts leave unaddressed the urgency of the problem of losses—and the need to employ every possible countermeasure, including high-tech options. If the technology can be considered as a limited tool (all the more so today when it is by no means a routine procedure), conservation- and restoration-minded communities might begin to scrutinize its potential utility, rather than dismissing it *tout court* for its shortcomings.

Many animals have already been cloned for a variety of purposes—mostly domestic and laboratory animals, like goats, cattle, pigs, mice, and cats among others. Because of the fascination exerted by this relatively novel development in biotechnology, we are likely to see more cloning, of both domestic and wild animals, down the pike. If high-profile publicity is a gauge, the proposal to clone extinct and endangered animals enjoys great popularity. Developments in this arena have been a mixed bag of success, failure, wishful thinking, future potential, and dubious motivation—but all bring home one point: that efforts to clone endangered and extinct species, for better or worse, are already with us and unlikely to go away.

The first endangered animal to be cloned, an Asian wild cow known as the Gaur, died within two days of birth. His death was not a setback for long. The European mufflon, an endangered Mediterranean wild sheep, was cloned a year later in Italy. Two clones of the Banteng, another endangered Asian wild cow, were created from cell-lines stored in the 1980s in the

San Diego Zoo; one of these animals survived and is a denizen of the zoo. There have been steps in the direction of cloning the Panda and the Asian Cheetah, but technical difficulties, political obstacles, and controversy have thwarted both endeavors to date. The Bucardo or Spanish Ibex, gone since 2000, may turn out to be the first extinct subspecies cloned from frozen tissue. The project to clone the extinct Thylacine (also known as Tasmanian Tiger and Tasmanian Wolf), from an alcohol-preserved specimen dating to 1866, was initiated in 1999 but quietly abandoned in 2005 as unfeasible (at least for the time being). South Korean scientist Hwang Woo-suk, internationally disgraced for fraudulent research on human embryonic stem cells, recently confessed buying "Mammoth" tissue samples from the Russian mafia and attempting to clone the Mammoth. One of the latest cases to appear in press involves Vietnam's antelope-resembling Saola—among the few mammals discovered in the twentieth century. The species found just ten years ago is already threatened with extinction, and, unable to breed the animals in captivity, some scientists want to clone them—indeed, have already unsuccessfully tried.

Both challenge and necessity in the domain of cloning endangered and extinct animals is applying the technology of "cross-species nuclear transfer" (or simply "cross-species cloning"). Opting for cross-species cloning—in which non-endangered species bear the brunt of the procedure—stems from self-evident objections to subjecting endangered animals to invasive procedures. Easily accessible cells of an endangered animal are used (for example, skin cells), the genome-bearing nucleus from such cells is extracted, and then the nucleus (or sometimes the whole cell) is injected into the enucleated egg-cell of a closely related, non-endangered species. From this chimera an embryo is coaxed into formation, which is then implanted into the womb of the surrogate, non-endangered animal.[8]

Cloning requires living cells, which is why the resurrection of extinct animals has proved elusive. But the application of the technology to endangered species is feasible and, as noted, already underway. The conservation rationale is to help prevent their extinction by boosting the numbers of animals (also an aim of captive propagation) and by maintaining their extant genetic diversity. Cloning of course cannot add genetic diversity, but by preserving cell-lines from as many animals of an endangered species as possible, its existing genetic variability can be placed in reserve for a future time. Should the numbers of an endangered species continue to decline, at least their present genetic profile (compromised though it already is) might be recovered. If such a species became extinct despite efforts to save it, freeze-preserved tissues might provide a fighting chance to bring it back.

Scientists like Robert Lanza, Oliver Ryder, and William Holt, who support cloning for conservation, do not necessarily call for immediate cloning ventures but have instead staked a precautionary position. They argue for the systematic stocking of "frozen zoos" as living databases for small and declining populations.[9] As Robert Lanza stated in an interview following the ill-fated cloning of the Gaur, "we wanted to send the message to the conservation groups that you should be protecting genetic diversity now. And although we still may not have the technology to do it efficiently, it is real. When an animal dies, all you have to do is freeze a few cells to preserve the genetics of the animal forever . . . Cloning is a tool to reintroduce genes that would otherwise be lost."[10]

Institutional action has paralleled such arguments, and recently gathered speed. In the United States, the Audubon Center for Research of Endangered Species (ACRES), the San Diego Zoo, and biotech company Advanced Cell Technology (ACT) have been at the forefront of preserving tissues of endangered species, and, to a limited extent, undertaking cloning experiments. The UK-based "Frozen Ark Project," inaugurated in 2004, involves the collaboration of numerous institutions world-wide—including the American Museum of Natural History, the San Diego Zoo, and the Laboratory for the Conservation of Endangered Species in Hyderabad, India—in building a global library for endangered animal cell-lines. The stated mission of The Frozen Ark is "to save DNA or frozen viable cells from endangered species before they go extinct. The DNA gives a vast amount of information about an animal's relationships, evolution, genetics, development, diseases, and ecology. If we act now we can rescue this information, or even the animals themselves. If not, there are no such hopes." While nowhere on its website does the organization explicitly advocate cloning—that possibility is clearly and intrinsically part of its enterprise.

Publications and interviews of scientists involved in cloning often reveal motives that are closely aligned with those of conservationists. And yet when it comes to cloning endangered and extinct animals, there is a dogged disconnect between reproductive biologists who undertake the experiments and conservation scientists who bear witness to the results. "Some conservation biologists have been slow to recognize the benefits of basic assisted reproduction strategies, such as in vitro fertilization, and have been hesitant to consider cloning," maintain Lanza and his colleagues.[11] Deep-seated distrust of high-tech solutions to ecological degradation understandably endures among the conservation-minded, who emphasize habitat "protection, management, and restoration" as the key for conserving species, as well as genetic diversity, ecological processes, and evolutionary potential.[12] Conservationists have consistently underscored that cloning is expensive, and that funds would better support conservation if they were funneled into habitat procurement and protection.[13] But cloning advocates have countered that a different type of patron tends to finance biotech, without resources being diverted from habitat conservation. On this view, "the sources of funding would not necessarily compete."[14]

While habitat protection is the crucial ingredient for conserving biodiversity, the jury is still out on whether cloning might serve as a restoration tool—a new technological spin on captive breeding. With the participation of conservation scientists, questions regarding potential habitat, the fate of cloned wild animals, and the reasoning for undertaking cloning projects (as well as the timing of such projects) would become prominent aspects of applying the technology to endangered or extinct species. Such cloning projects are not likely to be conceptualized and implemented for the benefit of the animals and their native habitats, as long as conservationists, ever-suspicious of hyperbolic biotech claims, hold cloning at arm's length. But by becoming actively involved, conservation scientists could steer cloning projects toward the goal of reintroducing animals to their available or restored habitats; indeed, as Dave Egan argues, cloning endeavors "may bring restoration [ecology] closer to its allied field of conservation biology."[15]

Without an explicit conservation intention and agenda, cloning extinct and endangered species will remain susceptible to experimentation for its own sake, or to the quest for the fame attending headline science. To serve the cause of conservation, the technology must be implemented in the context of a multidisciplinary effort in which cloning, itself, is an auxiliary part rather than the main event. Within a multidisciplinary team context all the pieces of the conservation puzzle, within which the assisted procreation of wild animals makes sense, can be addressed: habitat, behavior, reproduction, genetics, ecological interactions, and so on. A conservation paradigm must frame the cloning of endangered and extinct animals, if such projects are not to be driven by the ambitions of "boys with their toys" and "science for the sake of science," as the Director of the Tasmanian Conservation Trust wryly commented when questioned about the Thylacine cloning project.[16] Without a strong contingent of conservation scientists actively involved, cloned wild animals are far more likely to end up as displays in cages, human-created oddities of theme parks, or objects for advancing the careers of their makers.

Instead of highlighting the predictable contrast between the holistic approach of habitat conservation and the "laboratory gimmickry" of cloning,[17] conservationists might re-imagine cloning as a reproductive technology that can be leveraged to push for wild animal habitat. For example, tiger biologist Ullas Karanth dismisses the proposal of cloning tigers as "irrelevant," maintaining that "we are concerned about protecting habitats for them to live and not increasing their numbers."[18] What's more, he had nothing positive to say about the plan to clone the Asian Cheetah, given the absence of places for this critically endangered subspecies to live. While Karanth's denunciation of cloning endangered megafauna is well-founded—currently such projects seem more concerned with achieving a technical feat than genuinely serving the conservation of the species—an alternative tactic would be to press forward for Asian and Middle Eastern regions that might be ecologically restored for the reintroduction of the Asian Cheetah. Rather than denigrating cloning as glitzy, this same admittedly questionable feature can be exploited as opportunity for securing wilderness. In other words, cloning species that people have extinguished or decimated, and for which habitat can be restored and protected, can be a justifiable restoration strategy if it is tactically exploited: to negotiate habitat availability and to fulfill the moral and ecological need for restoration—of rescuing animals thoughtlessly destroyed and returning them to their ecological niches.

As misleading as hype about cloning can be, it can be turned into an advantage if habitat stipulations are successfully hitched to such endeavors. Just as charismatic animals are useful for conservation purposes as "umbrella species," providing popular grounds for protecting entire biotic communities, so the glamour of cloning makes it potentially serviceable as an "umbrella technology" for negotiating the restoration of places in which animals (boosted in numbers or brought back through cloning) can live. If cloning could be implemented as part of a conservation plan, subject to efforts to secure wild living spaces, then there is arguably no reason that the technology should not merit the support of the conservation community.

Unlike ambitions to resurrect animals of previous eras (like the Mammoth), the cloning of endangered and extinct species of the Holocene is far more ecologically sound and viscerally appealing, "if only because we might be able to care for those beings by returning them to their former or restored habitats."[19] Cloning animals destroyed by people has found both popular and scientific support because it taps into the need for restoration—restoration in the double sense of restoring justice and restoring the land.[20] While clones of endangered species have yet to be reintroduced to the wild, as David Quammen makes a point of noting, this is not necessarily an indicator of how things should, or will, stand in the future—and all the more reason for conservationists to become involved.

In discussions of cloning, mammals get most of the attention. It is regularly forgotten that frogs were cloned decades before the first mammal.[21] The potential of cloning charismatic megafauna like the Thylacine, Asian Cheetah, or Panda inevitably gets plenty of media coverage, while the possibility of cloning endangered or extinct amphibians has yet to make a headline. Ironically, however, frogs might presently receive the greatest benefit from cloning.[22] Costa Rica's Golden Toad has become a poster story of anthropogenic extinction. It has taught us that species are not safe from human impact even in protected natural areas, and it has served to call attention to the dire repercussions of climate change for biodiversity.[23] Would it not have been a sound provision if cell-lines from the Golden Toad had been preserved? By extension—how could the banking of cell-lines from the world's frogs be regarded as anything other than a rational safeguard, best undertaken immediately?

A final point in favor of cloning technology is to suspend *our* judgments, and allow future people to assess its applications. In this regard, Sarah Burnette of the Audubon Nature Institute's Center for Research of Endangered Species (AICRES) raises a valid question: "What if 100 years from now people finally figure out how to save the habitats, but there are no animals? Cloning is part of the answer."[24] Environmental ethicist Jeffrey Yule makes a cognate point: "If and when the human species gets to a point where the planet's many ecosystems have been restored sufficiently to support extinct species, it would be consistent with the tenets of conservation biology to consider restoring these species on a case-by-case basis."[25] However we reasonably censure cloning today—as technological fix or human artifact—we arguably owe future people to decide whether or not they want to use it.

As David Lowenthal reflects in this volume, "we cannot know what future generations may want, but we can anticipate what they may need to recover from some global calamity." Given humanity's incapacity to respond with needed alacrity to the biodiversity crisis—which in retrospect is bound to be seen as the main calamity of our time—the least we might do is step up the project of preserving cell-lines of endangered species from the whole animal kingdom: mammals, birds, reptiles, amphibians, insects, and so on. Future people might then decide for themselves between the better of the two "hyper-real" options—an Earth thinned of life or an Earth restored partially through cloning.

"With the cloning genie out of the bottle," in Egan's words, the cloning of endangered and extinct species will undoubtedly proceed apace. Is not restoration the appropriate rationale for such projects? Without an overarching conservation framework, the motives for cloning endangered and extinct animals are likely to remain nebulous, subject to political caprice, and driven by experimental curiosity or individual ambition. The fact that cloning is still in its infancy means that it should be approached with caution, if only for reasons of animal welfare.[26] At the same time, this early stage of cloning offers opportunity for conservation scientists to step into a forming picture, and help shape the ecological contexts within which cloning the extinct and endangered might be undertaken.

Any argument in favor of cloning must come to grips with the compelling concerns outlined in the beginning of this chapter. Endorsing the technology for conservation purposes must not be indiscriminate, nor vulnerable to grandiose, end-of-extinction illusions that a "technology-infatuated public" is often susceptible to.[27] And yet the specter of extinction provides abundant warrant for closely considering the potential usefulness of cloning, limited as it may be, rather than dismissing the technology as an untrustworthy technological fix. Human-driven extinction is spiritually and materially devastating; most measures to stop or reverse it are justified. Therefore if we lose species, have ourselves to blame, and cloning is the only way to bring them back—then let us use cloning by all means.

NOTES

1. See Cynthia Mills, "Second Chance," *Conservation in Practice* 7:4 (Oct.–Dec. 2006): 22–27.
2. E. O. Wilson, *The Creation: An Appeal to Save Life on Earth* (New York: Norton, 2006).

3. David Quammen, "Clone your Troubles Away: Dreaming at the Frontiers of Animal Husbandry," *Harper's* February 2005; David Ehrenfeld, *The Arrogance of Humanism* (New York: Oxford University Press, 1978).
4. Bill McKibben, *Enough: Staying Human in an Engineered Age* (New York: Henry Holt & Company, 2003); Gary Meffe, "Techno-Arrogance and Halfway Technologies: Salmon Hatcheries on the Pacific Coast of North America," *Conservation Biology* 6:3 (1992): 350–54.
5. David Ehrenfeld, "Transgenics and Vertebrate Cloning as Tools for Species Conservation," *Conservation Biology* 20:3 (2006): 723–32; Jeffrey Yule, "Cloning the Extinct: Restoration as Ecological Prostheses," *Common Ground* 1:2 (2002): 6–9.
6. Eric Katz, "Understanding Moral Limits in the Duality of Artifacts and Nature," *Ethics & the Environment* 7:1 (2002):138–45; Jack Turner, "The Wild and its New Enemies," in Ted Kerasote, ed., *Return of the Wild: The Future of our Natural Lands* (Wash., D.C.: Island Press, 2001).
7. Quammen, "Clone your Troubles Away."
8. Robert Lanza et al., "Cloning of an Endangered Species (*Bos gaurus*) Using Interspecies Nuclear Transfer," *Cloning* 2:2 (2000): 79–84; Sylvia Pagan Westphal, "Copy and Save," *New Scientist* (19 June 2004).
9. Oliver A. Ryder, "Cloning Advances and Challenges for Conservation," *Trends in Biotechnology* 20:6 (2002): 231–32; Oliver A. Ryder et al., "DNA Banks for Endangered Animal Species," *Science* 288:5464 (2000): 275–77, 2000.
10. Robert Lanza, "Second Chances: An Interview with Robert Lanza," *California Wild, The Magazine of the California Academy of Sciences* (Summer 2002).
11. Robert Lanza, Betsy Dresser, and Philip Damiani, "Cloning Noah's Ark," *Scientific American*, November 2000.
12. Ehrenfeld, "Transgenics and Vertebrate Cloning as Tools for Species Conservation;" Quammen "Clone your Troubles Away."
13. Sharon Begley, "Cloning the Endangered," *Newsweek*, 136:6 (16 October 2000): 56–57; Scott Weidensaul, "Raising the Dead," *Audubon* (May–June 2002): 58–66.
14. William V. Holt, Amanda R. Pickard, and Randall S. Prather, "Wildlife Conservation and Reproductive Cloning," *Reproduction, The Journal of the Society for Reproduction and Fertility* 127:3 (2004): 319.
15. Dave Egan, "Resurrection Ecology," *Ecological Restoration* 20:4 (2002): 237.
16. Quoted in Weidensaul, "Raising the Dead."
17. Quammen, "Clone your Troubles Away."
18. Interviewed in the *Deccan Herald*, www.deccanherald.com, October 3, 2005.
19. Egan, "Resurrection Ecology."
20. Eric Higgs, "What Is Good Ecological Restoration?," *Conservation Biology* 11:2 (1997): 338–48,; William R. Jordan III, *The Sunflower Forest: Ecological Restoration and the New Communion with Nature* (Berkeley: University of California Press, 2003).
21. Harry Griffin, "Cloning of Animals and Humans," in John Bryant, Linda Baggott la Velle, and John Searle, eds., *Bioethics for Scientists* (New York: John Wiley & Sons Ltd, 2002), 279–96.
22. Holt et al., "Wildlife Conservation and Reproductive Cloning."
23. Thomas E. Lovejoy and Lee Hannah, eds., *Climate Change and Biodiversity* (New Haven: Yale University Press, 2005); J. Alan Pounds, Michael P. L. Fogden, and Karen L. Masters, "Responses of Natural Communities to

Climate Change in a Highland Tropical Forest," in Lovejoy and Hannah, eds., *Climate Change and Biodiversity*, 70–74.

24. Amy Hembree, "Cloning is no Extinction Panacea," *Wired News*, www. wired.com, 13 February 2001.
25. Yule, "Cloning the Extinct."
26. Steven Best and Douglas Kellner, "Biotechnology, Ethics and the Politics of Cloning," *Democracy & Nature* 8:3 (2002): 439–65.
27. Ehrenfeld, "Transgenics and Vertebrate Cloning as Tools for Species Conservation."

26 NLIMBY

No Lions in My Backyard

C. Josh Donlan and Harry W. Greene

Lion: the fiercest and most magnanimous of the four footed beasts
—Samuel Johnson's Dictionary of the American Language (1775)

No lion shall be there,
Nor any ravenous beast shall go onto it,
They shall not be found there;
But the redeemed shall walk there
—Book of Isaiah (~2700 Before Present)

If they get near me, my family, friends or my property,
I'll be careful when I place the crosshairs on them,
And slowly squeeze the trigger of my Remington 300Ultra-Mag
—R. Weir (August 28, 2005, in response to the idea of lions in
North America)

What types of information should guide societies in their efforts to conserve and restore biodiversity? Should certain time periods in the past serve as reference points? And if so, what should those benchmarks be? As importantly, what types of information and experiences influence or bias our perspectives with respect to biodiversity conservation? These are important questions with wholesale implications for biodiversity and humanity, yet they are rarely discussed and thus this volume is particularly timely. These questions and their answers will inherently involve ecology, evolutionary biology, and the social sciences—but human behavior and psychology will also heavily influence them.

Given our deep and complex relationship with large animals[1], the explosive reactions from the scientific community, the media, and the public-at-large came as no surprise as they pounced on the 1700 words published in August 2005 in the journal *Nature* under the title, "Re-wilding North America."[2] In that short paper, along with ten co-authors, we fundamentally challenged current attitudes toward restoration and biodiversity conservation, and proposed exploring the potential costs and benefits of restoring large vertebrates and their biological roles to North America that went extinct roughly 13,000 years ago. Here we explore some of the human perspectives on ecological history and conservation by reflecting upon the suite of responses to our proposal of *Pleistocene Rewilding*. Rather than rebut criticisms, our purpose in this chapter is to explore how human sociology may play important and underappreciated roles in the way society makes decisions on the conservation and restoration of the natural world and its relationship with it.

WHAT IS PLEISTOCENE REWILDING?

Our proposal is based on a scientific and sociological framework that calls for restoration of missing ecological functions and evolutionary potential of lost North America megafauna, using extant conspecifics and closely related taxa as proxies. Proposed species included tortoises, horses, camels, cheetah, elephants, and lions—all native to the Americas until about 10,000 to 12,000 years ago. We argue that these taxa and others would contribute biological, economic, and cultural benefits to North America. Pleistocene Rewilding would be achieved, we argued, by carefully managed ecosystem manipulations whereby costs and benefits are objectively addressed on a case-by-case, locality-by-locality basis. The proposal generated a substantial response from the scientific community, media, and general public. In 2006, our longer paper on Pleistocene Rewilding elaborated on the ecological, evolutionary, economic, and esthetic justifications for restoring large vertebrates to North America, and discussed in detail potential costs and benefits as well as the major risks and challenges.[3]

In many ways our proposal was not novel. Some of our co-authors had earlier articulated similar ideas in less high-profile venues[4]; in particular, Paul Martin hinted at using species analogs over thirty years ago.[5] Others outside of North America have discussed and implemented restoration programs using closely related birds, reptiles, and mammals as proxies to replace extinct species from an ecological perspective.[6] A similar framework for insect conservation has been termed "resurrection ecology" by Robert Michael Pyle.[7] Until recently, these ideas and reintroductions, which have been largely limited to islands and parts of Siberia and Europe, have received little attention from either the scientific community or the general public. Our working group built on these previous ideas, and developed a cohesive framework for the use of ecological history and species analogs in biodiversity conservation. We widely reached the general public because we focused on large vertebrates—usually keystone species now largely absent—and, of course, published in a high-profile journal to which the media turns to for scientific reporting.

MEGAFAUNA AND THE MEDIA

The article in *Nature* received widespread, international attention in the print and radio media, and to a lesser extinct on television. Due to Associated Press coverage, the article appeared on the front page of many papers across the country, as well as receiving coverage in *USA Today*, *New York Times*, *The Economist*, *Christian Science Monitor*, National Public Radio, CBS News Radio, and the British Broadcast Service. The story reached newspapers and radio stations in Brazil, France, Germany, New Zealand, London, Australia, Canada, Africa, and India. Pleistocene Rewilding even attracted the talk shows, including ABC's *Good Morning America*, NBC's *Today Show*, and CNN's *Lou Dobbs Tonight*. The majority of the media coverage took place within a week of when the original article was published. In the following months, a number of full-length articles and editorials discussed the idea, including coverage in *Harper's Magazine*, *New York Times'* Editorial Page, and *New York Times Magazine*'s "Big Ideas of 2005."[8]

Media coverage of Pleistocene Rewilding ranged from scholarly to sensational and even comical (Figure 26.1). The idea of "cheetahs in your backyard" almost certainly played a major role in the widespread publicity. While some of the coverage took on the tone of a "Jumanji" sequel or a real-life Jurassic Park, much of the media took a balanced approach discussing the proposed pros and cons, quoting the proponents and the critics, while concurrently sliding in *Lions-eat-people* lines to garner the public's attention. Perhaps due to the media's lack of appreciation of ecological history, many articles focused on the potential benefits of Pleistocene Rewilding to endangered African megafauna rather than the heart of the proposal, which was the possibility of using large vertebrates to partially restore key biological processes to North America. This massive publicity raised the issue to a large portion of the American public of bringing bygone megafauna back to American soil. While many people probably failed to read details of the plan, it clearly piqued their interest and stirred some souls.

LIONS IN YOUR BACKYARD

During the height of the media attention, we received hundreds of unsolicited email correspondence (and a small amount of post mail) from citizens expressing enthusiasm, concern, and outrage. The response was overwhelming in terms of our capacity for rapid responses, and was generally bimodal in character: The majority who wrote was either excited and enthusiastic about Pleistocene Rewilding or adamantly and fiercely against it. Some emails and weblogs even resorted to personal attacks. The views voiced by those who wrote us fall into a few general categories that provide some insight, albeit perhaps biased ones, into our relationships with large animals.

Figure 26.1   Political cartoons regard the proposal of Pleistocene Rewilding. (Copyrights Cox & Forkum and John Trever)

Three general themes emerged from those who felt strongly against the idea of entertaining the reintroduction of large vertebrates to North America: elitism, imperialism, and fear. Elitism, we suspect, may have been expressed due to a combination of fear and a lack of appreciation for the ecological history of North America. While we know of no supporting data, the majority of North Americans are likely unaware that several species of elephants, horses, camels, and other megafauna roamed North America as recently as 10,000 years ago. Indeed in many ways, the press generated by Pleistocene Rewilding was one grand natural history lesson.

Many from Africa, as well some critics in the United States, accused us of conservation imperialism, a *let's go over to Africa, bring their animals back to the United States, and save them for the Africans* tone (Table 26.1). While our paper focused on the restoration of North America with captive animals already in the United States (the word captive was used five times in the two-page article) and made no mention of importing animals, this view was likely a result of inaccurate reporting by the media, an already height-ened sensitivity to western imperialism, and perhaps the second sentence of our article: "And now Africa's large mammals are dying stranded on a continent were wars are raging over scarce resources." Taken alone, this statement could be considered imperialistic, but it is accurate nonetheless. It is indisputable that many of Africa's megafauna are critically endangered and that wars are underway in Africa over resources.[9] Yet when this sen-tence is placed in context with the one preceding it—"North America lost most of its large vertebrate species . . . at the end the Pleistocene"—it takes on a different meaning, one that centers on the ecology and overall state of the world's remaining large vertebrates. Other critics expressed compelling concerns about detracting from Africa's tourism industry should the 100-year vision of Pleistocene Parks be realized in North America. However, available tourism data suggest that impact would be minimal to nonexis-tent.[10] Perhaps what is most interesting and ironically encouraging, is the strong connection between large animals and Africans. That deep connec-tion encourages a sense of stewardship for megafauna. If large trees had been our focus, we doubt there would have been such a strong response.

Fear drove many of the expressed concerns. Perhaps nothing makes us feel more human and vulnerable than a large predator that could kill and eat us. The reasons for this are as complex as they are historical, since our relationships with large animals extend back tens of thousands of years into the Pleistocene.[11] The detailed drawings of rhinoceros and lions in the Chauvet caves of France dated at >30,000 years ago are testament to this telling relationship, as is our ability to hunt and persecute large animals to the brink of extinction and beyond.[12] Whether such fears are justified or not, these perceptions influence public opinion and subsequently drive pub-lic policy. The negative economic impacts of large predators in the United States are often more myth than fact, and tourism-related deaths in South Africa attributed to lions, elephants, and other large mammals are rare.[13] Nonetheless, local human–predator conflicts (including deaths) are seri-ous problems in parts of Africa and India.[14] Mitigating for human–carni-vore conflict and elucidating the variety of benefits large predators bring to society is by no means a new conservation challenge. Progress in the United States is being made, as illustrated by reintroduction of wolves into

Table 26.1   A Selection of Correspondence Sent to the Authors in <u>Protest</u> to Pleistocene Rewilding

---

Elitist Response

You are proof that the state of higher education in America's Ivy League schools has sunk to a new low.

You probably do answer critical questions posed to you by other polite academics as you all sit smugly at the public funding trough and dream up dopey proposals like the instant one.

This is one of those good-hearted projects that tend to be suggested by people who have thrown in their lot with the animals to an extent that makes them incapable of understanding the views of members of their own species. Give it up, dear boy.

Of all the screwball, hairbrained ideas. You obviously live in an ivory tower and have no experience in the wild. I pray my tax dollars are not funding this stupid idea. In fact, I'm writing my congress woman to make sure it isn't.

Just a note to let you know that those of us who actually work for a living think you are a colossal asshat.

Imperialistic Reaction

Thank you for planning to rob Africa of her animals so you can beautify your barren great plains with her native animals!! Why come after the animals that support our tourism. Leave Africa and what is hers alone, thief!!!!!!!

The animals you are planning to use may be captives, but you are still stealing the identity of our land, and the backbone of our tourism industry. Our animals are unique to us and our landscape, they are among the last uniquely African possessions we have, it should always remain thus!!

Now here you come along, Mr. Genius, and plan to rob us . . . if America succeeds how many people will bother to go to Africa to see it. As an African I can tell you this, people don't come for a four star hotel treatment. You know what really burns me inside, you have robbed us before . . . now you want to take animals.

Fear of Large Animals

Turning loose wild animals in USA anywhere is moronic. You must not have any children or if you do you must think its ok that they will be lion food! You are a f*$#ing moron if you release killers in our homeland. I hope the cattle rancher guys shoot your ass or feed you to those lions if you release those killers into our ecosystem.

I want my kids to grow up where there are no lions, tiger, bears, rattlesnakes, alligators, crocks, or poison mosquitoes, or gang members either for that matter or anything else that would cause them harm! Learn to look at the big picture, let them go extent who cares as long as we and ours are safe!

If an Elephant ever comes tromping through my yard it'll get an ass full of buckshot.

I know we'll all appreciate it when are kids are eaten alive at a campsite, or when we get gobbled up while taking a hike.

If they get near me, my family, friends or my property then I'll be "really careful" when I place the crosshairs on them and slowly squeeze the trigger of my Remington 300Ultra Mag. Are you sane?

Table 26.2   A Selection of Correspondence Sent to the Authors in <u>Support</u> of
Pleistocene Rewilding

Love of Large Animals

I have just finished reading as many articles as I could find on the reintroduction of mega fauna to the United Sates. First, if this is even possible with the land greed we have in this country it would be a miracle but second, and the reason for this letter, if you are able to convince people to let this happen I will go back to school for what ever degree you would want me to have to help.

I want to congratulate you and your colleagues for this idea and I must confess that I had dreams about this for many years, since I first saw a fossilized cranium of Panthera atrox at a museum!

I wanted to write to say how wonderful I feel this approach is to conservation. I really hope the American people get behind this and you get support from the government too, I can only see benefits for the public, the local area (for jobs) and most importantly conserving rare animals in an environment that is going to keep them alive but also preserve their natural habits which they do not show in smaller parks & zoos.

As I'm not an expert in biology, my opinion doesn't carry much weight, but I think what you are proposing is a great idea. I've always wanted to go to Africa and see cheetahs and elephants in the wild, and now there is a possibility that I can do that in this country. I know the animals will be brought over for ecological purposes first, but it would still be cool to have "Cheetah Crossing" signs, or to see a pride of lions just hanging out while driving to the grocery store.

Response from the Ranching Community

I wasn't sure how far this had gone or if you were getting too many offers. This sounds intriguing, even if sabertooths and giant sloths can not be on the list . . . I might know of 47 square mile ranch in Central Arizona that might be interested.

I truly believe it is an amazing and promising concept. My family owns 26,000 acres in west Texas. Please advise if you think the private land concept is plausible. If so, I may be able to rally some landowners to the cause, if it can be economically feasible.

We're ranchers in Southeast Kansas. Let me know how we can help.

Support for a Proactive Positive Direction

As a physicist and long-time student of ecology, I love your proposal . . . Moreover, it has the virtues of positive action which is always more likely to succeed than complaint.

I'm a US Marine currently on tour in Al Anbar Province, Iraq. Since I was a child I have been amazed and in love with nature. I've always wanted to give something back to the earth that we take so much from as humans. Rewilding America on a large scale would be up there with the wonders of the ancient world in regards to achievement.

I just read your commentary in Nature and absolutely loved it. The lack of a historical context and almost static view of most conservation efforts today are really disappointing. I think that beyond the specific goals stated in the commentary that the discussion its generating might help to promote a view of conservation that looks back beyond 1492 and looks forward beyond this coming century.

continued

Table 26.2   continued

What a fantastic idea–giant safari parks in North America. I am so glad to hear
this being discussed. I hope in the future this puts an end to small overcrowded
zoos which encourage breeding, only to have these animals spend their lives in
small enclosures. it is not enough to keep these animals alive in zoos. we need to
look toward returning them to nature to freedom in the wild.

In spite of the tremendous barriers to this, you have a wonderful project! I know
that species transplantation is viewed with skepticism, but without a haven,
these species are "goners" in less that 50 years, except in captivity. But likewise,
anywhere threatened large species are in jeopardy, we have to protect them until
future ages can deal with the world respectively.

Yellowstone after decades of absence and persecution. Such bold human
actions, along with fear, are also testament to the enduring relationship and
fascination with large predators.

Three general themes also emerged from those who supported the idea of
Pleistocene Rewilding: the love of large animals, intrigue from the ranching
community, and support for a proactive conservation direction (Table 26.2).
We believe that the love for large animals likely stems from the same relationship that stirs our fears for those same creatures. Ancient rock art, cars, and
sports teams named after large mammals, as well as conservation programs
centered on large mammals are evidence of our fascination with charismatic
megafauna. Economic data lend credence to the cosmopolitan nature of this
allure: more people annually visit San Diego Zoo's Wild Animal Park for
catching glimpses of large mammals than all but a few U.S. National Parks.

The positive response from the ranching community came as somewhat
of a surprise, but in hindsight it could have been predicted for at least two
main reasons. First, many private ranches in Texas and elsewhere in the
western United States already contain large vertebrates such as cheetahs and
camels. While some of these megafauna are supported by ecological ratio-
nales, many are not (e.g., many African bovids and kangaroos). Further,
none of these ranches are managed in a manner proposed under Pleistocene
Rewilding, and most are managed solely for hunting enterprises. Nonetheless, ranchers are intimately familiar with large vertebrates on many different levels. Second, many ranchers are facing economic hardships and are
subsequently turning to alternative and complementary economic ventures,
including biodiversity conservation, to preserve their lifestyles and landscapes.[15] While the precise motives of the interested ranchers who contacted
us are unclear, we speculate that they include some blend of these issues.

And lastly there was much excitement demonstrated for the positive, pro-
active nature of Pleistocene Rewilding. This may turn out to be an important
point. Most of the environmental movement is too easily characterized as
"doom and gloom." It is difficult to captivate the general public and motivate change on a negative platform, especially when there is no immediate threat (perceived or otherwise, e.g., terrorism). Conservation biology is

largely consumed with refining the eulogy of nature. Eulogies, no matter how detailed, are bad strategies for a social movement. Moving away from mitigating extinction to pro-actively restoring ecosystems via the reintroduction of large vertebrates is arguably an exciting new platform for conservation biology.

ECOLOGICAL HISTORY, CONTEMPORARY PAROCHIALISM, AND THE SCIENTIFIC COMMUNITY

As with the media and general public, Pleistocene Rewilding precipitated a substantial response by the scientific community, including some gasps and groans. Some responses are arguably more legitimate and consistent than others. All hinge on matters of perspective. Many criticisms and concerns center on opportunity costs and uncertainty.[16] However, both are not unique to Pleistocene Rewilding; rather, we grapple with them with every conservation intervention. And contrary to some claims[17], we do not advocate Pleistocene Rewilding as a substitute or as a priority over ongoing conservation projects in Africa and North America.

Other concerns revolve around conservation benchmarks and the issue of nativeness.[18] While we acknowledge that "European" horses and "African" lions can be viewed as non-native to North America, such an argument could be considered inconsistent with other generally accepted conservation practices and is guilty of a parochial view of ecological history.[19] Rubenstein and colleagues claim that "modern day proxies species are wrong . . . different genetically from the species that occurred in North America during the Pleistocene;" yet they sidestep the question of why Bolson Tortoises, California Condors, and Peregrine Falcons are acceptable as proxies in some conservation programs?[20] Bolson Tortoises have been absent in the United States for over 10,000 years, and the remaining population in central Mexico surely differs genetically from the population that once roamed New Mexico. Was the Turner Endangered Species Fund wrong in restoring this critically endangered species to New Mexico? The last time California Condors soared over the Grand Canyon was in the late Pleistocene. Was the National Park Service misguided in reintroducing contemporary condor populations from California to Arizona? And was the Peregrine Fund wrong in reintroducing Peregrine Falcons from four different continents to the Midwest United States and parts of Canada as a proxy for the peregrines that once resided there and were extirpated by DDT in the 1960s? In each case the proxies certainly differed genetically from the extirpated population, and two of the three cases are already celebrated success stories with wholesale support from the conservation biology community.[21]

A similar rationale applies to horses and lions—each of which today occupies a fraction of its former distribution and was present in North America until the late Pleistocene. If Bolson Tortoises, Condors, and Peregrines can be appropriate proxies, why not lions? Could the real answer be that lions can eat us, rather than the objections more typically raised against our other exemplars? No one has argued against the introduction of the Bolson Tortoise because it is insufficiently native, or because it may promote unforeseen problems in disease ecology. Yet the "African Lion" is accurately labeled as such only in the sense that we might call *Mustela nigripes* the "Wyoming and South Dakota Ferret" subsequent to its extirpation throughout much of the Great Plains. Judging from molecular systematic assessments, conspecifics with the extant "African Lion" were found in the New World until the late Pleistocene, possibly as far south as Peru.[22] Of course lions then and now are not identical, but neither are the captive-bred northern Great Plains ferrets that were recently restored to Mexico identical with those that occurred at the southern range limit of *M. nigripes* a century ago.

Other critics have posited that global climate change since the late Pleistocene makes large vertebrate reintroductions to North America unviable.[23] While it is clear that North America's ecosystems are not the same today as they were 13,000 years ago, we hope dearly this would not preclude restoration activities since future anthropogenic climate change will likely dwarf those of the past. In fact habitats have been, are, and will continue to be dynamic on a time scale ranging from annual to millennia. But very few plants or small mammals went extinct during the late Pleistocene; rather, they migrated.[24] Thus, from a continental perspective, the major missing component of ecosystems today compared to the Pleistocene and prior is megafauna. Today's nature contains many anomalies because it misses what we can infer are critical cogs in the wheels—megafauna. Currently, we do not know if and where it is possible to reintroduce viable megafauna populations to North America. We suggest that it may be worth carrying out some scientific experiments to find out. A hypothesis-driven scientific approach under a cost-benefit framework will be critical if Pleistocene Rewilding is to be successfully implemented, particularly for addressing the greatest and most legitimate challenges such as disease ecology and the possibilities of unexpected ecological and social interactions.

Some biologists believe that Pleistocene Rewilding is simply not feasible in the world we live in today. Rubenstein and colleagues go as far to state that "We all remember 'Jurassic Park' . . . Pleistocene Rewilding of North America is only a slightly less sensational proposal."[25] Relapsed time since the late Pleistocene differs from that since the late Jurassic by a factor of about 15,000 (or about 150 million years), a phenomenal exaggeration of "slightly less" by any standard; moreover, the entire group to which dinosaurs pertain went extinct 60 million years ago, whereas mammoths, to which Asian elephants are more closely related than they are to extant African species, went extinct less than 4,000 years ago.[26] Many Pleistocene megafauna are extant today, and repatriating conspecifics or related proxies to areas where they went extinct about twice as long ago as the age of some living plants is simply not ecologically, evolutionarily, and "sensationally" equivalent to cloning giant extinct therapods (whose closest living relatives are birds) and then releasing them in modern environments.

In fact, Africa provides inspiration for the feasibility of Pleistocene Rewilding. The year following its founding, Kruger National Park was hardly the celebrated mainstay of southern African biodiversity it is today. In 1903, there were zero elephants, nine lions, eight buffalo, and very few cheetahs within the boundaries of the park. Due to vision and dedication of African conservationists, 7,300 elephants, 2,300 lions, 28,000 buffalo, and 250 cheetah would come to roam Kruger one hundred years later—as did 700,000 tourists bringing with them tens of millions of dollars.

A MEGAFAUNA RUNS THROUGH IT

Pleistocene Rewilding poses several questions that have thus far been largely ignored by the conservation community: What are our conservation benchmarks and why? Are we satisfied with the current trajectory of conservation biology, which is largely focused on slowing extinction rates and providing ever more detail for the obituaries of nature? Will we accept the risk of doing nothing (perhaps believing that doing nothing carries no risk), or will we attempt bold, carefully considered measures that sometimes fail?

We believe these questions and their answers are critical for the future of biodiversity and humanity. The answers and subsequent actions will largely dictate what kind of world future generations will live in and how much biodiversity will live alongside them. To answer those questions, we all (media, general public, the scientific community) must move past emotions and think deeply about our biases and our histories—and how they might affect the ways we think about and practice biodiversity conservation. Our parochial view of ecological history, fear, fascination, and our sense of place all influence our views on what is natural. It is not whether those views are good or bad that is most important; but rather, is it instructive to ask if they are consistent? Is 1492 an appropriate conservation benchmark for North America or is it racist? It appears that when we talk about the possibilities of proxies for restoration, we are relatively comfortable with the idea if the species is relatively small, benign, or out of sight, irrespective of "sameness" and the deepness of the benchmark (80 or 8,000 years). People are already restoring butterflies, island birds, and giant tortoises in ways that are consistent with Pleistocene Rewilding. Perhaps restoring megafauna may bring these deeper philosophical issues to the forefront of conservation, biology, and society.

NOTES

1. D. Quammen, *Monster of God: the man-eating predator in the jungles of history and the mind* (New York: W. W. Norton and Co., 2003); D. Peacock and A. Peacock, *The essential grizzly: the mingled fates of men and bears* (Guilford, Connecticut: The Lyons Press, 2006); P. Shepard, *Coming home to the Pleistocene* (Washington DC: Island Press, 1998); H. Kruuk, *Hunter*

and hunted: relationships between carnivores and people (Cambridge: Cambridge University Press, 2002).

2. C. J. Donlan et al., "Re-wilding North America," *Nature*, 436 (2005): 913–14.

3. C. J. Donlan et al., "Pleistocene Rewilding: An optimistic agenda for 21st century conservation," *American Naturalist* 168 (2006): 660–81.

4. P. S. Martin and D. A. Burney, "Bring back the elephants!," *Wild Earth* (Spring 1999): 57–64; D. A. Burney, D. W. Steadman, and P. S. Martin, "Evolution's second chance," *Wild Earth* (Summer 2002): 12–15.

5. P. S. Martin, "Pleistocene niches for alien animals," *Bioscience* 20 (1970): 218–22; P. S. Martin, "Wanted: a suitable herbivore," *Natural History* 78:2 (1969): 35–39.

6. I. A. E. Atkinson, "Introduced mammals and models for restoration," *Biological Conservation* 99:1 (2001): 81–96; D. W. Steadman and P. S. Martin, "The late Quaternary extinction and future resurrection of birds on Pacific islands," *Earth–Science Reviews* 61 (2003): 133–47; S. A. Zimov, "Pleistocene Park: Return of the mammoth's ecosystem," *Science* 308 (2005): 796–98; A. G. Jones, "Reptiles and amphibians," in *Handbook of ecological restoration*, M. R. Perrow and A. J. Davy, eds. (Cambridge University Press: Cambridge, 2002); D. B. Wingate, "Successful reintroduction of the yellow-crowned night-heron as a nesting resident on Bermuda," *Colonial Waterbirds* 5 (1982): 104–15.

7. R. M. Pyle, "Resurrection ecology: bring back the Xerces blue!," *Wild Earth* 10 (2000): 30–34.

8. Stolzenburg, W., "Where the wild things were," *Conservation in Practice* 7 (2006): 28–34; N. D. Kristoff, "Where deer and lions play," *New York Times* 33 (December 13, 2005); H. Nicholls, "Restoring Nature's Backbone," *PLoS Biology*, 4 (2006): e202.

9. S. Blake and S. Hedges, "Sinking the flagship: the case of forest elephants in Asia and Africa," *Conservation Biology* 18:5 (2004): 1191–1202; P. M. Gros, "The status and conservation of the cheetah Acinonyx jubatus in Tanzania," *Biological Conservation* 106:2 (2002): 177–85; C. André and J. Platteu, "Land relations under unbearable stress: Rwanda caught in the Malthusian trap," *Journal of Economic Behavior and Organization* 34 (1998): 1–47; J. Diamond, *Collapse: How societies choose to fail or succeed* (New York: Viking, 2004), 592.

10. C. J. Donlan et al., "Pleistocene Rewilding," 660–81.

11. D. Quammen, *Monster of God*; P. Shepard, *Coming home to the Pleistocene*.

12. P. Matthiessen, *Wildlife in America* (New York: Viking Press, 1989); D. Quammen, *Monster of God*.

13. K. M. Berger, "Carnivore-livestock conflicts: effects of subsidized predators control and economic correlates on the sheep industry," *Conservation Biology* 20 (2006): 751–61; D. N. Durrheim and P. A. Leggat, "Risk to tourists posed by wild mammals in South Africa," *Journal of Travel Medicine* 6 (1999): 172–79.

14. C. Packer et al., "Conservation biology: Lion attacks on human in Tanzania," *Nature* 436 (2005): 927–28; V. K. Saberwal et al., "Lion–Human conflict in the Gir Forest, India," *Conservation Biology* 8 (1994): 501–07.

15. N. Sayre, "Working wilderness: the Malpai Borderlands Group and the future of the western range" (Tucson: Rio Nuevo Press, 2006).

16. G. Chapron, "Re-wilding: other projects help carnivores stay wild," *Nature* 437 (2005): 318; M. A. Schlaepfer, "Re-wilding: a bold plan that needs native

megafauna," *Nature* 437 (2005): 951; E. Dinnerstein and W. R. Irvin, "Re-wilding: no need for exotics as natives return," *Nature* 437 (2005): 476.

17. G. Chapron, "Re-wilding: other projects help carnivores stay wild," *Nature* 437 (2005): 318; E. Dinnerstein and W. R. Irvin, "Re-wilding: no need for exotics as natives return," *Nature* 437 (2005): 476.

18. M. A. Schlaepfer, "Re-wilding," 951.

19. C. J. Donlan and P. S. Martin, "Role of ecological history in invasive species management and conservation," *Conservation Biology* 18:1 (2004): 267–69.

20. D. R. Rubenstein et al., "Pleistocene park: Does re-wilding North America represent sounds conservation for the 21st century," *Biological Conservation* 132 (2006): 232–38.

21. N. Snyder and H. Snyder, *The California Condor: A saga of natural history and conservation* (San Diego, CA: Academic Press, 2000); T. J. Cade and W. Burnham, eds., *Return of the Peregrine: A North American saga of tenacity and teamwork* (The Peregrine Foundation: Boise, Idaho, 2003).

22. N. Yamaguchi et al., "Evolution of the mane and group-living in the lion (Panthera leo): a review," *Journal of Zoology* 263 (2004): 329–42; J. Burger et al., "Molecular phylogeny of the extinct cave lion Panthera leo spelaea," *Molecular Phylogenetics and Evolution* 30 (2004): 841–49; B. Kurtén and E. Anderson, *Pleistocene Mammals of North America* (New York: Columbia University Press, 1980).

23. C. I. Smith, "Re-wilding: introduction could reduce biodiversity," *Nature* 437 (2005): 318.

24. S. K. Lyons, F. A. Smith, and J. H. Brown, "Of mice, mastodons and men: human-mediated extinctions on four continents," *Evolutionary Ecology Research* 6 (2004): 339–58.

25. D. R. Rubenstein et al., "Pleistocene park."

26. J. Krause et al., "Multiplex amplification of the mammoth mitochondrial genome and the evolution of Elephantidae," *Nature* 439 (2006): 724–27; H. N. Poinar et al., "Metagenomics to paleogenomics: large-scale sequencing of mammoth DNA," *Science* 311 (2006): 392–94.

Conclusions

27 Restoring Dirt Under the Fingernails

Eric Higgs

As evidence mounts for rapid changes in global climate systems, those of us who focus on restoring ecosystems sense the ground beneath us has given way, too. A field so tightly focused on setting clear goals for successful restoration now finds itself turning to multiple ecological trajectories and very long-term historical ranges of variation for inspiration. Talk of adaptation is growing, and restoration may perhaps become a way of anticipating and mitigating the effects of rapid ecological shifts: novel ecosystems and a world lightly attached to the past will be our tableau. This will be especially true for work that concerns the recovery of already threatened species, many of which will have their tenuous hold on survival further compromised. Then, of course, there are some who are whispering that restoration may in fact be an incoherent practice in a turbulent new nature. What is the benefit, so the argument goes, of fixing our sights on historical patterns and processes when these will be rendered immediately irrelevant or bypassed in the long run? This points to a turn away from history and perhaps from the very idea of restoration. In the future, we may well be turning to new (and some old) paradigms—recreation, regeneration, rehabilitation, reclamation, designer ecosystems, synthetic nature—to set conditions of engagement with natural processes.

This book arose from the inspired thinking of Marcus Hall, an historian. Most who attended the Transatlantic workshop in 2006 were inclined to history whether through the discipline of history or archaeology, anthropology and sociology. As always ought to be the case, the purpose of critical reflection is to ponder what might emerge from the dismantled superstructure of prior thinking, and this is exactly what was imagined for the workshop and the essays that arose from that gathering. The difficulty is that the lust for criticism of concepts can overrun what practitioners are doing on the ground. It was apparent to me that the simple effectiveness of present-day restoration gave way to fascinating conversations about the main stream and back eddies of restoration practice. My goal in this chapter is to recover a quotidian notion of restoration, and to set this understanding up against the extraordinary backdrop of portentous climate shifts. I will argue that historical knowledge will remain essential if not more important in coming years in deciding how to cope with ecosystems that have been degraded or destroyed. I suggest that humanists and human scientists need to respect the reality of natural processes and the practitioners who enact salutary interventions in ecosystems. We need more dirt under our fingernails as we explore the conceptual clumps and horizons.

HISTORICAL FIDELITY, RESILIENCE, AND RAPID CHANGE

For several years a local group has gathered in support of Beacon Hill Park and more particularly the degraded Southeast Woods. The Woods are a relict of the once common coastal Douglas Fir (*Pseudotsuga menzeizii*) ecosystem in the greater Victoria region. Settled in the 1840s, Victoria is the capital city of the Province of British Columbia and has steadily pushed back intact ecosystems through urban development. In spite of a reputation for environmental leadership in Canada, Victoria is afflicted with the same problems as most urban areas in the world: a precipitous drop in intact historically faithful ecosystems, smaller patches less well connected, and invasions of weedy species. Beacon Hill Park, a landmark in Victoria similar in significance to Victoria as Stanley Park is to Vancouverites or Central Park to New Yorkers, was established in the 1880s as an urban preserve and pleasure ground for leisure engagements. In the latter regard it has manifestly been successful as a location for public events, picnics, walkers and fanciers of fine ornamental gardens.

In recent years attention has shifted to restoring patches of biodiversity, including the showy springtime fields of common camas (*Camassia quamash*) and other native wildflowers. The largest intact ecosystem in the Park is the Southeast Woods, an area less than two hectares squeezed on two sides by major roadways, on the other two by Park facilities, and shot through with formal and informal walking trails. Coastal Douglas fir ecosystems, of which this is one, typically contain mature individual trees, some hundreds of years old, with a rich shrub layer and understory. Moisture is limited in Victoria and the proximity to ferocious winter storms off the Straight of Juan de Fuca give these forest remnants their distinctive character. The Woods face difficult challenges. First, a steady stream of weedy species including Scotch broom (*Cytisus scoparius*), English holly (*Ilex aquifolium*), English Ivy (*Hedera helix*), and spurge laurel (*Daphne laureola*) have occluded much of the understory species resulting in a loss of local vegetative diversity. Second, human traffic in all its various forms have worn away at the integrity of the forest. Informal braided trails, trampling of tree roots, vandalism, and litter are evident throughout. As one of the last wild ecosystems in downtown Victoria it is also a refuge for those whose activities lie at the margins of convention: the homeless, those seeking a haven for illegal drug use, and sexual activities. The Woods are in many respects a refuge from mainstream urban society.

Restoring this relict is a delicate walk along the line that divides ecological and cultural integrity in contemporary life. No amount of ecological do-gooding will be successful without addressing seriously the plight of people who use the Woods. Similarly, addressing social concerns without ecological awareness would erode whatever integrity is left of the Wood's ecological processes, patterns, and structures. A local volunteer community organization, Friends of Beacon Hill Park, has worked for three years at restoring projects. A graduate student at the University of Victoria, Jeff Ralph, was centrally involved in the efforts. Understanding the intricate connections of social and ecological well being, he wove his way through city committees and approval, worked with volunteers and unionized staff, and helped organized various restoration activities. Most Saturday mornings, volunteers meet at the Woods to remove invasive species.

A main ingredient in the restoration planning was the history of the site and the typical characteristics of coastal Douglas fir ecosystems. What did historical photographs, maps, and journals reveal about this remnant forest? What influences accounted for its particular configuration now, and what could reasonably be expected of a restoration effort? What species were missing and which ones could be reintroduced? What designs would work best to encourage a judicious balance of historical fidelity, ecological integrity, and social responsibility? These were the questions that occupied those involved with the project, although perhaps more clearly in retrospect than they appeared during the organic development of the project.

One sunny Saturday in October 2006, fifty students in my introductory ecological restoration class descended on the Park. Jeff Ralph, volunteers, and Park staff had organized three rotating work stations: one to pull ivy in an area previously treated; one ivy pull in a new area; and the third planting hundreds of native plants raised in a local nursery. The level of excitement was electric as students wielding loppers, pruners, and bare hands pulled and gathered several dozen massive piles of weeds from two of the sites. A camaraderie developed immediately, and the sounds of earnest conversations and laughter could be heard everywhere. At the planting station the results were startling. In two well-prepared plots where an old parking lot once stood, three waves of students planted native species including some which had been extirpated from the Woods.

There were three remarkable and diagnostic outcomes from the experience. First, many students commented that they felt a new sense of connection to the landscape. They understood it better by working with it, and would return to observe the health of their plantings. A bond of affection had formed. Second, many commented that the brief experience in the Woods opened their eyes to restorative challenges and possibilities elsewhere: from one specific experience they could extrapolate to other sites and possibilities. Finally, their connections placed them in an historical process, both in terms of realizing what had flowed from historical integrity to degradation to restoration and also by inaugurating their particular involvement (not all students, of course, will have a future connected to the site, but some will and that is a significant outcome).

This bond of affection needs to be considered in the face of changes that are predicted for the site in the wake of climate changes. Warmer, drier conditions will influence the Woods, perhaps pushing it away from its present coastal Douglas fir ecosystem toward a novel ecosystem. This latter possibility is likely given the small size of the site and the difficulty in achieving connectivity with other nearby ecosystems. My hunch is that those concerned with the site will accommodate such changes. There may be successive drought years or slowly rising temperatures that will produce shifts in the forest composition. The amount of diversity presently available should enable some species to flourish, including some rare ones, even as others struggle under the new conditions. The connections that exist for those volunteering at the site will create an adaptive relationship and through this make apparent both the importance of historical fidelity and ecological integrity. If anything, historical knowledge will provide guidance to those seeking an understanding of resilience at the site. Perhaps ecological restoration is ultimately a matter of restoring relationships between people and natural processes as it is about the restoration of natural processes.

IN DEFENSE OF HISTORY

There is growing concern, evident in some of the contributions to the present volume, about the propriety of using ecological restoration in the light of rapid climate change and a cultural disposition toward novelty and human-designed systems, including genetic modification. We need new approaches apropos of rapidly changing cultural and ecological conditions. Pleistocene rewilding, naturalized landscapes befitting the cultural character of a region, and designed systems for maximizing ecological services are emerging as paths that seem shiny compared with the rather dowdy traditional associations of restoration. Typically these newer approaches do not court historical knowledge.

As the winds of change pick up speed, we need to question carefully the role and value of historical knowledge in ecological restoration. If we push beyond simple instrumental accounts of history, as most historians of course do, then we find that history exerts a powerful countercurrent against simply sailing along. Elsewhere I argue that historical fidelity, the condition of being faithful to the historical knowledge if not to historical conditions per se, is one of four keystone concepts defining good ecological restoration. History exerts a particular discipline on the work of restorationists, and provides critical clues about how our practices should honor deep ecological processes.

I argue there are four reasons why history should become more, not less, important under rapid changing global climate systems. First, and perhaps most conventional, is the range of long-term variation that historical data provide us. Having focused much of our framework in northern temperate and montane ecosystems where I work on the last few hundred years, we may increasingly turn to and require palaeoecological knowledge back through the Holocene epoch and possibly beyond. Second, we will depend on reference ecosystems for models of how to restore or reconstruct ecosystems. That these references may come from further afield in anticipation of analogs that correspond with the rolling edge of an ecosystem's range, may well be the new reality. Third, the complexity and difficulty of understanding the history of an ecosystem enforces an important discipline on restorationists and as a consequence encourages a particular humility on our actions. Finally, history tempers our ambitions by showing how much folly is possible through hyperactive and heedless responses to change. Ecological sobriety may be our best practice in coming decades.

DIRT UNDER THE FINGERNAILS

My sense of the future of ecological restoration is that communities of well-intentioned citizens will continue to embrace the need for careful interventions to recover ecosystems that depend on historical knowledge for insight if not particular guidance. This view from the ground appears different from the prognostications built on global models of rapid change and readings of denatured lives in a technological world. The need to look backwards anchors our work, and through such attentiveness builds the kind of connections that emerged for the volunteers at Beacon Hill Park. The act of restoring—the palpable physical act of pulling invasives, building soil, or planting new species—instills a relationship with place and resists arrogating ecosystems to human control. In the end, restoration of ecosystems is always and also about the recovery of human connections with place.

Contributors

Mark B. Bain is associate professor of Systems Ecology in the Department of Natural Resources at Cornell University. His published works have focused on fish and macroinvertebrates in freshwaters and estuaries, ecosystem analyses and assessment, and restoration of species and habitats. His current research is on coastal ecosystems, ecology of sturgeon, restoration of waterways of New York, and pathogens in the Great Lakes.

Emily K. Brock is assistant professor of History at the University of South Carolina. Her current research explores the tensions at the intersection of the science of forestry and the demands of a growing lumber market in the early and middle twentieth-century Pacific Northwest.

James Bullock is a principal research scientist at the UK Centre for Ecology and Hydrology and a visiting professor at the Universities of Bournemouth and Liverpool. His research links human impacts on the environment with the conservation of species and communities at local and landscape scales. Currently he is carrying out research on large-scale species dynamics and the restoration of ecosystems.

David G. Casagrande is associate professor of environmental anthropology at Western Illinois University. He has studied human ecology of cultural groups in Mexico, Venezuela, and the United States. His research has been published in a variety academic journals and edited volumes. He was formally editor-in-chief of the *Journal of Ecological Anthropology*. He is currently using resilience theory to integrate human responses to flooding with ecological restoration along the Mississippi River.

Eileen Crist is associate professor in Science and Technology in Society at Virginia Tech where she teaches courses in environment, science, and technology. She has published a number of refereed and popular papers. She is author of the book *Images of Animals: Anthropomorphism and Animal Mind* (Temple University Press, 2000), which investigates the implications of behavioral science for the question of animal mind. She is also coeditor of *Scientists Debate Gaia* (MIT Press, 2004) and *Gaia in Turmoil* (MIT Press, 2009). Her recent research focuses on environmental science, environmental ethics, and biodiversity conservation.

Althea Davies is based in the School of Biological and Environmental Sciences, University of Stirling, Scotland. Her interest in the long-term history of upland ecosystems is reflected in a recent environmental history project on interactions between plant diversity, land-use, and market incentives in upland Scotland over the last 400 years. She is currently continuing the aims of her recent fellowship to make long-term ecology and management more accessible to upland policymakers and managers and a more integrated part of ecology.

Jan E. Dizard is a sociologist and the Charles Hamilton Houston Professor of American Culture at Amherst College (USA), where he has taught since 1969. Among his publications are *The Minimal Family* (with Howard Gadlin), *Going Wild: Hunting, Animal Rights and the Contested Meaning of Nature*, and *Mortal Stakes: Hunters and Hunting in Contemporary America*. He is currently working on restoration as one aspect of how we socially construct nature.

C. Josh Donlan is the director of Advanced Conservation Strategies, whose purpose is to provide innovative, self-sustaining, and economically efficient solutions to environmental challenges by building cross-sector synergy and integrating biological, economic, technological, and socio-political threats and opportunities. His research and writing has received widespread attention, including the *New York Times Magazine*'s Big Ideas of 2005 and The Best American Science and Nature Writing of 2008. Josh was named "25 of 2005 Saving the Planet" by *Outside Magazine*.

Jenifer E. Dugan is an associate research biologist at the Marine Science Institute of the University of California, Santa Barbara. She is a coastal marine ecologist with wide-ranging research interests and expertise in sandy beach ecosystems. She and Anita Guerrini, both investigators with the Santa Barbara Coastal Long Term Ecological Research program (U.S. National Science Foundation), are collaborating on a multidisciplinary book on the ecological history of a southern California wetland.

James Feldman is assistant professor of Environmental Studies and History at the University of Wisconsin Oshkosh. His book, *A Storied Wilderness: Nature, History, and the Rewilding of the Apostle Islands* will be published by the University of Washington Press.

Harry W. Greene is professor in the Department of Ecology and Evolutionary Biology at Cornell University. His honors include the University of

California, Berkeley's Distinguished Teaching Award and the American Society of Naturalists' Edward Osborne Wilson Naturalist Award, and his *Snakes: the Evolution of Mystery in Nature* made the New York Times' list of "100 Most Notable Books." He teaches biology for non-majors at Cornell and is finishing a book about the personal value of natural history.

Paul H. Gobster is research social scientist with the USDA Forest Service's Northern Research Station in Chicago. His current research examines people's perceptions of natural areas restoration and management, the interface of aesthetic and ecological values in landscape, and the design and provision of urban green spaces to encourage healthy lifestyles.

Matthias Gross is senior research scientist in sociology at the Helmholtz Centre for Environmental Research (UFZ) in Leipzig, Germany. His most recent book, *Ignorance and Surprise,* is forthcoming with MIT Press. His recent research focuses on ecological restoration in post-industrial settings and the revitalization of contaminated sites.

Anita Guerrini is Horning Professor in the Humanities and Professor of History at Oregon State University. Her teaching and research fields include history of science and medicine, and environmental history. She and Jenifer Dugan, both investigators with the Santa Barbara Coastal Long Term Ecological Research program (U.S. National Science Foundation), are collaborating on a multidisciplinary book on the ecological history of a southern California wetland.

Marcus Hall is currently senior lecturer of environmental sciences at the University of Zurich where he teaches courses in environmental history and environmental studies. Before moving to Europe, he was assistant professor of history at the University of Utah where he held the Environmental Humanities Research Professorship. His publications include *Nature and History in Modern Italy* (2010, co-edited with Marco Armiero) and *Earth Repair: A Transatlantic History of Environmental Restoration* (2005), winner of the Downing Book Award from the Society of Architectural Historians. He is currently researching the background and promise of salvage in conservation.

Keith Harrison is a research associate at Sheffield Hallam University and a Senior Researcher with Hallam Environmental Consultants Ltd. He has undertaken research on the mapping of wetland change using archives and the application of GIS mapping techniques.

Eric Higgs is professor and director of environmental studies at the University of Victoria, Canada. He teaches courses in ecological restoration, technology and culture, and environmental issues. He is author of *Nature By Design: People, Natural Process, and Ecological Restoration* (2003). From 2001 to 2003 he was chair of the Society for Ecological Restoration International.

Kathy Hodder is senior lecturer in Conservation Ecology at the University of Bournemouth, UK. She is currently conducting research linking policy and governance with biodiversity conservation and predicting development impacts on wildlife.

Nobusuke Iwasaki is a geographer and researcher at the National Institute for Agro-Environmental Sciences in Tsukuba, Japan, where he develops GIS models to evaluate landscape diversity that supports rural biodiversity and land-use functions.

Jozef Keulartz is associate professor of Applied Philosophy at Wageningen University, the Netherlands. He has been appointed special chair for Environmental Philosophy at the Radboud University Nijmegen, the Netherlands. He has published in different areas of science and technology studies, social and political philosophy, bioethics, environmental ethics and nature policy. Keulartz is member of the scientific council of the European Centre for Nature Conservation (ECNC) and member of the Netherlands Commission on Genetic Modification (GOGEM).

David Kidner worked as a process design engineer in the petroleum industry before turning to social science with a PhD in psychology from London University. For the past three decades he has taught critical social science and environmental philosophy in Britain and the United States, and is currently at Nottingham Trent University. He is the author of *Nature and Psyche: Radical Environmentalism and the Politics of Subjectivity* (SUNY Press, 2001).

David Lowenthal, emeritus professor of geography, University College London, formerly secretary of the American Geographical Society, has been a Fulbright, Guggenheim, Leverhulme, and Landes Fellow and is a Fellow of the British Academy. Among his books are *West Indian Societies, Geographies of the Mind, The Past Is a Foreign Country, Landscape Meanings and Values, The Politics of the Past, The Heritage Crusade and the Spoils of History, George Perkins Marsh: Prophet of Conservation*, and *Paysage du temps sur le paysage*.

Daniel McCool is professor of Political Science, and director of the Environmental Studies program, at the University of Utah. He has written or edited seven books, primarily on environmental issues or American Indian policy. He is currently writing a book on the politics of river restoration in the United States.

Timo Myllyntaus, professor of Finnish history at the University of Turku, Finland, is also the secretary general of ICOHTEC (International Committee for the History of Technology) and the regional representative of the Nordic countries at the European Society for Environmental History. With Mikko Saikku he edited *Encountering the Past in Nature: Essays in Environmental History* (2001) and with Margrit Müller, *Pathbreakers: Small European Countries Responding to Globalisation and Deglobalisation* (2008).

Ian Rotherham is reader at Sheffield Hallam University and Director of the Tourism and Environmental Change Research Unit. He has written over 250 papers and articles and a number of books. His research includes landscape history, the economics of landscape change, issues of invasive alien species, and aspects of tourism development. He has been awarded a number of international research prizes.

William D. Rowley holds the Griffen Chair in Nevada and the West in the History Department at the University of Nevada, Reno. Now finishing a two-volume history of the U.S. Bureau of Reclamation and the development of western water resources, his previous books include studies of western range regulation and American agricultural issues.

Chris Smout is Professor Emeritus at the University of St. Andrews and Historiographer Royal for Scotland. He has served as Deputy Chairman of Scottish Natural Heritage, 1991–1997. Among his books are the Ford Lectures at Oxford, published as *Nature Contested; Environmental History in Scotland and Northern England since 1600* (Edinburgh UP, 2000), *A History of the Native Woods of Scotland 1500–1920*, (with Alan Macdonald and Fiona Watson, Edinburgh UP, 2005) and *Exploring Environmental History* (Edinburgh UP, 2009). He is currently working on the environmental history of the Firth of Forth.

David Sprague is an anthropologist and senior researcher at the National Institute for Agro-Environmental Sciences in Tsukuba, Japan. He supervises historical GIS projects with the objective of proposing landscape and biodiversity indicators for evaluating rural land-use change.

Mairi J. Stewart is research fellow at the UHI Centre for History, based at North Highland College, Scotland, where she is principal researcher on a social history of twentieth-century forestry. She has worked on a range of environmental history projects in Scotland. Among her publications are two chapters in T. C. Smout, ed., *People and Woods in Scotland: A History* (2003). She is joint organizer of the Scottish Woodland History Discussion Group.

David Tomblin is a PhD Candidate in the Department of Science and Technology in Society at Virginia Tech. His dissertation research explores the role of ecological restoration in the political resurgence of the White Mountain Apache Tribe in the second half of the twentieth century.

Miguel Vasquez is professor of Anthropology at Northern Arizona University in Flagstaff, where he teaches courses on Development; Ethnographic Methods; Community, Technology, & Values; and Sustainability in the Southwest: Lessons from Puebloan Culture. He works closely with diverse community organizations in northern Arizona, focusing on local and indigenous sustainability in face of the impacts of globalization. He was awarded the 2008 Presidents Faculty Award at Northern Arizona University.

Frans Vera is personal advisor of the Director of the National Agency for Nature Management and Forestry (Staatsbosbeheer) and temporary lecturer at the Wageningen University, the Netherlands. In his *Grazing Ecology and Forest History*, he challenges the theory of the closed canopy forest as the baseline for the natural vegetation on places where trees can grow in the lowlands of both Europe and the eastern United States. The main part of his work concerns reestablishing natural processes with large indigenous ungulates in European nature reserves.

Laura A. Watt is assistant professor of environmental studies and planning at Sonoma State University (USA), where she teaches a variety of courses in environmental history, politics, and planning. She is currently completing a book with the University of California Press on the history and management of a protected working landscape at the Point Reyes National Seashore.

Lynne M. Westphal, PhD, is Project Leader and Research Social Scientist with the US Forest Service's Northern Research Station in Chicago. She manages the *People and Their Environments* research work unit. Recent research investigated the simultaneous ecological and economic revitalization of a rustbelt landscape, farming community ideas for future farm landscapes that effectively incorporate ecosystem services, and assessments of empowerment outcomes from urban greening projects.

Nicki J. Whitehouse is lecturer in Palaeoecology, Queen's University Belfast, Northern Ireland. With broad interest in late Quaternary environmental change, Whitehouse's expertise lies in the analysis of sub-fossil beetles from a variety palaeoenvironments. Much of her recent work has been concerned with examining early Holocene landscape and woodland structure in response to natural and human-induced change and biotic changes associated with the transition to agriculture.

Index